T0324765

Springer Texts in Mechanical Engineering

Frederick F. Ling
Series Editor

Warren R. DeVries

Analysis of Material Removal Processes

With 94 Figures

Springer-Verlag
New York Berlin Heidelberg London Paris
Tokyo Hong Kong Barcelona Budapest

Warren R. DeVries
Department of Mechanical Engineering
Rensselaer Polytechnic Institute
Troy, NY 12180-3590
USA

Series Editor
Frederick F. Ling
Director, Columbia Engineering Productivity Center
and
Professor, Department of Mechanical Engineering
Columbia University
New York, New York 10027-6699
and
William Howard Hart Professor Emeritus
Department of Mechanical Engineering,
Aeronautical Engineering and Mechanics
Rensselaer Polytechnic Institute
Troy, New York 12180-3590
USA

Library of Congress Cataloging-in-Publication Data
DeVries, W. R. (Warren Richard)
Analysis of material removal processes / Warren R. DeVries.
p. cm. – (Springer texts in mechanical engineering)
Includes bibliographical references and index.
ISBN 0-387-97728-7
1. Machining. I. Title. II. Series.
TJ1185.D489 1991
671.3'5 – dc20 91-30501

Printed on acid-free paper.

Production managed by Hal Henglein; manufacturing supervised by Robert Paella.
Camera-ready copy prepared by the author.
Printed and bound by R.R. Donnelley & Sons, Harrisonburg, VA.
Printed in the United States of America.

9 8 7 6 5 4 3 2 1

ISBN 0-387-97728-7 Springer-Verlag New York Berlin Heidelberg
ISBN 3-540-97728-7 Springer-Verlag Berlin Heidelberg New York

Series Preface

Mechanical engineering, an engineering discipline born of the needs of the industrial revolution, is once again asked to do its substantial share in the call for industrial renewal. The general call is urgent as we face profound issues of productivity and competitiveness that require engineering solutions, among others. The Mechanical Engineering Series (MES), published by Springer-Verlag, which is intended to address the need for information in contemporary areas of mechanical engineering, is well known to the scientific community.

This series, which has included important monographs and graduate level textbooks, has now been bifurcated into two subsets. While research monographs will continue to be published within the MES, textbooks will be included in a new series entitled Springer Texts in Mechanical Engineering (STME). STME has the same editorial board as MES and both series will, in fact, simultaneously appeal to the scientific community, with the former stressing educational aspects (student textbooks), and the latter emphasizing research achievements (research monographs).

Both MES and STME are conceived as comprehensive series which will cover a broad range of concentrations important to mechanical engineering graduate education and research. We are fortunate to have a distinguished roster of consulting editors on the advisory board, each an expert in one of the areas of concentration. The names of the consulting editors are listed on the first page of the volume. The areas of concentration are: applied mechanics; biomechanics; computational mechanics; dynamic systems and control; energetics; mechanics of materials; processing; thermal science; and tribology.

Professor Wang, the consulting editor for processing, and I are pleased to present the text *Analysis of Material Removal Processes* by Professor DeVries as part of Springer Texts in Mechanical Engineering.

New York, New York *Frederick F. Ling*

Preface

Material removal processes - machining and grinding in this book - are an integral part of a large number of manufacturing systems. They may be either the primary manufacturing process or an important part of preparing the tooling for other processes like forming or molding. In recent years industry and educational institutions have concentrated on the material removal system, perhaps at the expense of the process. This book concentrates on material removal processes; in particular, analysis and how it can suggest ways to control, improve or change the process. This knowledge is more important with automated computer controlled systems than it has ever been before, because quantitative knowledge is needed to design and operate these systems. Practical experience is another important ingredient, but it is difficult to quantify. This experience resides in skilled people that are becoming more difficult to find, e.g., machinists, process planners, and tool designers. As a result, engineers will need to provide this knowledge. This book provides the technological knowledge about material removal needed for setting up, operating or understanding the reasons behind experience based "rules of thumb."

Analysis of Material Removal Processes is a textbook for junior and senior undergraduates and beginning graduate students in engineering. At Rensselaer, "Metal Cutting" is taught annually as an elective to about 40 students. Most of these students are mechanical engineers, but students from materials engineering, industrial engineering and even electrical engineering have taken the course. This level and type of student has the background in mechanics, materials, heat transfer and mathematics that is assumed in the text. The book is also suitable for professionals in industry with responsibilities for planning, improving or designing material removal processes. The perspective of the book is that process planning is the manufacturing design activity that complements product design. Topics cover the modeling and mechanisms of machining technology that support the analysis that goes into process planning. The traditional machining topics of mechanics, cutting edge materials and wear, and machining and grinding processes are all aimed at providing the analysis needed for planning a material removal process. Simple analytical and, where necessary, empirical models are used to predict forces, bulk temperatures, surface roughness, and estimate unit processing times. Separate chapters cover both machining economics and machine tool stability. A new innovation is a chapter that is devoted to metrology, experimental design, tolerance and quality; topics that are important in providing empirical data for new processes and measuring the performance of operating processes. Computer solutions, either by programs or spread sheets, are encouraged by example problems in the text and computer code included in the Appendix. From the material covered in the text and the nearly 300 references in the Bibliography, an engineer should be able to understand and develop elementary models of machining and grinding process, and use them to suggest improvements and innovations for improved quality and productivity.

There are a few things about the text that initially may catch the attention of both an instructor and a student. First, there are no pictures of machines or machine tools. This is intentional. Machine tools and technology are always changing, but the analysis really doesn't. A visit to an instructional laboratory, a factory or shop, or a trade magazine are probably the best place to see machine tools or up to date pictures of them. Second, no English units are used in the text except to say what the common English units are for things like cutting speeds or feed rates. This was an attempt to force me to think in terms of SI units, after growing up with English units. The numbering of equations is also a little hard to follow unless you know the idea behind the numbering scheme. The equations are not always numbered consecutively, and many have letters. For example in Chapter 5, all equations that appear as Eq (5.1?) refer to surface roughness for edges where the nose or corner radius is zero. This means that Eqs (5.1a and b) for a generic roughness are close together but Eqs (5.1c and d) for the roughness in turning appear about thirteen pages later, with a number of other equations defined in between. Equations that are for essentially the same thing have the same number, modified by a letter.

In preparing this text, students that have taken "Metal Cutting" from me at Rensselaer really deserve thanks and acknowledgement. Their puzzled looks, questions and comments led to improvements in the organization and coverage in this text. They also were quick (and correct) to point out the impossible homework and quiz problems that have since be corrected and show up as examples and problems in the text. A number of graduate teaching assistants have worked with me over the years and their comments and suggestions have also been incorporated. Mr. Mengyik Yeong has been particularly helpful in the last stages of preparing the manuscript. Drafting the manuscript for *Analysis of Material Removal Processes* was done while I was on sabbatical at the University of California at Berkeley where my hosts were Dave Dornfeld in the Mechanical Engineering Department and the Forest Products Laboratory. Ms. Hollis McEvilly has taken care of most of the word processing, formatting and other details of manuscript preparation. She has done an outstanding job, and more importantly, she has been patient even when I haven't; her efforts are greatly appreciated.

July, 1991 Warren R. DeVries

Contents

List of Figures

List of Tables

Nomenclature

Symbol	Dimensions	Definition
A	L^2	area
a	L/rev	feed
b	L	chip width, width of cut or radial depth of cut
C	F-t/L	damping coefficient
C?		corrections for ratio of normal to power force components with subscripts: r - rake, w - wear and c - chip thickness
c	F•L/T•M	specific heat capacity
D	L	workpiece diameter
d	L	depth of cut or axial depth of cut.
E	F/L^2	elastic or Young's modulus
E	F/L^2	specific energy for material removal, with subscripts: c - cutting and g - grinding
F	F	force
f	t^{-1}	frequency (Hz)
G		grinding ratio
G	F/L^2	shear modulus
H_B	F/L^2	Brinell hardness
H_V	F/L^2	Vickers hardness
$H_{R?}$		Rockwell Hardness where ? indicates the scale, e.g., A, B or C
h	L	uncut chip or cut thickness
h_c	L	chip or cut thickness
h_{eq}	L	equivalent chip thickness in grinding
h_{avg}	L	average chip thickness
h_{max}	L	maximum chip thickness
K	grits/L	circumferential spacing of active grits
K?		specific cutting energy corrections with subscripts: r - rake, w - wear and c - chip thickness
KT	L	crater wear depth
k	F•L/T•L•t	thermal conductivity
k'		wear coefficient
k_c	F/L	cutting stiffness
k'_c	F/L^2	specific cutting pressure
$k_{dyn}(s)$		dynamic transfer function
k_{eq}	F/L	static equivalent stiffness of machine or setup
k_g	F/L	grinding stiffness
k_l	F/L	static stiffness of contact
L	L	length
L	L	workpiece length
l_s	L	length of the shear plane
l_c	L	contact length in machining or grinding

M	M	mass
N	t^{-1}	rotational speed (revolutions)
P	F•L/t	power
Pe		Peclet number
q	F•L/t	heat flow
R	?	range
R	L	radius
R_a	L	arithmetic or center-line average surface roughness
R_q	L	root mean square or rms surface roughness
R_t	L	maximum peak to valley surface roughness
s	t^{-1}	Laplace variable
s_g	L/grit	feed per active grit
s_z	L/edge	feed per tooth or chip load
$T_{??}(\bullet)$		rotation matrix in the ?? plane
T	F•L	torque
t	t	time
V	L^3	volume
VB	L	flank wear land width
v	L/t	velocity vector
v_c	L/t	chip velocity or speed
v_f	L/t	feedrate
v_s	L/t	peripheral speed of grinding wheel
v_R	L/t	cutting speed for reference tool life
v_w	L/t	peripheral speed of workpiece in grinding
X	L	reference axis of major feed motion
x	L	relative motion along X- axis
Y	L	reference axis of minor feed motion
y	L	relative motion along Y- axis
Z	L^3/t	volumetric removal rate without subscript for machining and for grinding with subscripts: w - workpiece and s - wheel wear rate
Z	L	reference axis of spindle rotation
z	L	relative motion along Z- axis
α		clearance angle, with subscripts: n - normal, f - side, p - back, o - orthogonal, b - base
β		friction angle ($\tan^{-1}(\mu)$)
γ		rake angle, with subscripts: n - normal, v- velocity, e - effective, f - side, p - back, o - orthogonal, g - geometric.
γ		shear strain
$\dot{\gamma}$	t^{-1}	shear strain rate
ε		tensile strain
ε		porosity
ζ		damping factor or ratio
Θ		dimensionless temperature

θ	θ	temperature with subscripts: o - ambient, c - chip, r - rake, w - workpiece in grinding and s - surface in grinding
κ	L^2/t	thermal diffusivity
κ_r		tool cutting edge angle
κ_r'		tool minor cutting edge angle
Λ_s	$L^3/F \cdot t$	abrasive removal parameter (volume rate per unit force)
Λ_w	$L^3/F \cdot t$	metal removal parameter (volume rate per unit force)
λ		inclination angle in oblique cutting
μ	?	mean
μ		friction coefficient
ρ	M/L^3	density
σ	F/L^2	normal stress
σ	?	standard deviation
τ	F/L^2	shear stress
τ	t	time constant or time delay
ϕ		shear plane angle
ω	t^{-1}	frequency or rotational speed (radians)

1 Introduction

Material removal processes - machining and grinding in this book - are an integral part of a large number of manufacturing systems. They may be either the primary manufacturing process or an important part of preparing the tooling for other processes like forming or molding. In recent years industry and educational institutions have concentrated on the material removal system, perhaps at the expense of the process. This book concentrates on material removal processes. In particular, analysis and how it can suggest ways to control, improve or change the process. This knowledge is more important with automated computer controlled systems than it has ever been before, because quantitative knowledge is needed to design and operate these systems. Practical experience is another important ingredient, but it is difficult to quantify. This experience resides in skilled people that are becoming more difficult to find, e.g., machinists, process planners, and tool designers. As a result, engineers will need to provide this knowledge. This book provides the technological knowledge about material removal needed for setting up, operating or understanding the reasons behind experience based "rules of thumb."

1.1 The Design Problem In Material Removal - Process Planning

Since this book is for engineers, and all engineers do design work, either synthesis or analysis, one way of looking at the coverage of this book is from a design perspective. Typically someone studying material removal will concentrate on designing a process to produce a product. The builder of machine tools needs to know about the loads, rates and dynamics of the process to design a product - the machine tool. The design of a process to produce a product is still the main reason for the analysis of material removal processes and has a special name: *process planning*.

Process planning is a design problem because for each part there are numerous constraints and more than one solution. With practical machining processes, the analysis will sometimes be approximate, but is aimed at helping answer design questions like:

- Is machining the most appropriate method considering: volume and batch size, accuracy and finish, materials and cost constraints?
- What type(s) of surfaces have to be generated, e.g., internal or external surfaces of revolution, vertical or horizontal flat surfaces, tapers or contours?
- What type(s) of technology should be used, e.g., tried and true versus new and unknown, and where should the work be done, "in house" or "farmed out"?

There are also a number of constraints to the planning problem: technological constraints called specifications, economic constraints on cost or cycle time and practical constraints such as shop or plant capacity to meet delivery dates. Examples of these constraints related to material removal processes are:

- Product specifications such as accuracy, tolerance, or surface finish affect the choice of process; the tighter the geometric specifications, the more likely the need for machining or grinding.
- Specifications on the material to be processed lead to questions like: If there is previous experience with a material, is it easy or difficult to process, or if it is a new material, is there any data on how to process it?
- Cost, particularly unit cost, may be the economic constraint in some situations. However, if the process plan is for a station on a transfer line or a system coupled to other processes, the processing time for a single stage may be the bottle neck for the system. The cost or tool life for a single process is subordinate to the system. Then knowing what is technologically possible is a way to meet the constraints imposed by the system.
- Practical limitations that a particular facility has to deal with are specific machine tools with limits on power, stiffness, type, controls, etc.

The solution of this design problem involves selecting and parametrically specifying a number of design variables. This book deals with material removal processes so the restricted set of design or process planning variables includes:

- The process(es) needs to be planned to meet the product specifications and the other planning constraints. This means making a decision on processing in one step, using the same process but in several steps, i.e., roughing and finishing, or determining that technologically several processes are needed to meet the constraints imposed by product specifications, i.e., rough machining followed by finish grinding.
- For a chosen processing path, expendable tools, cutting fluids, and jigs and fixtures have to be selected and specified. These are often items that require a large amount of lead time, so making the correct decision as early as possible is important.
- Selections of machining conditions, e.g., feed, speed and depth of cut, is part of preparing NC programs for complex parts. Knowing how to change these machining conditions is often the most immediate way of making a machine tool that is chattering operate in a stable region without having to purchase a new machine tool or re-design the workholding system.
- If a process is on-going, can it be done better or be improved using new materials, technology or controls? Models and analysis are needed for this type of planning to extend beyond the limits of experience and empirical data.

Solutions to these planning or design problems have major components of experience, empirical data and engineering analysis. The latter two components will be the major emphasis of this book, but to be successful it is important to find someone that can add the experience component.

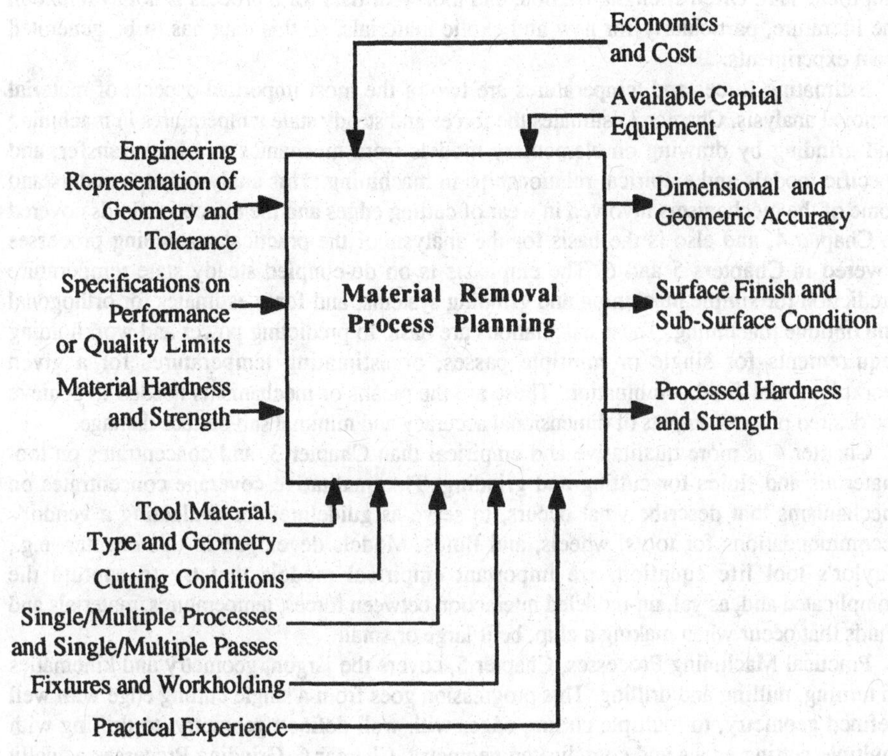

Figure 1.1 Representation of the inputs, constraints, outputs and controls in process planning for material removal.

1.2 Analysis of Material Removal for Process Planning

Figure 1.1 illustrates some of the considerations that go into material removal process planning: This level of process planning is the focus of this book. A planning decision to use material removal, rather than forming, joining or molding, has already been made. Figure 1.1 uses the IDEF standard block diagram [LeClair82]. The format for the diagram is: the inputs to the planning process are on the left, the outputs are on the right, constraints are on the top, and the means or mechanisms to transform the inputs to the outputs are on the bottom. Later this representation summarizes each chapter, but here it is a way to summarize the coverage of this book.

Material removal processes can change the size and shape of a variety of materials and produce accurate and geometrically complex parts with a high quality surface finish. Chapter 2 is a short tutorial on *metrology*, *tolerances*, *quality*, and *experimental design*, all important in designing with reasonable tolerances, checking process capability and developing empirical data. For example, metrology becomes important in assessing the process planning outputs of dimensional and geometric accuracy, as well as surface finish. Tolerance and quality go together. Tolerance is an input specification that usually determines what process to use, and quality control serves as a way to use the output of a material removal process to track and improve process capability. Experimental design, or how to collect and analyze data in an efficient way, covers collecting and modeling empirical data. Often strength, friction, and tool wear data for a process is not available in the literature, particularly for new and exotic materials, so this data has to be generated from experiments.

Estimating forces and temperatures are two of the most important aspects of material removal analysis. Chapter 3 estimates the forces and steady state temperatures in machining and grinding by drawing on elementary models from mechanics and heat transfer, and specific models and empirical relationships in machining. This analysis helps understand some of the mechanisms involved in wear of cutting edges and the effects of fluids covered in Chapter 4, and also is the basis for the analysis of the practical machining processes covered in Chapters 5 and 6. The emphasis is on de-coupled steady state temperature prediction for simple machining and grinding systems, and force estimates for orthogonal and oblique machining. These calculations are basic to predicting power and workholding requirements for single or multiple passes, or estimating temperatures for a given workpiece-tool-fluid combination. These are the means or mechanisms needed to achieve the desired process outputs of dimensional accuracy and minimal sub-surface damage.

Chapter 4 is more qualitative and empirical than Chapter 3, and concentrates on tool materials and fluids for cutting and grinding. The qualitative coverage concentrates on mechanisms that describe what occurs, to serve as guidelines for evaluating a vendor's recommendations for tools, wheels, and fluids. Models developed in this chapter, e.g., Taylor's tool life equation, are important empirical models that try to capture the complicated and, as yet, un-modeled interaction between forces, temperatures, materials and fluids that occur when making a chip, be it large or small.

Practical Machining Processes, Chapter 5, covers the jargon, geometry and kinematics in turning, milling and drilling. This progression goes from a single cutting edge with well defined geometry, to multiple cutting edges with well defined geometry, to drilling with multiple cutting edges and complicated geometry. Chapter 6, Grinding Processes actually takes this progression a step further by considering a myriad of cutting edges with statistically defined geometry. Both chapters emphasize estimating the static forces and theoretical finish produced by a process when selecting machining conditions. Two ways to estimate forces are used: the mechanics approach of Chapter 3 and empirical techniques based on specific cutting or grinding energy and power law models.

Chapter 7 concentrates on what is traditionally termed machining economics. This chapter deals with optimizing unit processes based on three criteria: unit time, cost and profit. Writing the unit time, cost and profit equations for selected processes, and showing the form of the resulting solutions is the emphasis of this chapter. These solutions are alternatives to selecting machining conditions based only on technological considerations and constraints.

The final chapter, Chapter 8 - Modeling Machining Vibration for Stability Analysis, is where it is acknowledged that machining is more than kinematics and mechanics; it includes dynamics and vibration. Both time domain and frequency domain models for describing the motion of a cutting edge and grinder are developed. The dynamic model is the tool used to analyze the effect of a machine tool's structure on chatter vibrations that occur only during the machining or grinding process. Unless the limitations of a particular machine tool are known during the planning process, the machining conditions selected using optimized static criteria still may not allow a part to be produced because of vibration. Simple stability models or stability charts provide a short term solution to getting a process to run. The modeling suggests how to improve a machine or workholding design.

No mention was made of using the analysis techniques in this book to change the inputs shown in Figure 1.1. Traditionally, that has been the approach in bringing a product design from concept through manufacture. Today this approach is beginning to change. Outputs or preliminary process plans feed back to the design system, changing process inputs to accommodate technology advances or cost constraints. Also, little has been said about the form in which geometric and tolerance information enters the planning process. Traditionally this has been through engineering drawings, but over time it is becoming more common for this information to be generated, transferred and modified on Computer Aided Design (CAD) systems. As it becomes more common to transfer design information electronically, it also becomes more important for an engineer to be able to give fast and timely analysis on how to manufacture a design. Only when the manufacturing analysis link is complete can the benefits of Design for Manufacture or Simultaneous Engineering be fully realized. This book will help provide quantitative answers to the material removal part of this loop.

1.3 References

[Chang90] Chang, Tien-Chien, (1990), *Expert Process Planning for Manufacture*, Addison-Wesley, Reading, MA.

[GalKni86] Gallagher, C.C., and W.A. Knight (1986), *Group Technology Production Methods in Manufacture*, Ellis Horwood, Ltd. Chichester, England.

[LeClair82] LeClair, Steven R., (1982), "IDEF the Method, Architecture the Means to Improved Manufacturing Productivity," SME Technical Paper #MS82-902, SME.

[Koren83] Koren, Y., (1983), *Computer Control of Manufacturing Systems*, McGraw-Hill, New York.

[PreWil77] Pressman, R. S., and Williams, J. E., (1977), *Numerical Control and Computer-Aided Manufacturing*, John Wiley, New York.

2
Metrology, Quality Control and Data Analysis

This chapter covers the topics of metrology, tolerancing, quality control and experimental design before discussing the different ways machining changes the shape of a part in other chapters. Tolerances on dimensions and surface finish are used for comparing different processes, and when the tolerances or surface finish specified are small, this usually leads a process planner to choose machining or grinding processes. Metrology, the science of measurement, plays an important role in checking if specifications are met, and is needed to develop experimental data for analyzing a process. The most efficient use of experimental data occurs not by chance, but when experiments are designed and planned.

When manufacturing more than one item, there is always some variation between parts and components. A design engineer's job is to specify how much of this variation can be tolerated without affecting the functional performance and safety of a product. This is how tolerance and specifications on performance become constraints in Figure 2.1 that represents the coverage of this chapter. Tracking down the source of the variation, knowing what is inherent to the process and what part of it can be improved is part of designing the process, particularly the design of a quality control system to monitor the material removal process. To do this in a quantitative way, measurements are needed and have to be analyzed so they become the means to monitor the performance of a machining or grinding process. Of course, some of the uncertainty comes from incomplete knowledge of the materials or processes, so experiments and empirical models have to be developed to fill in these knowledge gaps. In machining and grinding processes incomplete knowledge of materials is the rule rather than the exception, so there are many places where experimental design methods can profitably be employed.

Figure 2.1 indicates four of the topics covered in this chapter: metrology, tolerancing, quality or process control, and experimental design. Because each is usually covered in a separate course, their coverage here is limited to a brief discussion and a total of seven examples. Check the references at the end of the chapter when you need more details.

2.1 Metrology

Metrology is the science of measurement. Characteristics used to specify measurement instruments for experiments and process or quality control are defined. While there are numerous ways to make process measurements, specifying instrument performance is an important part of process development and planning. A way to make these specifications is illustrated with a simple statistical model that demonstrates the metrology rule of thumb that *the instrument should be ten times better than what you want to measure.*

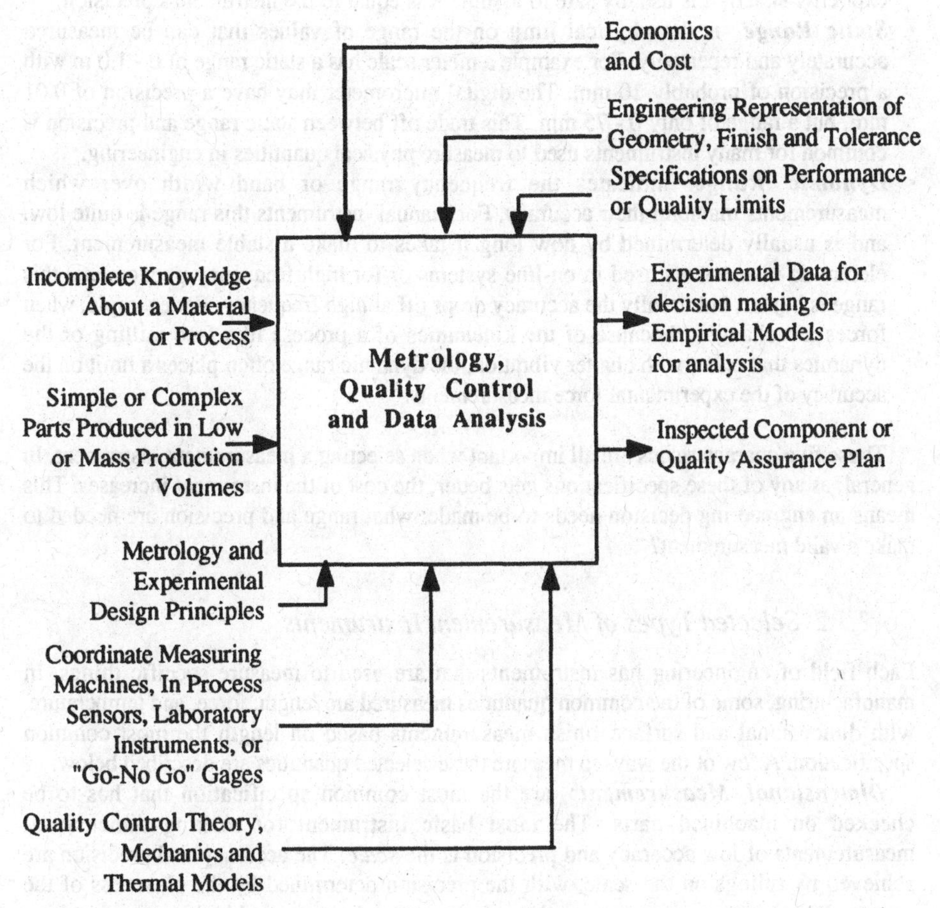

Figure 2.1 The inputs, constraints, outputs and controls for metrology, quality control and data analysis in a machining system.

2.1.1 Definitions of Basic Metrology Terms

Some common metrology terms that appear in specifications for measurement devices, either for manual inspections or for on-line computer control systems, are:

- ***Accuracy*** is the degree of conformity to a standard or true value. For example, since the international standard for length measurement is the meter, if an instrument reads a meter, its accuracy would be in comparison to this international standard.
- ***Precision*** defines the degree of refinement with which a measurement can be stated. This can usually be determined by inspecting the instrument and finding out the number of digits displayed or the "least count". For example on a meter scale the least count might be 10 mm, or for a digital micrometer the number of decimal places on the liquid crystal display indicates the precision.
- ***Repeatability*** is the range of variation in repeated measurements. This value usually

is less than the precision of the instrument. As a result, if a repeatability value is not explicitly stated, it is usually safe to assume it is equal to the instrument's precision.

- ***Static Range*** is the physical limit on the range of values that can be measured accurately and repeatedly. For example a meter scale has a static range of 0 - 1.0 m with a precision of probably 10 mm. The digital micrometer may have a precision of 0.01 mm, but a range of only 0 - 75 mm. This trade off between static range and precision is common for many instruments used to measure physical quantities in engineering.
- ***Dynamic Range*** indicates the frequency range or band width over which measurements maintain their accuracy. For manual instruments this range is quite low, and is usually determined by how long it takes to make a stable measurement. For electronic instruments used in on-line systems or for high frequency phenomena, this range is higher, but usually the accuracy drops off at high frequency. For example, when forces are changing because of the kinematics of a process like face milling or the dynamics that occur with chatter vibration, the dynamic range often places a limit on the accuracy of the experimental force measurements.

These five characteristics are all important when selecting a measurement instrument. In general, as any of these specifications gets better, the cost of the instrument increases. This means an engineering decision needs to be made: what range and precision are needed to make a valid measurement?

2.1.2 Selected Types of Measurement Instruments

Each field of engineering has instruments that are used to measure specific things. In manufacturing, some of the common quantities measured are length, force, and temperature, with dimensional and surface finish measurements based on length the most common specification. A few of the ways to measure these selected quantities are described below.

Dimensional Measurements are the most common specification that has to be checked on machined parts. The most basic instrument for making dimensional measurements of low accuracy and precision is the *scale*. The accuracy and precision are achieved by rulings on the scale, with the precision determined by the closeness of the rulings. Direct comparison of the scale with the part to be measured is the way of making the measurement.

Micrometers are probably the most common manual method for making dimensional measurements in manufacturing. Accuracy is determined by the pitch of the threads used to adjust the jaws of the micrometer. The precision is determined by the ability to divide a rotation of the thread using a scale on the body of the instrument. Micrometers can routinely achieve a precision of 0.1 mm, and by adding a vernier, a scale which further subdivides the rotation of the thread, the precision can be increased by a factor of ten to 0.01 mm.

For setting standards or laboratory work, accurately machined blocks called *gage blocks* are used. By selecting different combinations of these blocks and stacking them together, the resulting stack will be accurate to within ±0.0001 mm. Using the gage blocks as a standard, measurement is made by direct comparison.

Another dimensional measurement method used primarily for its precision is the fringe counting *interferometer*. Counting interference fringes produced by light waves from a reference beam and the part to be measured is the principle of operation. A stable monochromatic light source, typically a laser, is used. The precision is determined by the wavelength of the light source, usually on the order of 0.1 μm.

These methods are measurement principles used to make dimensional measurements. *Coordinate measuring machines* (CMM's) use these principles along with transducers to digitally display or record dimensional measurements. They also make it easier to measure geometric features like flatness, parallelism, hole center locations and other features of a machined or ground part that need more than a single measurement to be defined. This way of specifying geometry is referred to as geometric dimensioning and tolerancing [GeoMet82], and requires planning in selecting reference datums in design, manufacture and inspection.

Surface Measurement usually involves geometric features on two scales: the large scale undulations that have to do with waviness and flatness, and the small scale undulations that are usually termed roughness. This geometry of a part is usually measured on a scale an order or magnitude less than what would be used to specify the precision of a dimensional measurement.

Flatness and waviness are usually measured using light interference patterns. A way to do this is with a monochromatic light source and optical flats. To work correctly the optical flats must be parallel and flat to an accuracy of less than the wavelength of light used. When light is shown through the optical flat as it rests on the surface to be measured, a fringe pattern develops. Each fringe corresponding to a change in elevation, due to waviness or flatness deviations in the surface, of one wavelength of light.

On a smaller scale, *roughness* is usually measured by a stylus that moves across the surface. The instrument is such that the vertical displacement of the stylus is amplified, both mechanically and electronically, to give a time varying signal that represents the micro changes in the surface plus any waviness. The waviness is usually removed by using a high pass filter; the break frequency of the filter is termed the *wavelength cutoff* often designated by λ_c. Wavelength cutoffs are usually selected from the progression: 0.075 mm, 0.25 mm, 0.75 mm, 2.5 mm or 7.5 mm. If a value isn't specified, $\lambda_c = 0.75$ mm is probably the most common cutoff. The smaller cutoffs are about the same as the spacing between feed marks so these settings will filter out feed mark contributions to surface roughness.

After applying the cutoff, the positive and negative deviations from the waviness profile z(x) are used to compute the roughness. The measurements are made over a specific traverse length L, which should be about 3 times the wavelength cutoff or $L > 3\lambda_c$. The American National Standards Institute (ANSI) standard for roughness is defined as [SurTex78]:

$$R_a = \frac{1}{L}\int_0^L |z(x)|\, dx, \qquad (2.1)$$

and is often termed the Arithmetic Average (AA) or Center Line Average (CLA) roughness. Methods to estimate the theoretical value of R_a as a function of cutting edge geometry and cutting conditions are given in Chapters 5 and 6. Another measure of roughness more prevalent in Europe than the U.S.A., is

$$R_q = \sqrt{\frac{1}{L}\int_0^L z^2(x)\, dx} \qquad (2.2)$$

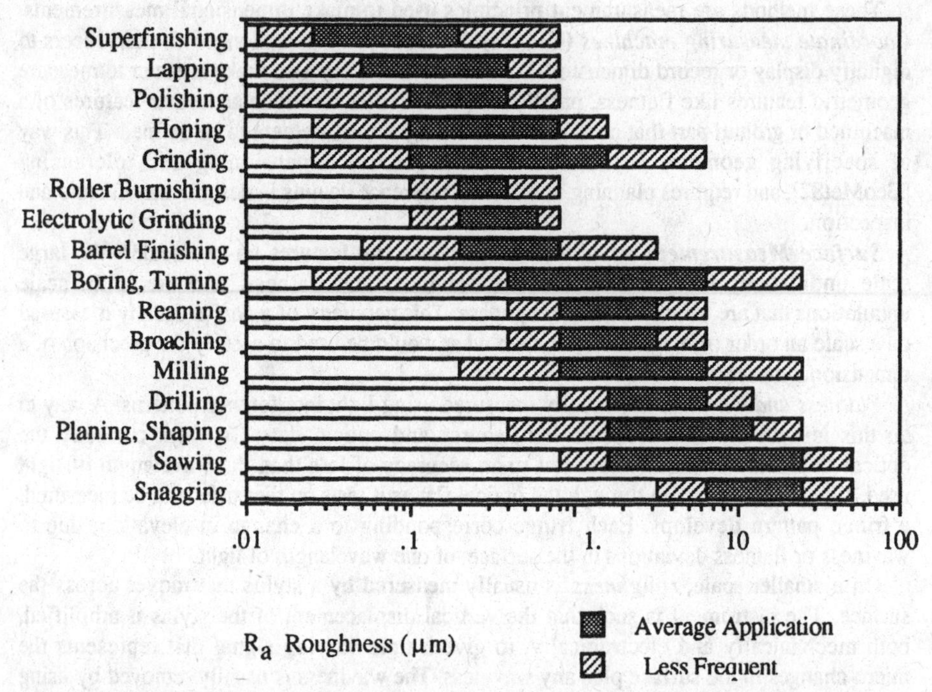

Figure 2.2 Ranges of R_a surface roughness in machining and grinding based on [SurTex78].

and is termed the root mean square (RMS) roughness. R_q is exactly the same as the RMS voltage in electrical engineering or the standard deviation in statistics. The third common method of measuring the roughness of a surface is in terms of the extremes in a length L of a profile. The maximum peak to valley height

$$R_t = \max[z(0)...z(L)] - \min[z(0)...z(L)] \tag{2.3a}$$

considers only the single maximum and minimum in a profile of length L. Methods to estimate R_t are also given in Chapters 5 and 6. Because R_t deals with extremes, for real surfaces it can be an erratic estimate of the roughness. R_z, which takes an average of n of these extremes, is often used for real surfaces.

$$R_z = \frac{1}{n}\sum_{i=1}^{n}\left(i^{th} \max[z(0)...z(L)] - i^{th} \min[z(0)...z(L)]\right) \tag{2.3b}$$

As a guide to the range of R_a surface roughness that are possible, Figure 2.2 provides some guidelines for common machining and grinding processes.

Equations (2.1) - (2.3) are the most common ways to quantify the roughness of a surface. They only consider micro scale height deviations on a surface, not the spacing between peaks or valleys. As a result, methods that use the correlation length or spectrum of a profile are showing up on commercial instruments, c.f., [SurTex78] and [Pet*etal*79].

Force Measurement is usually done indirectly by either measurement of the pressure or stress acting on a known area (piezo electric transducers) or by measurement of small displacements (strain gage technology). Devices that measure forces in these ways are usually termed *load cells* or *dynamometers*. Their importance is primarily in experimental work, such as measuring loads when determining stress-strain behavior of materials or verifying forces predicted in machining. There are practical machining systems that incorporate force measurement in a control system to obtain uniform quality in a machining process. Another practical way that the major force component, later referred to as the power force component F_p, is measured is using a *power meter*. As will be shown in Chapter 3, and later in Chapter 5, the product of this force and the speed is the major contributor to the power required in machining and grinding. Using a power meter to measure F_p usually lacks the accuracy and dynamic range of a load cell or dynamometer, but has the practical advantage of not interfering with the process. As a result it is a popular method for control or process monitoring.

In subsequent chapters on the mechanics of machining and grinding, calculations are aimed at estimating the forces measured by dynamometers. This comparison of experiment and calculations is a well accepted way of checking a model. In other instances, the problem may be to calculate a micro-property like stress, from a macro-measurement of force with a dynamometer.

Whenever these force measurements are made in machining, the design specifications of interest always include: range and resolution, stiffness so that the measurement doesn't affect the process, insensitivity to elevated temperatures like those encountered in machining, and cross-talk between components is of concern when multiple force or torque components are measured with one dynamometer. For machining dynamics experiments, the dynamic range of the device is of great importance.

Temperature Measurement, like force measurement, is of primary concern in experimental investigations and when controlling a process. For example, the temperatures in metal removal strongly influence how long a cutting tool will last. In grinding applications, much of the energy provided turns into heat, which may flow into the workpiece causing it to distort or affect the workpiece's subsurface strength or fatigue properties.

Most temperature measurement methods are indirect. For example, specially treated materials that change color at a specific temperature are used to infer the maximum temperatures reached on a workpiece. Detecting the radiation given off by bodies at high temperatures is a non-contact, indirect temperature measurement method. However, the most common temperature measurement method uses thermocouples. The thermocouple junction is formed by either a pair of wires mounted at the point of measurement, or the interface between say, a workpiece and cutting tool. In either case, reliable temperature measurement methods are difficult to make and require a great deal of skill. For example, mounting a thermocouple can be difficult, particularly if the machining operation is milling or grinding, because in addition to mounting the thermocouple, the signal has to be taken off a rotating spindle. The work-tool thermocouple technique, on the other hand, works fairly well with rotating spindles, but is an aggregate temperature at an unknown point.

2.1.3 Simple Model for Measurement Variation

Instruments are used to quantify variation in a part or changes in a process, so ideally their readings should be unaffected by the instrument itself. However as the term repeatability implies, all instruments introduce some variation to a measurement. The problem is to decide what specification to make for an instrument, so that the instrument's contribution to the total variation in a part or process is "small."

To analyze this problem, a simple statistical model of the variance in a measurement σ_m^2 is expressed as a function of the variance of the instrument σ_i^2 and the variance of the part or process σ_p^2. Now if the two sources of variation in the measurement are not related or are independent, then the variances of these sources are additive

$$\sigma_m^2 = \sigma_i^2 + \sigma_p^2 \quad (2.4a)$$

More commonly, we deal with standard deviations, i.e., the square root of the variance, so

$$\sigma_m = \sqrt{\sigma_i^2 + \sigma_p^2} \quad (2.4b)$$

Eq (2.4) leads to a justification for a "rule of thumb" common in selecting a measurement instrument: "the instrument should be ten times better than what you want to measure." Better in this case means that if you want to measure a part with a standard deviation of σ_p, then you should select an instrument with a $\sigma_i = 0.1\ \sigma_p$. Applying this rule of thumb, from Eq (2.4b) the measurements read will be

$$\sigma_m = \sqrt{(0.1\ \sigma_p)^2 + \sigma_p^2}$$

$$= \sigma_p \sqrt{1.01} \approx \sigma_p\ 1.005,$$

or the measured variation σ_m, is accurate to within 0.5% when compared to the "true value" of σ_p.

There are two common ways to determine a numerical value for σ_i when selecting an instrument. The easiest and most direct way is when the variation in the instrument is specified by its repeatability, and usually is written as a plus-minus tolerance. If the repeatability is not in the list of specifications, the precision of the instrument can often be used instead. Either way, an agreed upon convention is that the standard deviation is one third the tolerance limit, i.e., σ_p = repeatability/3. Of course if σ_p is given in the specification, use it directly.

Example 2.1 - Selecting a Micrometer

Suppose that a micrometer has to be selected to measure the variation in a part that has a standard deviation $\sigma_p = 0.01$ mm. Two micrometers are available, one without a vernier and the other with a vernier. The precision of the first micrometer is ±0.03 mm and, because of the vernier, the second micrometer has a precision of ±0.003 mm and a higher cost.

Estimate the measured standard deviation expected when using each micrometer to measure parts where σ_p is known to be 0.01 mm.

Solution: Assume that the accuracy of each micrometer is equivalent, in this case less than 0.003 mm. Because there is no other data available to estimate σ_i, assume that the precision can be used to estimate of the repeatability tolerance. This means that σ_i is one third the precision for each micrometer. Using Eq (2.4) to estimate the measurement standard deviations, results for the two micrometers are:

Without vernier

$$\sigma_i = \frac{0.03 \text{ mm}}{3} = 0.01 \text{ mm}$$

$$\sigma_m = \sqrt{(0.01)^2 + (0.01)^2} = 0.014 \text{ mm}$$

With vernier

$$\sigma_i = \frac{0.003 \text{ mm}}{3} = 0.001 \text{ mm}$$

$$\sigma_m = \sqrt{(0.01)^2 + (0.001)^2} = 0.01005 \text{ mm}$$

2.2 Tolerancing

Tolerancing is a way of specifying acceptable limits on part variation that will not affect the functional performance, safety or assembly of a part or component. There are two practical reasons for being interested in tolerances and how they interact:

- If tolerance limits are made too small, manufacturing costs can increase drastically. Tolerance specifications in many cases will determine what method of manufacture will be used. A process planner knows that small tolerances usually means selecting expensive machining or grinding processes.
- When tolerance limits are too large, it may be impossible to have interchangeable parts or stock spare parts. The lower processing costs associated with large tolerances are usually replaced by costs associated with 100 percent inspection and selective assembly.

When this tradeoff between small and large tolerance limits is done intelligently, making the best use of material removal processes technology, a firm gets a cost and quality advantage.

While tolerances apply to geometric forms like parallelism or concentricity [GeoMet82] or [Spotts83], this section only deals with dimensional tolerances. Figure 2.3 gives an indication of the range of dimensional tolerances that some selected material removal processes can achieve. The lower tolerance limits in this figure are always more expensive to produce than the higher limit. Typically tolerances are specified in terms of a nominal or average value and the acceptable positive and negative deviations from this value.

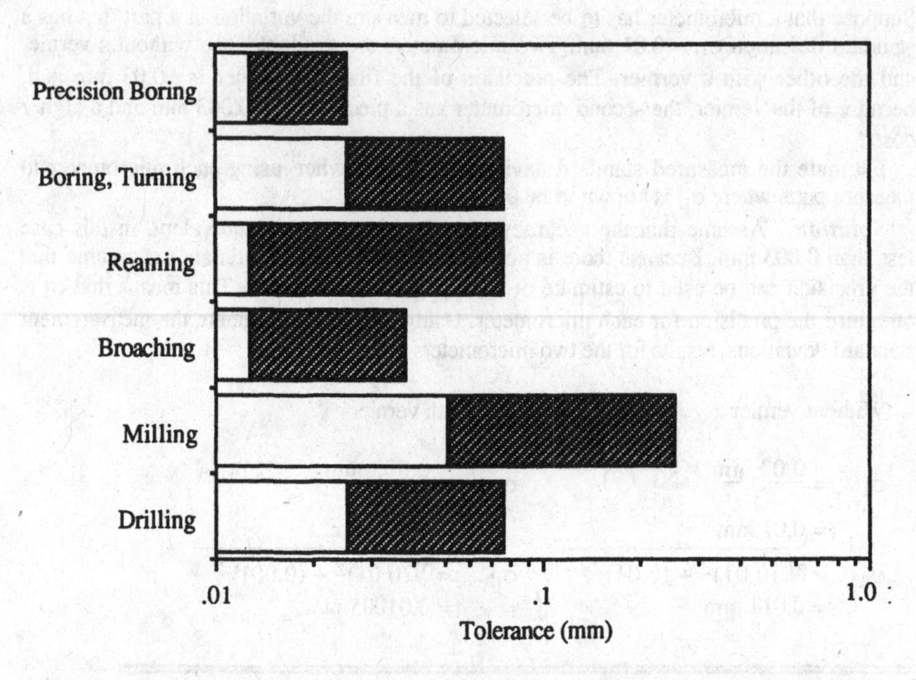

Figure 2.3 Ranges of dimensional tolerances for some machining processes based on [Spotts83].

These specifications may be of the form:

$$\text{nominal value}\,{}^{-\text{ lower tolerance}}_{+\text{ upper tolerance}}$$

if the upper and lower tolerance limits are not equal. If the tolerance limits are symmetric or *bilateral* they may be represented as:

nominal value ± bilateral tolerance.

2.2.1 Probability Model for Statistical Tolerancing

Geometric and statistical tolerances are the main approaches to specifying tolerable product variation. Geometric tolerances are more conservative, but also more costly to manufacture.

Statistical dimensional tolerancing with symmetric or bilateral limits can be thought of as a way of specifying the mean and standard deviation of a probability distribution. The nominal value in the tolerance specification gives the mean

$$\mu = \text{nominal value}. \tag{2.5}$$

Tolerance limits give the standard deviation of the distribution. For symmetric bilateral tolerances, the agreed upon convention relating tolerances and standard deviations is

Table 2.1 The Standard Normal Distribution

$$P[z \le z_0] = \int_{-\infty}^{z_0} \frac{1}{\sqrt{2\pi}} e^{-w^2/2} dw \quad \text{and} \quad P[z \le -z_0] = 1 - P[z \le z_0]$$

z_0	$P[z \le z_0]$	z_0	$P[z \le z_0]$	z_0	$P[z \le z_0]$
0.00	0.500	1.10	0.864	2.05	0.980
0.05	0.520	1.15	0.875	2.10	0.982
0.10	0.540	1.20	0.885	2.15	0.984
0.15	0.560	1.25	0.894	2.20	0.986
0.20	0.579	1.282	0.900	2.25	0.988
0.25	0.599	1.30	0.903	2.30	0.989
0.30	0.618	1.35	0.911	2.326	0.990
0.35	0.637	1.40	0.919	2.35	0.991
0.40	0.655	1.45	0.926	2.40	0.992
0.45	0.674	1.50	0.933	2.45	0.993
0.50	0.691	1.55	0.939	2.50	0.994
0.55	0.709	1.60	0.945	2.55	0.995
0.60	0.726	1.645	0.950	2.576	0.995
0.65	0.742	1.65	0.951	2.60	0.995
0.70	0.758	1.70	0.955	2.65	0.996
0.75	0.773	1.75	0.960	2.70	0.997
0.80	0.788	1.80	0.964	2.75	0.997
0.85	0.802	1.85	0.968	2.80	0.997
0.90	0.816	1.90	0.971	2.85	0.998
0.95	0.829	1.95	0.974	2.90	0.998
1.00	0.841	1.960	0.975	2.95	0.998
1.05	0.853	2.00	0.977	3.00	0.999

$$\sigma = \frac{\text{tolerance}}{3}. \tag{2.6}$$

Each measurement is considered a random variable x, that has associated with it a probability density, f(x). The density function assigns a probability or expected frequency of occurrence to each range of x values. The most common and versatile density function is the Gaussian or Normal density function given by

$$f(x) = \frac{1}{\sigma\sqrt{2\pi}} e^{\frac{-(x-\mu)^2}{2\sigma^2}}, -\infty < x < \infty.$$

While other density functions can be used to model dimensional variation, this book will use the Gaussian or Normal density almost exclusively. A shorthand notation for specifying that a random variable x, has a Normal density function with mean μ and standard deviation σ is

$$x \sim N[\mu,\sigma^2].$$

The probability that the random variable x, is less than or equal to a specified value x_0, is found by writing and evaluating

$$P[x \leq x_0] = \int_{-\infty}^{x_0} f(x)\, dx. \tag{2.7}$$

To simplify the integration of the Gaussian density function for arbitrary values of μ and σ, a transformation is usually applied to get a random variable that has a mean of zero and a variance of one. This allows values for probabilities to be tabulated in this Standard Normal form, as in Table 2.1. The transformation is,

$$x \sim N[\mu,\sigma^2] \rightarrow z = \frac{x-\mu}{\sigma} \rightarrow z \sim N[0,1]. \tag{2.8}$$

Probabilities like those in Eq (2.7) can be found by applying Eq (2.8) and looking up $P[z \leq z_0]$ in Table 2.1.

Example 2.2 - Probability Model for a Tolerance Specification.

Assume that plates like those shown in Figure 2.4 have to be face milled to a tolerance specification on its thickness of 10.00 ± 0.30 mm.

What is the assumed probability density for a plate thickness, and what is the probability that the thickness will be within the tolerance limits?

Solution: Using the agreed upon convention in Eq (2.6), the standard deviation is

$$\sigma = \frac{0.30 \text{ mm}}{3} = 0.10 \text{ mm}.$$

Without any other information about the distribution of these dimensions, it is reasonable to assume that the thicknesses have a Normal distribution with $\sigma = 0.10$ mm and $\mu = 10.00$ mm. Then the short hand notation for this dimension is $x \sim N[10.00, 0.10^2]$, with the sketch of the density shown in Figure 2.4. This answers the first part of the question.

To determine the probability that the thickness will be within tolerance, the desired probability is

$$P[\text{Single plate within tolerance}] = P[9.70 \text{ mm} \leq x \leq 10.30 \text{ mm}].$$

It can most easily be found using the standard transformation of Eq (2.8) and Table 2.1 to find the probability. The transformation is

$$P[9.70 \text{ mm} \leq x \leq 10.30 \text{ mm}] \leftrightarrow z = \frac{x - 10.00}{0.10}, \text{ or } P[-3 \leq z \leq 3].$$

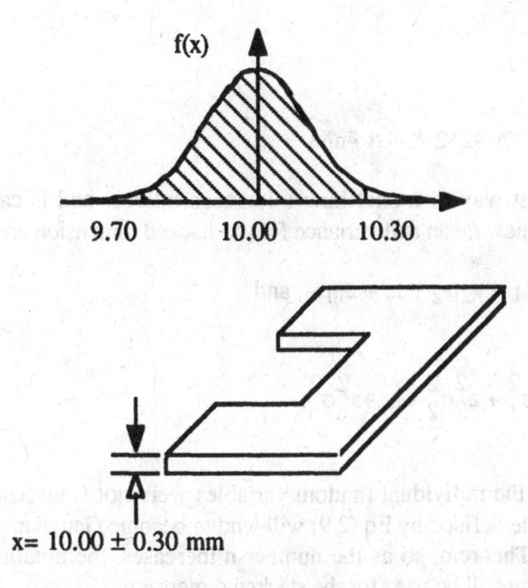

Figure 2.4 Thickness tolerance on a single plate.

The shaded area in Figure 2.4 is the desired probability, and it can be evaluated using symmetry and the values in Table 2.1 by realizing that

$$
\begin{aligned}
P[-3 \leq z \leq 3] &= P[z \leq 3] - P[z \leq -3] \\
&= P[z \leq 3] - (1 - P[x \leq 3]) \\
&= 0.999 - (1 - 0.999) \\
&= 0.998
\end{aligned}
$$

This probability can be considered a frequency, which means that 499 out of 500 plates would be within tolerance; a high probability. This is also the reason for agreeing that the tolerance values should be 3 times the standard deviation, as given by Eq (2.6).

2.2.2 Modeling Tolerance Stack Up and Interference

The type of statistical model used to model the tolerance of a single plate in Example 2.2, can be used to answer questions such as: If a number of components with a known tolerance are put together, what is the tolerance of the overall assembly? Or, given the nominal dimensions and tolerance specifications of two parts that must fit together in an assembly, what are the chances that the assembly either will or will not fit together?

One way to look at this problem is in terms of combinations of randomly selected components. Mathematically these components are termed independent random variables. For example, the components in an assembly can be considered independent random variables x_i, each with a nominal dimension or mean value μ_i and variance σ_i^2 determined from the tolerance. When these elements are assembled in a simple linear way, the "stacked" dimension is a new random variable y that can be represented by,

$$y = a_1x_1 + a_2x_2 + ... + a_nx_n. \tag{2.9a}$$

This is the simplest way to model how tolerances interact and is called an elementary tolerance link. The new mean and variance for the stacked dimension are given by

$$\mu_y = a_1\mu_1 + a_2\mu_2 + ... + a_n\mu_n, \text{ and} \tag{2.9b}$$

$$\sigma_y^2 = a_1^2\sigma_1^2 + a_2^2\sigma_2^2 + ... + a_n^2\sigma_n^2. \tag{2.9c}$$

Moreover, even if the individual random variables were not Gaussian to begin with, the new random variable defined by Eq (2.9) will tend to become Gaussian. This is predicted by the Central Limit Theorem, so as the number n increases, the notation used for Normal random variables can still be used for the stacked dimension

$$y \sim N[\mu_y, \sigma_y^2].$$

From this simple idea, the stack up of tolerances or the associated interferences can be determined. If, for example, the stack up of tolerances is to be modeled, the coefficients a_i, $i = 1,2 ... n$, in Eq (2.9) are positive, typically one. Similarly, if interferences are the difference between dimensions, at least one coefficient in Eq (2.9) is negative.

Example 2.3 - Stack Up of Tolerances.
If the metal plates in Example 2.2, are stacked 4 high to make a sub-assembly, what is the nominal thickness and the tolerance of the stack?

Solution: From Example 2.2, each of the 4 plates has a thickness distribution

$$x_i \sim N[10.00, 0.10^2], i = 1,2,3,4.$$

If the four plates needed for a sub-assembly are picked at random, i.e., no selective assembly, then the assumption of independence is reasonable. For this stack up assembly, the coefficients in Eq (2.9) are all ones, and from Eq (2.9b) the nominal thickness of the stack is

$$\mu_y = 10.00 + 10.00 + 10.00 + 10.00 = 40.00 \text{ mm} = \text{nominal thickness}$$

and the variance of the stack can be calculated using the variance of each plate in Eq (2.9c),

$$\sigma_y^2 = (0.10)^2 + (0.10)^2 + (0.10)^2 + (0.10)^2 = 4(0.10)^2.$$

From the standard deviation $\sigma_y = 0.20$ mm, the tolerance is $3\sigma_y = 0.60$ mm.
Combining these results, the the tolerance specification for the stacked plate is 40.00 ± 0.60 mm as shown in Figure 2.5.

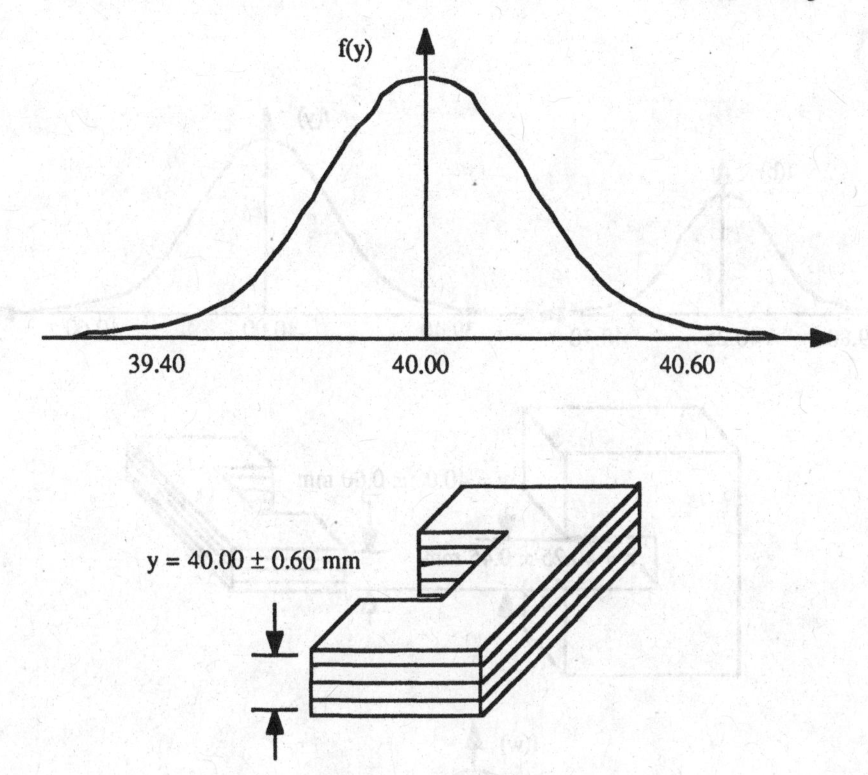

Figure 2.5 Thickness tolerance for a stack of 4 plates.

As an aside, analyzing the accuracy of gage blocks is done in exactly the same way. This means each gage block must be extremely accurate and temperatures must be controlled to be able to maintain the accuracy of ±0.0001 mm quoted in Section 2.1 for a stack of gage blocks.

Example 2.4 - Interference fit.

Suppose that the stacks in Example 2.3 have to fit into brackets made with a specification of 40.25 ± 0.45 mm as shown in Figure 2.6. If the sub-assembly stacks and brackets are selected at random, what is the probability of the assembly not fitting together?

Solution: The assembly will not fit together if the stack is thicker than the bracket. The difference between the stack thickness and bracket opening is called *clearance* when it is positive and it is termed *interference* when this difference is negative. The solution is based on finding the probability of interference.

To solve this problem using Eq (2.9), the first step is to define a model for the difference between the bracket and stack sub-assembly dimensions. If b is a random variable to denote the bracket opening, then from Eq (2.6), the standard deviation of the bracket is $\sigma_b = 0.45/3 = 0.15$ mm, so the short hand notation for b is

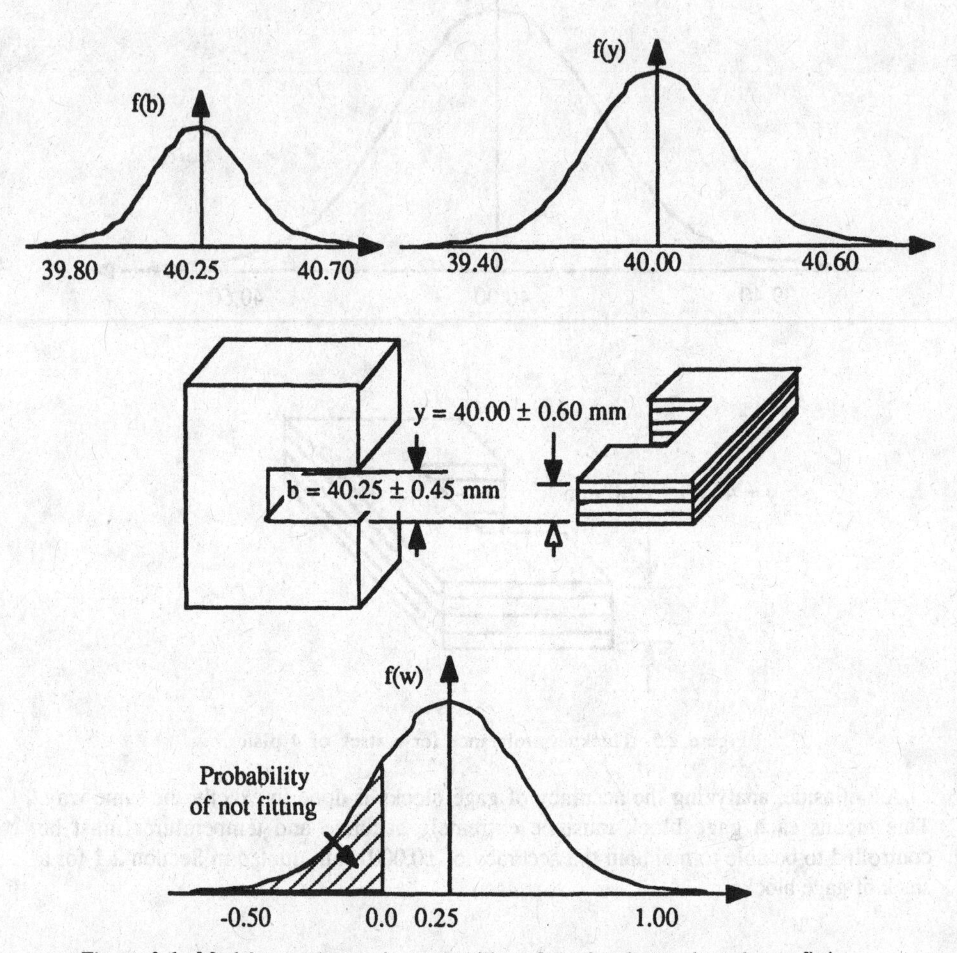

Figure 2.6 Model to estimate the probability of the bracket and stack not fitting.

$$b \sim N\,[40.25,(0.15)^2].$$

The model for the interference or clearance can be defined as the difference between the random variable for the bracket opening and the random variable for the stack thickness. This new variable w, can be written

$$w = b - y.$$

This is the same form as Eq (2.9), where $a_1 = 1$ and $a_2 = -1$, so that:

$$\mu_w = \mu_b - \mu_y = 40.25 - 40.00$$
$$= 0.25 \text{ mm},$$
$$\sigma_w^2 = \sigma_b^2 + \sigma_y^2 = (0.20)^2 + (0.15)^2$$

$$= (0.25)^2, \text{ or}$$
$$w \sim N[0.25,(0.25)^2].$$

To find the probability that this assembly will not fit is equivalent to finding the probability that the difference is negative, i.e., that there is interference. This is given by

$$P\,[\text{interference}] = P[w < 0]$$

and corresponds to the shaded area in Figure 2.6. This can be computed using Eq (2.8) to transform w to z so that Table 2.1 applies.

$$P\,[w < 0] = P[z = (\frac{w - 0.25}{0.25}) < -1]$$
$$= 1 - P[z \leq 1] = 1 - 0.841$$
$$= 0.159$$

or over 15% of the assemblies would not fit together with the specified tolerances. This is an unacceptably large probability and points out why a product designer tends to favor small tolerances to avoid this problem. At this point, the best way to address the problem is to decide which processes, those for making the plates or the bracket, could best be used to decrease the probability of the assemblies not fitting.

2.3 Quality Control

Quality is often measured in terms of how well a part, sub-assembly or final product adheres to design tolerances and specifications. These specifications become constraints in process planning. Quality control methods have two important functions. Initially these methods document the inherent variation in a process, sometimes referred to as the *process capability*, to be sure it is less than the design tolerance specifications. Once it has been determined that it is technically possible to meet tolerance specifications, quality control is a way to monitor and improve the performance of a process [DeV*etal*91] [Duncan74].

2.3.1 Quality Control Methods

Today few production machining or grinding processes have direct feedback control of the two most often specified attributes of a part: dimensions and surface finish. As a result, samples of the product are taken, measured, and based on this data a decision is reached as to whether or not the product meets quality standards. If agreed upon standards are not met, the real engineering work begins; trying to find the source of the problem and correcting it. It should be evident that quality control is a manual version of feedback control with sampling replacing sensors, humans replacing both the control law and the electro-mechanical actuators, and the quality control technique providing the filtered control signal.

Because quality control involves decision making based on samples, it relies on the theory of probability and the application of statistics: probability for decision making models and statistics to analyze data used with these models. To catch defects as quickly as possible, simple methods of analysis have been devised that use simple calculations, charts and graphs to provide manual feedback as quickly as possible.

While a complete coverage of quality control methods is not possible in this book, the issues involved in designing a quality control plan can be summarized:

- Deciding the nature of what has to be controlled such as is it a "go" or "no go" type *attribute* or is it a variable that has *continuous* values.
- Selecting and agreeing upon the method for implementing the control, considering the chances of passing bad parts (*consumer's risk*) and the chances of rejecting good parts (*producer's risk*).
- A sampling plan must be devised that considers the size and frequency of sampling. Design of the sampling plan should include specifying the metrology system.
- Finally, the methods for correcting and improving the process have to be outlined, which is often associated with looking for what are termed *assignable causes*. For machining and grinding processes, some of the ways to correct or improve the process are based on the principles and technology covered in subsequent chapters.

2.3.2 $\bar{X}$ and R Charting as a Representative Quality Control Method

Control charting is supposed to make monitoring a process simple, ideally so that a machine operator can control and improve the process. While the theory behind the chart may be involved, the calculations to use the chart should be simple to perform.

The $\bar{X}$ and R charts were developed for this purpose as a way of making sure that the nominal values and tolerances in Section 2.2 are met during manufacture. The $\bar{X}$ or *average chart* is used to track nominal values and the R or *range chart* tracks the tolerance. These charts are for continuous, rather than "go" or "no go" type variables. Both charts have a target value and upper and lower control limits (UCL and LCL) that are used to indicate when the process has changed, initiating a search for assignable causes. When determining these targets and limits, the following assumptions are made:

- The process is in *statistical equilibrium*. Good parts are being produced. This is a key part of the process because this is where the initial process capability is determined.
- Because $\bar{X}$ is the average of several samples, the Central Limit Theorem assures that $\bar{X}$ is approximately Gaussian, even if the individual measurements used to compute the average are not.
- The range of the sample measurements R, is used to estimate the standard deviation of the measurements, simplifying calculations. A branch of statistics called *order statistics* provides the theoretical basis.

Other factors that must be determined are n, how many samples should be used to compute $\bar{X}$ and R, and how often to sample. These are economic decisions because:

- As n increases, the process can be monitored better. But the cost of a measurement increases too, because of the need for direct labor or extensive metrology equipment.

Table 2.2 Coefficients for 3-σ $\overline{X}$ and R Charts

Number of Observations in a Sample	Factors for Control Limits		
n	A_2	D_3	D_4
2	1.880	0	3.267
3	1.023	0	2.575
4	0.729	0	2.282
5	0.577	0	2.115
6	0.483	0	2.004
7	0.419	0.076	1.924
8	0.373	0.136	1.864
9	0.337	0.184	1.816
10	0.308	0.223	1.777
11	0.285	0.256	1.744
12	0.266	0.284	1.716
13	0.249	0.308	1.692
14	0.235	0.329	1.671
15	0.223	0.348	1.652
16	0.212	0.364	1.636
17	0.203	0.379	1.621
18	0.194	0.392	1.608
19	0.688	0.404	1.596
20	0.180	0.414	1.586
21	0.173	0.425	1.575
22	0.167	0.434	1.566
23	0.162	0.443	1.557
24	0.157	0.452	1.548
25	0.153	0.459	1.541

- **Determining the frequency of sampling (once per hour, once per shift or once per day) will be a trade off between the time spent making measurements and how quickly a problem can be detected and corrected.**

Calculations of $\overline{X}$ and R are simple. For a sample of size n, the following calculations have to be made.

$$\overline{X} = \frac{1}{n}\sum_{i=1}^{n} x_i = \text{the average or x-bar} \qquad (2.10a)$$

$$R = \max(x_1, x_2, .. x_n) - \min(x_1, x_2, .. x_n) = \text{the range} \qquad (2.11a)$$

When setting up the chart, these calculations should be made and plotted to make sure that the process is in equilibrium. Then it is possible to determine the limits on the control chart.

$\bar{\bar{X}}$ target value for the averages. Most of the time this is the nominal value that is part of the dimensioning and tolerancing for a design. Less frequently it is the average of the previous $\bar{X}$'s if the process in is equilibrium.

$\bar{R}$ target value for the range. It is the average of the previous R's taken with the process in equilibrium.

Once the target values are established, control limits can be added. The control limit will depend on constants like those given in Table 2.2 for constructing 3-σ UCL and LCL limits for both $\bar{X}$ and R charts. Then the control limits for tracking the nominal size are

$$LCL_{\bar{X}} = \bar{\bar{X}} - A_2 \bar{R} \text{ and} \tag{2.10b}$$

$$UCL_{\bar{X}} = \bar{\bar{X}} + A_2 \bar{R}. \tag{2.10c}$$

Similarly, the control limits for the R chart are

$$LCL_R = \bar{R} D_3 \text{ and} \tag{2.11b}$$

$$UCL_R = \bar{R} D_4. \tag{2.11c}$$

Remember that the constants in Table 2.2 are for a particular sample size and include the correction that is needed to use $\bar{R}$ to estimate the standard deviation for $\bar{\bar{X}}$.

Using the $\bar{X}$ and R charts is aimed at determining changes in either the nominal dimension indicated by $\bar{X}$, or the tolerance indicated by R. If changes are detected, something should be done to exploit the "good" changes or correct "bad" changes in the process. Figure 2.7 is for charts where the first 10 samples of size 4 were used to set up the chart because the process appears to be in statistical equilibrium.

After these first ten samples are illustrations of conditions that could require action. Cases in Figure 2.7 that warrant corrective action and their possible causes are listed below.

- In the period 12 - 13 both $\bar{X}$ and R are within the control limits, but $\bar{X}$ is beginning a trend above $\bar{\bar{X}}$. This is termed a *run* and should be watched since it may indicate a shift in the mean or nominal value. If these measurements are turned shaft diameters, the run could be due to the nose wear on the cutting edge, causing the diameters to increase.

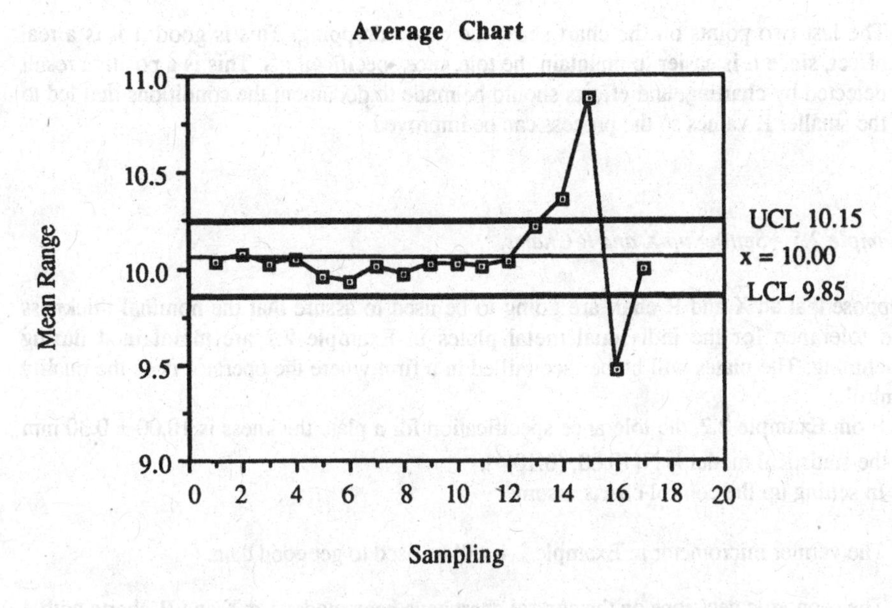

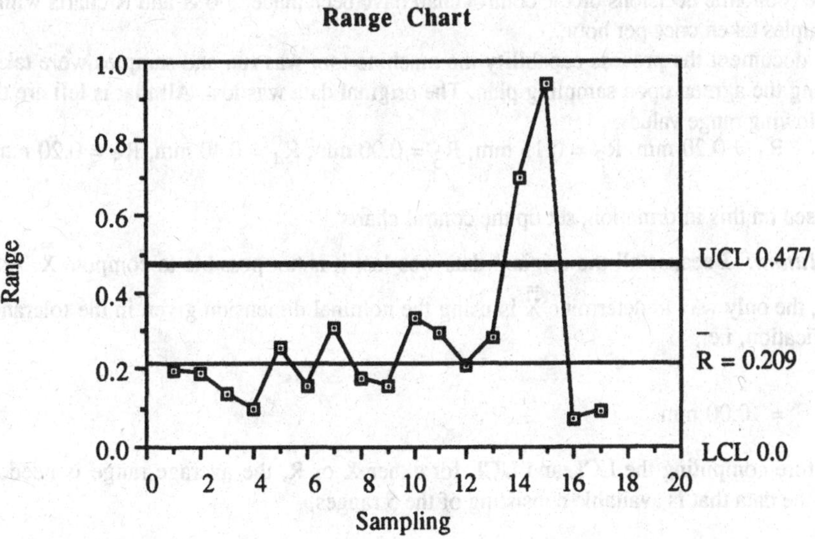

Figure 2.7 Examples of $\bar{X}$ and R charts where the first 10 samplings were used to establish the process capability and the changes that the chart detects.

- **While $\bar{X}$ is out of the control limits at points 14 and 15, it is more significant that R is outside the limits for 2 samples. If the process is under control, it is highly unlikely that both $\bar{X}$ and R would be out of their control limits at the same time. Possible causes are a new and inexperienced operator, or some changes in the process that make it difficult to control.**

- The last two points on the chart show the range dropping. This is good if it is a real effect, since it is easier to maintain the tolerance specifications. This is a positive result detected by charting, and efforts should be made to document the conditions that led to the smaller R values so the process can be improved.

Example 2.5 - Setting up $\bar{X}$ and R Charts.

Suppose that an $\bar{X}$ and R chart are going to be used to assure that the nominal thickness and tolerance for the individual metal plates in Example 2.2 are maintained during machining. The plates will be be face milled in a firm where the operator does the quality control.

From Example 2.2, the tolerance specification for a plate thickness is 10.00 ± 0.30 mm or the statistical model is $N[10.00, (0.10)^2]$.

In setting up the control charts assume:

- The vernier micrometer in Example 2.1 will be used to get good data.
- The economic decisions on the control chart have been made: 3-σ $\bar{X}$ and R charts with 4 samples taken once per hour.
- To document the process capability the machine tool was run and samples were taken using the agreed upon sampling plan. The original data was lost. All that is left are the following range values:

  $R_1 = 0.20$ mm, $R_2 = 0.15$ mm, $R_3 = 0.00$ mm, $R_4 = 0.40$ mm, $R_5 = 0.20$ mm.

Based on this information, set up the control charts.

Solution: Because all the original data was lost it is not possible to compute $\bar{\bar{X}}$. As a result, the only way to determine $\bar{\bar{X}}$ is using the nominal dimension given in the tolerance specification, i.e.,

$$\bar{\bar{X}} = 10.00 \text{ mm.}$$

Before computing the LCL and UCL for either $\bar{X}$ or R, the average range is needed. From the data that is available consisting of the 5 ranges,

$$\bar{R} = \frac{[0.20 + 0.15 + 0.00 + 0.40 + 0.20]}{5} = 0.19 \text{ mm}$$

The control limits are constructed for a sample size of n = 4, from Table 2.2 $A_2 = 0.729$ so

$$UCL_{\bar{X}} = 5.00 \text{ mm} + 0.729\ (0.19 \text{ mm}) = 5.14 \text{ mm and}$$

$$LCL_{\bar{X}} = 5.00 \text{ mm} - 0.729\ (0.19 \text{ mm}) = 4.86 \text{ mm.}$$

For the upper and lower control limits on R, $D_3 = 0$ and $D_4 = 2.282$ so

$UCL_R = 2.282\ (0.19\ \text{mm}) = 0.43\ \text{mm}$ and
$LCL_R = 0.0\ (0.19\ \text{mm}) = 0.00\ \text{mm}$.

A side issue is the question of whether this chart, which measures the performance of the process, indicates that the process is capable of meeting the plate tolerance specification. The short answer is yes. An $\bar{R} = 0.19$ mm for $n = 4$ is consistent with a $\sigma_x = 0.10$ mm. This is because it can be shown that $\bar{R}/4 < \sigma < \bar{R}/3$ is approximately true [Duncan74]. The empirical evidence from setting up the chart is that the process can actually do better than the tolerance specification.

2.4 Experimental Design Methods for Empirical Data

Experimental design is the planning done before running experiments to simplify the analysis that is done after running the experiments. The objective is always to carry out experiments: to provide the data needed to estimate parameters for a known model, to quickly screen variables that are not significant from subsequent experiments, or to develop a new empirical model. Running an experiment is expensive and time consuming, so running experiments without planning them is wasteful.

There are a number of books on experimental design listed at the end of this chapter, for example [Box*etal*78], [Phadke89] or [Taguch87]. This section will be a short introduction to the topic with two examples used to cover the basics of experimental design. The examples will illustrate common instances where experimental design techniques are used: when experiments are needed to find the parameters of a known model, and when experiments are needed to screen several variables in the process of building an empirical model.

Example 2.6 - Parameter Estimation When A Model is Known

Many times data is not available for new materials that have to be processed. Then experiments have to be run to gather data for one of the empirical models needed to analyze the process. For example, a model used in later chapters relates the average shear strength τ_s, needed to cause metal flow to the average thickness of a chip h_{avg}. Many times this is modeled by a power law equation of the form

$$\tau_s = \tau_0 \left(\frac{h_{avg}}{h_R} \right)^a . \tag{2.E1}$$

τ_0 and a are the parameters of this empirical model that have to be determined from experiments in which only the average chip thickness is changed and "all other things are kept constant". Note that in this empirical model, like several that will appear later, a specified parameter h_R, has been introduce to "keep the units clean" by assuring that the quantity raised to the power a is a pure number. Another interpretation of h_R is that it is the value of h_{avg} that results in a shear stress of τ_0. In this case where the chip thickness is measured in millimeters, $h_R \equiv 1$ mm.

While the power law model is not linear when its used, by taking a log *transformation* of Eq (2.E1), the model becomes

$$\log\left(\frac{\tau_s}{\tau_0}\right) = \log(\tau) - \log(\tau_0) = a \log\left(\frac{h_{avg}}{h_R}\right) \text{ or } \log(\tau_s) = \log(\tau_0) + a \log\left(\frac{h_{avg}}{h_R}\right) \quad (2.E2)$$

After transformation this is equivalent to a linear model, and if the *error* associated with each experimental ε_i is included, the model form is

$$Y_i = Y_0 + a X_i + \varepsilon_i. \quad (2.12a)$$

The *linear least squares method* is aimed at finding estimates of $\hat{Y}_0$ and $\hat{a}$ that minimize

$$\sum_{i=1}^{n} \varepsilon_i^2 = \sum_{i=1}^{n} (Y_i - \hat{Y}_0 - \hat{a} X_i)^2 \quad (2.13)$$

(If each ε_i is *independent:* - the errors are not systematic - and has a Gaussian or Normal distribution, then the least squares method is also what is termed a *maximum likelihood* method). The solution, is well known for the two parameters shown and is often implemented on programmable calculators. A more general solution can be written using matrix notation, if the Y_i's on the left side, and the variables associated with the parameters and the errors on the right side of Eq (2.12a), are written as

$$\mathbf{Y} = \begin{bmatrix} Y_1 \\ Y_2 \\ . \\ . \\ Y_n \end{bmatrix}, \mathbf{X} = \begin{bmatrix} 1 & X_1 \\ 1 & X_2 \\ . & . \\ . & . \\ 1 & X_n \end{bmatrix} = [\mathbf{X_0} \mid \mathbf{X_1}] \text{ and } \mathbf{E} = \begin{bmatrix} \varepsilon_1 \\ \varepsilon_2 \\ . \\ . \\ \varepsilon_n \end{bmatrix}. \quad (2.14a)$$

Note that boldface variables are vectors or matrices, and boldface variables like $\mathbf{X_0}$ with a subscript represent a column vector that is adjoined with $\mathbf{X_1}$ to form the matrix $\mathbf{X}$. The solution is found by differentiating the squared error Eq (2.13) with respect to the model parameters. Setting the resulting linear equations equal to zero leads to the solution

$$\hat{\beta} = \begin{bmatrix} \hat{Y}_0 \\ \hat{a} \end{bmatrix} = (\mathbf{X}^T\mathbf{X})^{-1}(\mathbf{X}^T\mathbf{Y}). \quad (2.15a)$$

In addition to this solution, a valid question is how good are the answers or what *confidence intervals* can be used to bracket the true underlying values. If it is valid to assume that the errors are independent and, to a lesser extent, that they are Gaussian, the *sample variance* s^2, can be calculated. This is done by using estimated values as if they were the "true" values

$$s^2 = \frac{\sum_{i=1}^{n} (Y_i - \hat{Y}_0 - \hat{a} X_i)^2}{n-2}. \tag{2.16a}$$

Equation (2.16a) is an *estimate of the variance of the error when measuring* Y_i. (The theoretical value can be thought of as σ_m^2 in Section 2.1.3.) The *variance-covariance matrix for the estimated parameters* is

$$\mathrm{Var}[\hat{\beta}] = \begin{bmatrix} \mathrm{Var}[\hat{Y}_0] & \mathrm{Cov}[\hat{Y}_0 \, \hat{a}] \\ \mathrm{Cov}[\hat{a} \, \hat{Y}_0] & \mathrm{Var}[\hat{a}] \end{bmatrix} = (X^T X)^{-1} s^2. \tag{2.16b}$$

Elements of the $\mathrm{Var}[\hat{\beta}]$ matrix can be used to put confidence limits on the estimated parameters.

While no calculations have been made yet and Eqs (2.15 - 16) are often covered under the heading of regression or curve fitting, some of the main ideas common to all experimental design have in fact been covered:

- There is an underlying model, Eq(2.E1) in the form of a power law. This means some assumptions have already been made on what are important variables; in this case only h_{avg}.
- A transformation, the logarithmic one in Eq (2.E2), was needed to look at the problem and get it into the form of Eq (2.12). Often with factorial designs, linear transformations of continuous variables or simple coding transformations for qualitative variables are used.
- The term independent was use to describe the ε_i's. As a practical definition, independence means that there are no systematic errors in the experiments. To assure this, the experiments should be run with the same batch of material by the same person, and the order of experiments should be randomized to block out any time effect. (Anyone that has done experiments knows that often these precautions are difficult to implement.)
- The "design" part of experimental design is choosing the **X** matrix in Eq (2.14) so it provides data for the underlying model that is useful. A minimum experimental design requirement is that $(X^T X)$ in Eqs (2.15-16) is not singular; ideally it is diagonal. Assuring this is determined by both the number of experiments ($n \geq 2$ for this model) and the levels of the experimental variables. If inference will be made about how good the estimated parameters are, the n-2 term in Eq (2.16a) dictates the need for more than two experiments.

Now for the numerical example that uses these techniques to find $\hat{Y}_0$ and $\hat{a}$. The form of the model is known, so designing the experiment is a case of selecting the number of levels consistent with the model - a minimum of two are required for this model which is linear after the log transformation. Also, to check on the goodness of the estimated parameters, extra experiments are needed. A replicated experiment will be run so that confidence intervals can be constructed for the two estimated parameters.

The two levels are selected to cover a region of average chip thicknesses where the power law relationship between chip thickness and shear strength is expected to apply. *This practical knowledge is what the experimenter must bring to experimental design.* In this case the two levels selected are h_{avg} = 0.075 and 0.200 mm. Table 2.3 shows these values, the measured strength τ_s, and the transformed values that make up the **X** matrix and the **Y** vector. This table also illustrates the notation used subsequently: lower case x's and y's are used to denote the experimental data before transformation, while the upper case values apply after transformation.

Table 2.3 Experimental Data for Example 2.6 (h_R = 1 mm)

x h_{avg}(mm)	y τ_s(MPa)	X_0	X_1 $\log(h_{avg}/h_R)$	Y $\log(\tau_s)$	E
0.075	612	1	-1.125	2.787	-0.013081
0.200	517	1	-0.699	2.713	0.014772
0.075	650	1	-1.125	2.813	0.013081
0.200	483	1	-0.699	2.684	-0.014772

The matrices needed to estimate the transformed parameters are calculated from the data in Table 2.3.

$$(\mathbf{X}^T\mathbf{X}) = \begin{bmatrix} 4.000 & -3.648 \\ -3.648 & 3.508 \end{bmatrix} \text{ and } \mathbf{X}^T\mathbf{Y} = \begin{bmatrix} 10.997 \\ -10.072 \end{bmatrix}.$$

Solving for the transformed parameters by Eq (2.15a) and the sample standard deviation from Eq (2.16a) gives:

$$\begin{bmatrix} \hat{Y}_0 \\ \hat{a} \end{bmatrix} = \begin{bmatrix} 4.833 & 5.026 \\ 5.026 & 5.511 \end{bmatrix} \begin{bmatrix} 10.997 \\ -10.072 \end{bmatrix} = \begin{bmatrix} 2.533 \\ -0.237 \end{bmatrix} \text{ and}$$

$$s^2 = \frac{\sum_{i=1}^{4} (Y_i - (2.533 - 0.237\, X_i))^2}{2} = (0.019731)^2$$

From Eq (2.16.b), the variance covariance matrix is

$$\begin{bmatrix} \mathrm{Var}[\hat{Y}_0] & \mathrm{Cov}[\hat{Y}_0 \hat{a}] \\ \mathrm{Cov}[\hat{a}\, \hat{Y}_0] & \mathrm{Var}[\hat{a}] \end{bmatrix} = \begin{bmatrix} 0.0019 & 0.0020 \\ 0.0020 & 0.0021 \end{bmatrix}.$$

With these estimated values, the lower and upper confidence intervals for the estimated parameters can be constructed using

Table 2.4 The Student t-Distribution

$$P[z \le z_0] = \int_{-\infty}^{z_0} \frac{\Gamma[(r+1)/2]}{\sqrt{\pi r}\,\Gamma[r/2]\,(1+w^2/r)^{(r+1)/2}}\,dw \quad \text{and} \quad P[z \le -z_0] = 1 - P[z \le z_0]$$

Degrees of Freedom r	$P[z \le z_0]$				
	0.900	0.950	0.975	0.990	0.995
1	3.078	6.314	12.706	31.821	63.657
2	1.886	2.920	4.303	6.965	9.925
3	1.638	2.353	3.182	4.541	5.841
4	1.533	2.132	2.776	3.747	4.604
5	1.476	2.015	2.571	3.365	4.032
6	1.440	1.943	2.447	3.143	3.707
7	1.415	1.895	2.365	2.998	3.499
8	1.397	1.860	2.306	2.896	3.355
9	1.383	1.833	2.262	2.821	3.250
10	1.372	1.812	2.228	2.764	3.169
11	1.363	1.796	2.201	2.718	3.106
12	1.356	1.782	2.179	2.681	3.055
13	1.350	1.771	2.160	2.650	3.012
14	1.345	1.761	2.145	2.624	2.977
15	1.341	1.753	2.131	2.602	2.947
16	1.337	1.746	2.120	2.583	2.921
17	1.333	1.740	2.110	2.567	2.898
18	1.330	1.734	2.101	2.442	2.878
19	1.328	1.729	2.093	2.539	2.861
20	1.325	1.725	2.086	2.528	2.845
21	1.323	1.721	2.080	2.518	2.831
22	1.321	1.717	2.074	2.508	2.819
23	1.319	1.714	2.069	2.500	2.807
24	1.318	1.711	2.064	2.492	2.797
25	1.316	1.708	2.060	2.485	2.787
26	1.315	1.706	2.056	2.479	2.779
27	1.314	1.703	2.052	2.473	2.771
28	1.313	1.701	2.048	2.467	2.763
29	1.311	1.699	2.045	2.462	2.756
30	1.310	1.697	2.042	2.457	2.750

$$\hat{\beta} \quad \pm [t_\alpha(\text{d.f}) \text{ or } z_\alpha] \times \sqrt{\text{Var}[\hat{\beta}]}. \qquad (2.16c)$$

With this relatively small sample size it is unreasonable to use the Normal distribution. Instead, the Student's t-distribution in Table 2.4 should be used. (As an aside, the question often comes up, when is n large enough so that the Normal distribution table can be used instead of the Student's t-Table. A pragmatic answer is when the degrees of freedom reach 30, since most t-Tables end at that point.)

For a 95% confidence interval with 2 degrees of freedom, the denominator used to compute the sample standard deviation, the t-value required is $t_2(1-\alpha/2=0.975) = 4.303$. The diagonal elements of the variance covariance matrix are used to compute the confidence intervals

$$\hat{Y}_0 \pm 4.303 \sqrt{\mathrm{Var}[\hat{Y}_0]} = [2.346, 2.719] \text{ and } \hat{a} \pm 4.303 \sqrt{\mathrm{Var}[\hat{a}]} = [-0.437, -0.038].$$

Because the confidence interval for $\hat{a}$ does not include zero, it is significant, but barely. Since $\hat{a}$ is negative, τ_s tends to increase as the average chip thickness decreases, and Chapter 3 discusses mechanisms that explain this empirical result. Applying the inverse transformations

$$\hat{\tau}_0 = 10^{2.533} = 341 \text{ (MPa) and } \hat{a} = -0.237,$$

so that the empirical model relating average chip thickness to shear strength in machining is

$$\tau = 341 \text{ (MPa)} \left(\frac{h_{avg}}{1 \text{ mm}}\right)^{-0.237}$$

Example 2.7 - Two Level Factorial Design for Screening Experiments

In Chapter 4 the empirical tool life equation proposed by F.W. Taylor will be covered in detail. The most common form of this equation represents tool life T, as a function of only the cutting speed v,

$$\left(\frac{T}{T_R}\right) = \left(\frac{v}{v_R}\right)^{-1/n}.$$

The generalized tool life equation includes two other cutting conditions: feed a and depth d. This model is:

$$\left(\frac{T}{T_R}\right) = \frac{K}{\left(\frac{v}{v_R}\right)^{1/n} \left(\frac{a}{a_R}\right)^{1/n_1} \left(\frac{d}{d_R}\right)^{1/n_2}}.$$

The reason for favoring the simplified form is that the effects of feed and depth have been found to be minor. However, with an empirical equation, the only way to arrive at such a conclusion is by an experiment; a *screening experiment* to determine which variables have little effect on a response like tool life, so that they can be neglected in subsequent analysis.

A full 2^3 *factorial design* will be used for the screening experiments in this example, but first a word about this notation. The 2 indicates that 2 levels will be used, the exponent 3 indicates that 3 variables will be considered (v, *a* and d), and the fact that $2^3 = 8$ means that this will be done in a total of 8 experiments. The statistics that can be estimated from these 8 experiments are: $\binom{3}{0}$ average+$\binom{3}{1}$ main effects+$\binom{3}{2}$ two factor interactions $\binom{3}{3}$ three factor interaction = 8. (The terms in parentheses are the binomial coefficients.) This means that the underlying *model* for this experimental design is:

$$Y = \beta_0 X_0 + \beta_1 X_1 + \beta_2 X_2 + \beta_3 X_3 + \beta_{12} X_1X_2 + \beta_{13} X_1X_3 + \beta_{23} X_2X_3$$

$$+ \beta_{123} X_1X_2X_3 + \varepsilon \qquad (2.12b)$$

where the X_i's are *coded* or *transformed* variables. Usually the dependent variables x_i, are coded to two levels, -1 and +1. For quantitative variables, this can be done by using experience or knowledge of the process to select high and low values that cover the range that should be explored by the experiment. The coding transformation is

$$X_i = \frac{x_i-(x_{i,high}+x_{i,low})/2}{(x_{i,high}-x_{i,low})/2}, \qquad (2.17)$$

which translates the data based on the average, and normalizes it based on the average difference. Referring back to Eq (2.12b), the β's reflect the importance of each term. Because of the coding, if the magnitude of a coefficient is large the effect is important, and when a coefficient is small the effect or interaction probably can be neglected.

The experimental design matrix for a 2^3 factorial design is usually written as shown in Table 2.5. This table also shows the high and low levels selected based on knowledge of machining technology as representative of conditions where the tool life model would be needed. The transformations used to code and scale the feed, speed and depth are from Eq (2.17). The pattern for generating the rows and columns should be obvious, with the order shown termed the "Yates order," after an algorithm that was designed to simplify hand calculations. (Another way of looking at the design is if $-1 \rightarrow 0$ and the matrix is rotated about the left edge, the coded variables simply are counting from 0 to 2^3-1 in binary.) The important thing about using this design for running experiments is that it is an *orthogonal design*, i.e., $X_i \bullet X_j = 0$ for $i \neq j$ and $X_i \bullet X_i = 2^3$ for i = 1, 2 and 3. Orthogonality allows several variables to be changed at once, while still being able to determine the effects of a single variable.

One reason that a two level factorial designed experiment is useful in screening applications can be seen by looking at the tool life values in the measured response column. Tool life, as will be explained in Chapter 4, is simply the cumulative cutting time before an edge wears out, so the "better" the tool life the longer the experiment. In Table 2.5 the range of tool lives is from 1080 s to 10346 s, with the total experiment taking nearly 12 hours of machining time. In a case like this where the experiments are long or tedious to run, the maximum amount of information should be extracted from the experimental data. Designed experiments assure this.

The analysis of this designed experiment is given in Table 2.6, where 5 columns have been added to the design matrix in Table 2.5 . The first column is simply all +1's like the $\mathbf{X_0}$ vector in Example 2.6. The elements in the other 4 columns are generated by multiplication of the elements of the vectors $\mathbf{X_1}$, $\mathbf{X_2}$ and $\mathbf{X_3}$ that generate or form the basis for this design. Once all the vectors, each of which is unique for this full factorial design, are generated, the effects listed in the last row of Table 2.6 can be calculated. For example, the effect of v, the speed is

$$(\mathbf{X_1} \bullet \mathbf{Y})/8 = -3765$$

Table 2.5 2^3 Factorial Design and Transformation Equations Data from [Wu64a] and [Wu64b]

$\mathbf{X_1}$	$\mathbf{X_2}$	$\mathbf{X_3}$	$\mathbf{Y}$
v_{low} = 1.68 m/s v_{high} = 3.36 m/s	a_{low} = 0.25 mm/rev a_{high} = 0.56 mm/rev	d_{low} = 1.24 mm d_{high} = 2.56 mm	Tool Life T (s/edge)
$\dfrac{v-(v_{high}+v_{low})/2}{(v_{high}-v_{low})/2}$	$\dfrac{a-(a_{high}+a_{low})/2}{(a_{high}-a_{low})/2}$	$\dfrac{d-(d_{high}+d_{low})/2}{(d_{high}-d_{low})/2}$	
-1	-1	-1	9646
+1	-1	-1	2218
-1	+1	-1	9940
+1	+1	-1	1627
-1	-1	+1	10346
+1	-1	+1	2110
-1	+1	+1	7218
+1	+1	+1	1080

which means that a unit increase of the coded variable (about 1 m/s for v) would cause the tool life to decrease by about one hour (-3765 s). Interactions are handled the same way, i.e., the interaction between feed and depth is

$$(\mathbf{X_2 X_3} \bullet \mathbf{Y})/8 = -483.$$

This negative interaction indicates that either increasing *or decreasing* f and d simultaneously can decrease the tool life, but this decrease is only about 12% of the

Table 2.6 Analysis of the Experimental Results in Table 2.5

	Original Design Vectors			Vectors Generated As Element By Element Products				
X_0 Average	X_1 v Effect	X_2 *a* Effect	X_3 d Effect	X_1X_2 v *a*	X_1X_3 v d	X_2X_3 *a*d	$X_1X_2X_3$ v *a* d	Y T s/edge
+1	-1	-1	-1	+1	+1	+1	-1	9646
+1	+1	-1	-1	-1	-1	+1	+1	2218
+1	-1	+1	-1	-1	+1	-1	+1	9940
+1	+1	+1	-1	+1	-1	-1	-1	1627
+1	-1	-1	+1	+1	-1	-1	+1	10346
+1	+1	-1	+1	-1	+1	-1	-1	2110
+1	-1	+1	+1	-1	-1	+1	-1	7218
+1	+1	+1	+1	+1	+1	+1	+1	1080
5388	-3765	-557	-335	152	171	-483	373	

amount expected with a unit increase in speed. Examining the other effects and interactions, clearly the cutting speed v, has the biggest effect on tool life. This is followed, in order, by f and the two factor interaction between f and d, however these are all less than 15% of the effect of v. This analysis of the 2^3 factorial designed experimental results supports the conclusion that the simplified Taylor tool life expression is adequate for most situations.

Another detail relates to the underlying model and orthogonal properties of this design. Applying the least squares method to estimate the parameters in Eq (2.12b) is easy because of the design and the coded data. Referring to Eq (2.14a) and (2.15a), because

$$X = [X_0 \mid X_1 \mid X_2 \mid X_3 \mid X_1X_2 \mid X_1X_3 \mid X_2X_3 \mid X_1X_2X_3] \text{ then} \tag{2.14b}$$

$$X^TX = 2^3\, I \text{ and } (X^TX)^{-1} = 2^{-3}\, I$$

$$\hat{\beta} = \begin{bmatrix} \hat{\beta}_0 \\ \hat{\beta}_1 \\ \cdot \\ \cdot \\ \hat{\beta}_{123} \end{bmatrix} = \frac{1}{2^3}(X^TY) = \frac{1}{2^3}\begin{bmatrix} X_0 \bullet Y \\ X_1 \bullet Y \\ \cdot \\ \cdot \\ X_1X_2X_3 \bullet Y \end{bmatrix} \tag{2.15b}$$

This is an application of factorial design to handle the problem of screening. Taking time to design a set of experiments maximizes the information and minimizes the number of experiments to accomplish the task. However, this design has limits. What they are and how they can be overcome are listed below.

- There is not enough data to put confidence intervals on anything because every test condition was used to estimate something; the design was *saturated* with all the degrees of freedom used up. To estimate confidence intervals several *center point experiments* that correspond to $\{X_1, X_2, X_3\} = \{0, 0, 0\}$, can be used to compute the sample standard deviation.

- For 3 or 4 variables, running 8 or 16 experiments may not be a burden for screening, but if the number of variables gets much greater it is often better to use other experimental designs. An alternative is to use *two level fractional factorial designs* that are based on the full factorial designs in this example.
- Other experimental designs have to be used if a model includes terms like X_1^2. Two levels are not enough, and one approach is to use a *star design*.

2.5 References

[AllStu84] Allocca, John A., and Allen Stuart, (1984), *Transducers: Theory and Applications*, Reston Publishing Co., a Prentice-Hall Company, Reston, VA.

[ASTME67] American Society of Tool and Manufacturing Engineers, (1967), *Handbook of Industrial Metrology*, Prentice Hall, Englewood Cliffs, NJ.

[BS113472] British National Standard BS1134, (1972), "Assessment of Surface Texture," British Standards Institute, London.

[Bjørke89] Bjørke, Øyvund, (1989), *Computer Aided Tolerancing, Second Edition*, ASME Press, New York.

[Box*etal*78] Box, George E.P., William G. Hunter, J. Stuart Hunter, (1978), *Statistics for Experimenters: An Introduction to Design, Data Analysis and Model Building*, Wiley, NY.

[DeV*etal*91] DeVor, R.E., T.H. Chang, J.W. Sutherland, (1991) *Statistical Quality Design and Control: Contemporary Concepts and Methods*, MacMillan, New York, N.Y.

[Doebel83] Doebelin, Ernest O., (1983), *Measurement Systems: Applications and Design*,McGraw Hill Book Co., New York.

[Duncan74] Duncan, Acheson J., (1974), *Quality Control and Industrial Statistics, Fourth Edition*, Richard D. Erwin, Inc, Homewood, IL.

[GeoMet82] Geo-Metrics II, (1982), "Dimensioning and Tolerancing," *ANSI/ASME Standard*, Y14.5M, pp. 35-52.

[Pet*etal*79] Peters, J., P. Vanherck and M. Sastrodinoto, (1979), "Assessment of Surface Topology Analysis Techniques," *CIRP Annals, v28/2*.

[Phadke89] Phadke, Madhav Shridhar, (1989), *Quality Engineering Using Robust Design*, Prentice Hall, Englewood Cliffs, NJ.

[Spotts83] Spotts, M.F., (1983), *Dimensioning and Tolerancing for Quality Production*, Prentice-Hall, Inc., Englewood Cliffs, NJ.

[SurTex78] *Surface Texture, Surface Roughness, Waviness and Lay*, (1978), American National Standard Institute B46.1-1978., Published by ASME, United Engineering Center, 345 East 47th Street, New York, NY 10017.

[Taguch86] Taguchi, Gen'ichi, (1986), *Introduction to Quality Engineering: Designing Quality Into Products and Processes,* English Translation by Asian Productivity Organization, Tokyo.

[Taguch87] Taguchi, Gen'ichi, (1987), *System of Experimental Design: Engineering Methods to Optimize Quality and Minimize Costs,* Don Clausing, Technical Editor, (English translation by Louise Watanabe Tung), UNIPUB/Kraus International Publications, Dearborn MI and American Supplier Institute.

[Wu64a] Wu, S.M., (1964a), "Tool-Life Testing by Response Surface Methodology - Part I," *ASME Transactions, Journal of Engineering for Industry,* Series B, Vol. 86, pp. 105-110.

[Wu64b] Wu, S.M., (1964b), "Tool-Life Testing by Response Surface Methodology - Part II," *ASME Transactions, Journal of Engineering for Industry,* Series B, Vol. 86, pp. 110-116.

2.6 Problems

2.1. One of the trade offs in designing quality control charts is selecting the number of samples n. As was explained in Section 2.3, when the coefficients in Table 2.2 are used to set up 3-σ $\overline{X}$ and R charts, the sample size is taken into account in the coefficients A_2, D_3 and D_4. This problem is aimed at showing how the charts are affected by different sample sizes. The data is for diameter measurements made on a process that appears to be in statistical equilibrium based on Figure 2.8. Measurements listed in Table 2.7 are grouped so that charts can be set up for n = 4 and n = 6. The tolerance on this diameter is 25.00 ± 0.30 mm.

a. Determine the upper control limits for both $\overline{X}$ and R charts when n is 4 and 6.
b. Which UCL's change by more than 10 percent when n is increased from 4 to 6, the one for $\overline{X}$, R or both?
c. Determine if the process is capable of meeting the tolerance specifications for the diameters. (Because the process is in statistical equilibrium, you should be able to estimate the sample variance for these measurements quite precisely.)

Table 2.7 Measurements to Compare Effect of Sample Size on Control Charts

Sample	Diameter measurements (mm) for n = 4				Extra measurements for n = 6.	
1	24.85	25.11	25.13	24.88	24.94	25.05
2	25.03	24.85	24.96	25.07	24.98	24.95
3	25.00	24.86	24.97	25.09	24.76	24.99
4	25.07	25.07	25.22	24.87	24.94	25.06
5	25.00	24.95	24.88	24.89	25.06	25.00

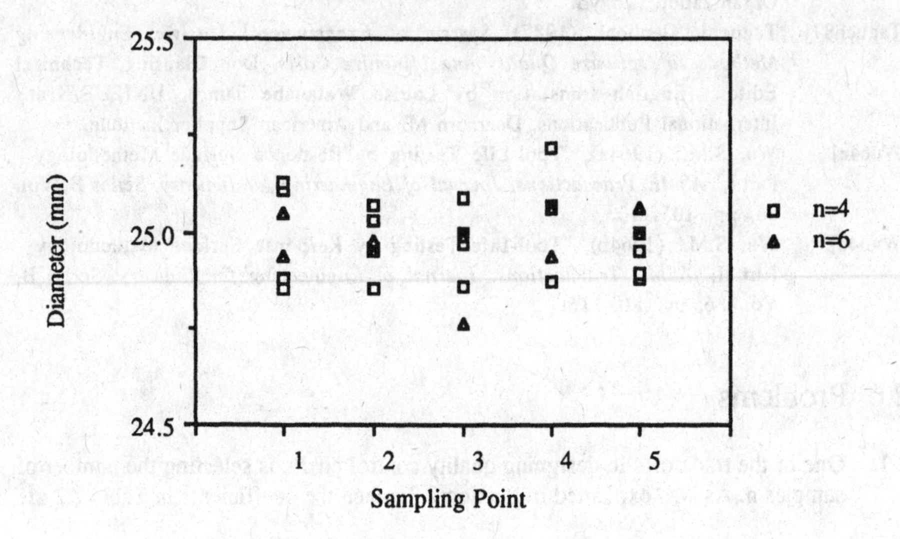

Figure 2.8 Plot of measurements for Problem 2.1.

2.2 Use the data in Table 2.5 to estimate n, n_1 and n_2 in the generalize tool life equation talked about in Example 2.7, i.e.,

$$\left(\frac{T}{T_R}\right) = \frac{K}{\left(\frac{v}{v_R}\right)^{1/n}\left(\frac{a}{a_R}\right)^{1/n_1}\left(\frac{d}{d_R}\right)^{1/n_2}}.$$

Assume T_R = 60 s/edge and you should specify the values of v_R ,a_R and d_R that you used to estimate the parameters. (Hint: This can be done quite simply by choosing the correct transformation(s) since this data came from a factorial design experiment.)

3
Mechanics and Thermal Models for Machining

Predicting forces and temperatures are two of the most common reasons for modeling machining processes. Both of these quantities are needed for practical reasons related to planning a machining operation. Forces exerted by the cutting edge on the workpiece can cause deflections that lead to geometric errors and difficulty meeting tolerance specifications. Reaction forces of the workpiece on the cutting edges can, if large enough, cause catastrophic failure of the cutting edge. The product of the force and velocity vector are used to predict power requirements for sizing a new machine tool or for predicting the production rates possible with an existing machine tool. Temperature predictions are used to estimate how the process of machining or grinding affects a cutting edge's useful life or how elevated temperatures can change the mechanical properties of the workpiece. Both of these considerations are important for economical operation of the process or the safety and performance of the machined product.

As Figure 3.1 shows, when making force and temperature calculations, material and thermal properties of both the cutting edge and the workpiece are needed. The geometry and orientation of the cutting edge are variables that have to be determined in a detailed process plan because they affect process performance. Selecting cutting conditions - the feed, cutting speed and depth of cut - are other important planning variables. Cutting conditions are important because they determine the chip cross section and the velocity that the workpiece material approaches the cutting edge. Methods for making force and temperature calculations are based on mechanics, heat transfer and tribology.

3.1 Classification of Chip Formation

Models for the mechanics and thermal aspects of cutting and grinding always require some idealizations. Three basic classifications of chip formation have been recognized over the years based on: the geometry of the cutting edge(s), the chip morphology and the zone used to model plastic flow. Figure 3.2 presents a graphical outline of these different ways the chip formation problem can be classified.

3.1.1 Types Based on the Cut Geometry

The three major ways the cut geometry can be classified are: orthogonal, oblique and three dimensional cutting.

Orthogonal cutting uses a single cutting edge, but the most important characteristic is that the velocity of the workpiece material v, is orthogonal to this cutting edge. It is assumed that the chip flows up the surface of the wedge shaped edge with the chip velocity v_c parallel to the incoming velocity v. This case has been the one where most of the modeling work has been done. This means that all the analysis can be done in the plane

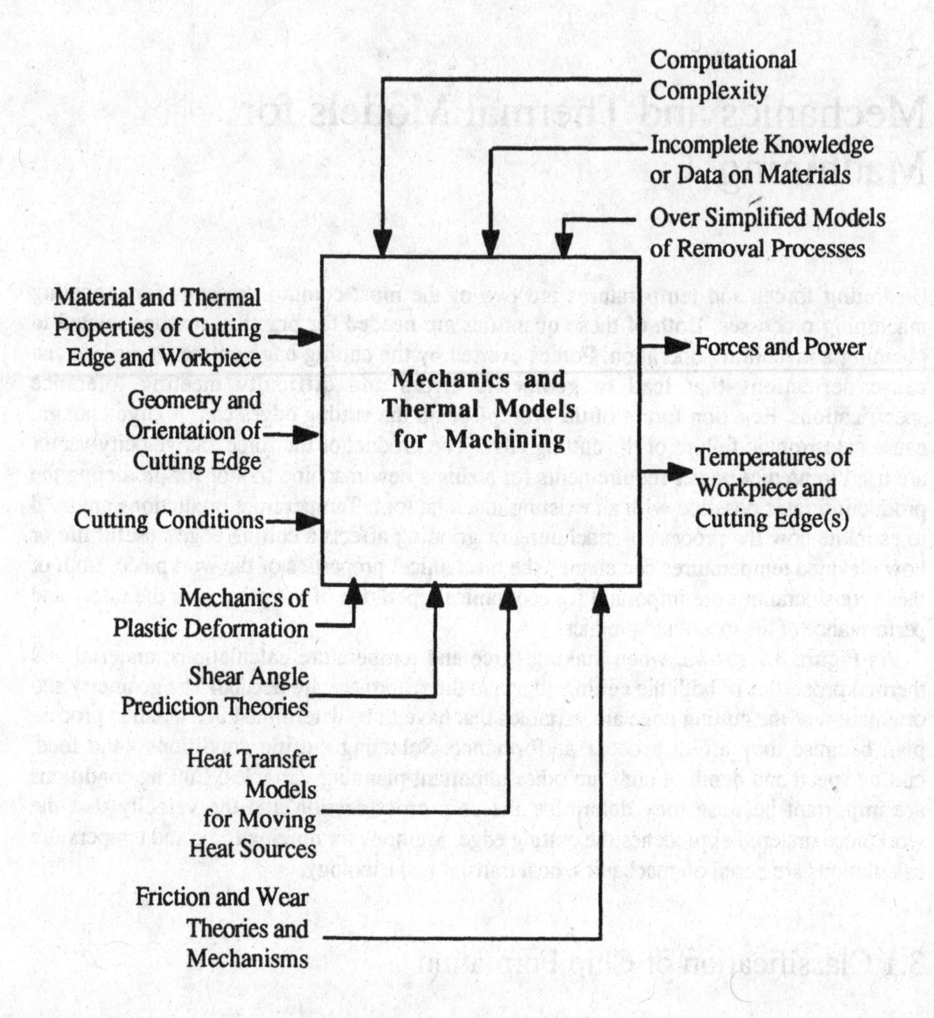

Figure 3.1 Inputs, Constraints, Outputs and Mechanisms in Developing Mechanics and Thermal Models for Machining.

that contains the two velocity vectors. This will be the case shown most often. It should be realized that very few practical machining processes meet the requirements of orthogonal cutting exactly. Cases that can be modeled as orthogonal machining include: turning a tube, cutoff or parting operations, some cases of shaping or planning, and slotting operations with special milling cutters.

Oblique cutting, like orthogonal cutting, only uses one cutting edge. However, as the term oblique implies, the velocity of the workpiece material v, is oblique or inclined to the cutting edge at an angle λ. This has several implications. One is that the direction of the chip velocity v_c, is no longer parallel to the cutting speed. This means that the angle that the chip makes as it flows up the face of the cutting wedge has to be determined. The mechanics of oblique cutting can be analyzed as a two dimensional problem if the correct reference frame is chosen. Then the three dimensional force vector can be found by applying

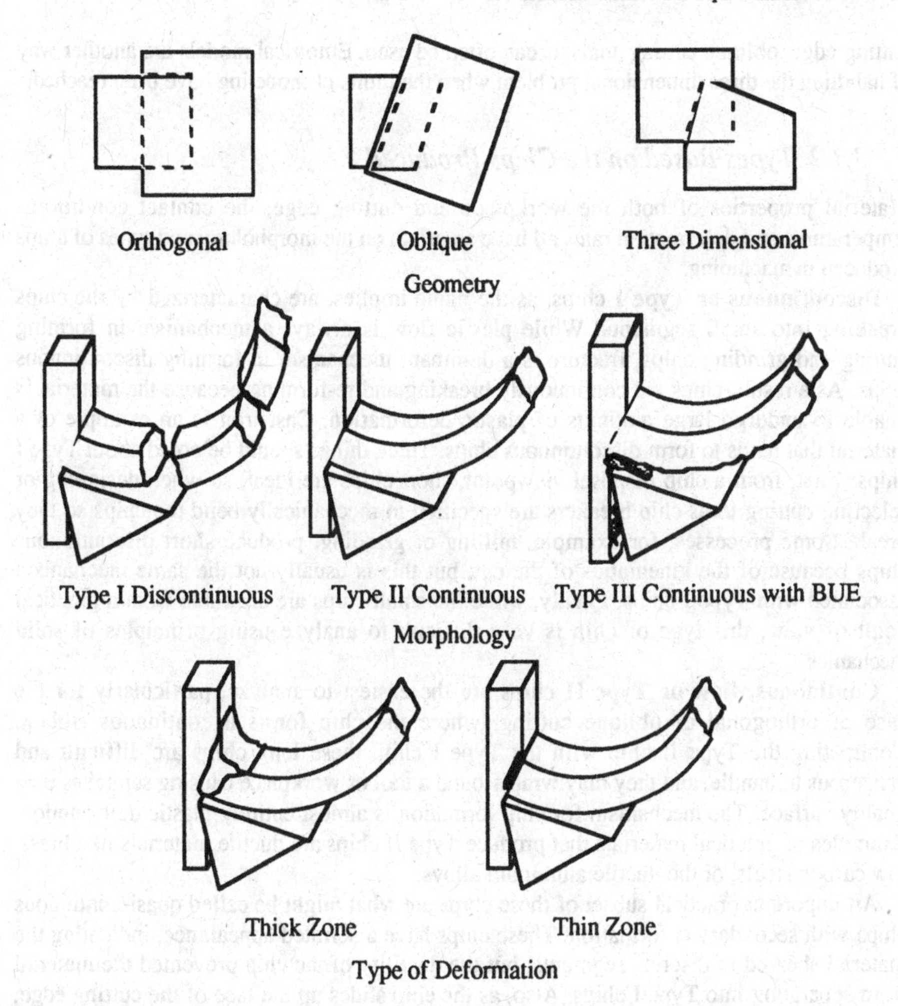

Figure 3.2 Classification of Chip Formation Based on Geometry, Morohology and Type of Plastic Deformation.

a rotational transformation. Cuts with helical end mills, face mills or many turning tools are usually analyzed as oblique machining.

Three dimensional cutting differs from both orthogonal and oblique cutting because cutting is done by multiple edges, rather than a single edge. These multiple edges are usually referred to as the major and minor cutting edges. The nomenclature to identify these edges for practical machining processes is covered later in Chapter 5. The common characteristic of all these processes is that the analysis is inherently three dimensional. The problem cannot be transformed to a two dimensional problem by selecting a reference frame. This is because at least two edges are cutting simultaneously and the deformation of the workpiece material interacts along these edges. So far, there are no analytical or closed formed solutions to either the mechanics or thermal problem in three dimensions, but there are several numerical approaches based on the finite element method [CarStr88], [Lee84]. As might be expected, most practical machining and grinding processes should be classified as three dimensional cutting. Fortunately, when most of the cutting is done on the major cutting edge, oblique cutting analysis can often be used. Empirical models are another way of handling the three dimensional problem when the limits of modeling have been reached.

3.1.2 Types Based on the Chips Produced

Material properties of both the workpiece and cutting edge, the contact conditions, temperatures and deformation rates all have an effect on the morphology and types of chips produced in machining.

Discontinuous or Type I chips, as the name implies, are characterized by the chips breaking into small segments. While plastic flow is always a mechanism in forming cutting and grinding chips, fracture is a dominant mechanism in forming discontinuous chips. As a result, chips are continuously breaking and re-forming because the material is unable to undergo large amounts of plastic deformation. Cast iron is an example of a material that tends to form discontinuous chips. Three things should be noted about Type I chips: First, from a chip disposal viewpoint, short chips are ideal, so when designing or selecting cutting tools chip breakers are specified to mechanically bend the chips so they break. Some processes, for example, milling or grinding, produce short discontinuous chips because of the kinematics of the cut, but this is usually not the same mechanism associated with Type I chips. Finally, while the small chips are desirable from a practical point of view, this type of chip is very difficult to analyze using principles of solid mechanics.

Continuous, flow or Type II chips are the easiest to analyze, particularly for the case of orthogonal or oblique cutting, where the chip forms a continuous ribbon. Contrasting the Type II chip with the Type I chip, these long chips are difficult and dangerous to handle, and they may wrap around a tool or workpiece causing scratches on a quality surface. The mechanism for chip formation is almost entirely plastic deformation. Examples of practical materials that produce Type II chips are ductile materials like brass, low carbon steels, or the ductile aluminum alloys.

An important practical subset of these chips are what might be called quasi-continuous chips with secondary deformation. These chips have a serrated appearance, indicating the material sheared in discrete segments, but the ductility of the chip prevented the material from separating into Type I chips. Also, as the chip slides up the face of the cutting edge, the friction forces resists the motion causing deformation on this contact surface. This quasi-continuous behavior has two ramifications that will be discussed later. The first is that this segmented appearance will be highly idealized in Section 3.2 to estimate strains in making a chip. The second ramification will be in Chapter 8, where quasi-continuous chip formation is proposed as a mechanism that initiates machine tool chatter.

Continuous chips with built up edge or Type III chips have plastic deformation as their basic mechanism, just like Type II chips. In fact the secondary deformation mentioned with regard to the quasi-continuous chip is more severe, with high interface temperatures and pressures sufficient to cause some of the chip material to weld to the cutting edge. Additional chip material builds up on the previously welded material, hence the name built up edge (BUE). As the BUE grows, it tends to become unstable because the moment created by the forces acting on the BUE exceeds the strength of the edge cross-section. The BUE breaks off, often taking a small amount of the cutting edge with it and gouging the workpiece surface in the process. The entire process can then repeat. Both the wear of the cutting edge and the roughened workpiece surface are reasons why Type III chips should be avoided. Fortunately, for most materials this cyclic process is most prevalent at cutting speeds below the ones used in practical machining operations.

Also, in some cases the BUE may actually form so as to increase the angle at the tip of the cutting wedge, which can actually decrease forces; something to remember when designing a diagnostic machining system based on force measurement.

3.1.3 Types Based on Deformation Zone

Plastic deformation is particularly important for Type II and III chips. A basic assumption needed before predicting forces or temperatures is a model of the deformation zone.

Thick zone or shear *zone* models assume that shear is the main mechanism for chip formation, but this deformation occurs in a three dimensions *volume* or *region* ahead of the cutting edge. This probably is the way that all deformation occurs. Based on videos of the zone in front of the cutting wedge, at low cutting speeds, say less than 1 m/s, it is clear that the deformation is in a region bounded by two planes that intersect at the cutting edge. Several researchers have modeled the deformation zone in this way to solve the cutting mechanics problem [Oxley63], [Oxley89] or [Trent77]. As the speed increases, the included angle between these two planes tends to decrease so that it appears that this volume collapses to a single plane.

Thin zone or shear *plane* models appear to be a reasonable way to model the deformation in oblique and orthogonal machining processes at the speeds where these processes normally operate. Shear deformation is assumed to occur along a single plane called the *shear plane*. As one can imagine, a model that assumes an infinitely thin plane separating a region of no plastic deformation from a region with a finite amount of deformation presents some theoretical difficulties, for example shear discontinuities. Even with this theoretical limitation, the thin zone model is the one used in most mechanics and thermal calculations for orthogonal or oblique cutting. At the cutting speeds where most machining is done, it is the simplest model that gives a closed form solution and reasonable predictions.

3.2 Mechanics Models for Machining

The ways of classifying chip formation in Section 3.1 serve as ways of classifying the assumptions used to analyze chip formation. Almost all analysis of the mechanics of machining, including those presented here, consider Type II or continuous chips. The most straight forward analysis is with the thin zone model, which fortunately is appropriate for most practical machining operations. Two variations to the analysis will be in terms of cutting edge geometry: orthogonal and oblique machining.

3.2.1 Orthogonal Machining

Because of the assumption of orthogonal machining, analysis can be done in the plane represented by Figure 3.3. A single cutting edge forms a line coincident with the z axis, going out of the plane of the figure. The nomenclature for the thin zone orthogonal cutting with continuous chip is aimed at describing three things: the workpiece or uncut chip, the chip, and the cutting edge.

With the thin zone or shear plane model, a well defined boundary is assumed between the workpiece and the chip, termed the shear plane. This plane has an edge of length l_s inclined at an angle ϕ_o - the orthogonal shear plane angle - that is measured relative to the

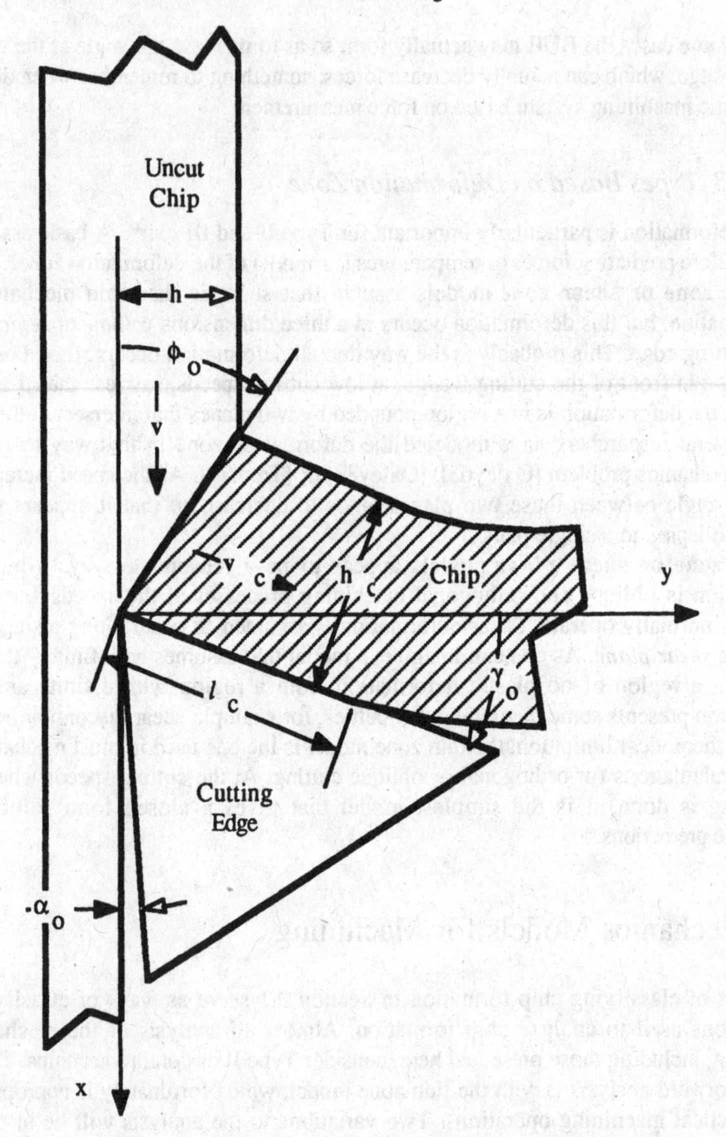

Figure 3.3 Chip Geometry for Orthogonal Cutting.

velocity v of the workpiece material as it approaches the cutting edge. The geometry of the uncut chip is characterized by its width b, which by assumption does not change as the chip is formed, and the uncut chip thickness h. Subsequently when practical machining processes are considered in Chapter 5 , it will become clear that b and h are functions of the depth, feed and edge geometry that are selected when planning a machining operation. The chip and forces to produce it will depend on the boundary defined by the shear plane angle ϕ_o. As a result, predicting ϕ_o is one of the continuing areas of research in machining, c.f., Table 3.1. Another important chip boundary is the length of contact l_c, which defines the length of the sliding interface between the chip and the cutting edge. This contact length is a key geometric parameter for the temperature calculations in Section 3.3. These boundaries

Table 3.1 Some Shear Plane Angle Prediction Models [Shaw*etal*53]		
Model	Reference	Based on
$\phi_o = 45 - 0.5\,(\beta - \gamma_o)$	Ernst-Merchant [ErnMer41]	Minimizing force
$\phi_o = 0.5\cot^{-1}K - 0.5\,(\beta - \gamma_o)$	Merchant [Mercha45a & b]	Minimizing force and empirical coefficient K
$\phi_o = 45 - \quad (\beta - \gamma_o)$	Lee-Shaffer [LeeSha51]	Slip line theory
$\phi_o = 45 - \quad \beta + 0.5\,\gamma_o$	Stabler[Stable51]	
$\phi_o = 45 - 0.5\tan^{-1} 2\mu + 0.5\,\gamma_o$	Hucks[Hucks52]	
$\phi_o = 0.5\cot^{-1}K - 0.5\tan^{-1} 2\mu + 0.5\,\gamma_o$	Hucks[Hucks52]	
$\phi_o = 50 - 0.8\,(\beta - \gamma_o)$	Palmer-Oxley [PalOxl59]	Aggregate solution for shear zone model

define the chip of width b and thickness h_c that separates from the workpiece as it moves at the chip velocity v_c parallel to the rake face of the cutting edge.

The final geometry is that of the cutting edge, and again the subscript "o" is used to denote the geometry for orthogonal cutting. For chip formation mechanics, the orthogonal rake angle γ_o is most important. It is measured *in* the plane that contains the two velocity vectors v and v_c and is measured *from* the normal to the velocity vector v, with a positive γ_o shown in Figure 3.3. Glancing ahead to Table 3.1, γ_o shows up in all the shear angle prediction equations. The orthogonal relief or clearance angle, α_o is also measured *in* the plane that contains the two velocity vectors v and v_c, but it is measured *from* the velocity vector v to what is termed the flank or relief surface of the cutting edge. As the terms relief or clearance imply, this angle is to maintain a clearance between the cutting edge and the newly generated surface. Remember that a process planer can affect the cutting process by choosing γ_o and α_o.

Geometric and kinematic relationships are needed before estimating the forces to make a chip. As a starting point, a mechanism for forming the chip has to be proposed. The common one is the forces needed to move the workpiece material into the cutting edge, producing a stress field. The assumed mode of plastic flow is by shearing, and with a thin zone model the shearing is confined to a plane. Figure 3.4 illustrates this idealized process, where the dashed lines in Figures 3.4a and b mark out an undeformed segment of arbitrary thickness Δw. The solid segment in Figure 3.4b shows how the material must deform to produce a Type II continuous chip. As a segment moves up the rake face of the cutting edge, a shear strain γ produced by the shear stress τ_s, is needed as a boundary of the chip rotates to match the solid line segment with the cutting edge. Because of this relative motion along the shear plane, in addition to the cutting speed v and the chip velocity v_c, there must be a velocity of the chip relative to the shear plane, the shear velocity, denoted by $v_{c/s}$. If the chip is incompressible and the density is unaffected by the deformation process, then based on continuity the velocity relationships can be summarized as:

$$\mathbf{v} = \mathbf{v}_{c/s} + \mathbf{v}_c. \tag{3.1}$$

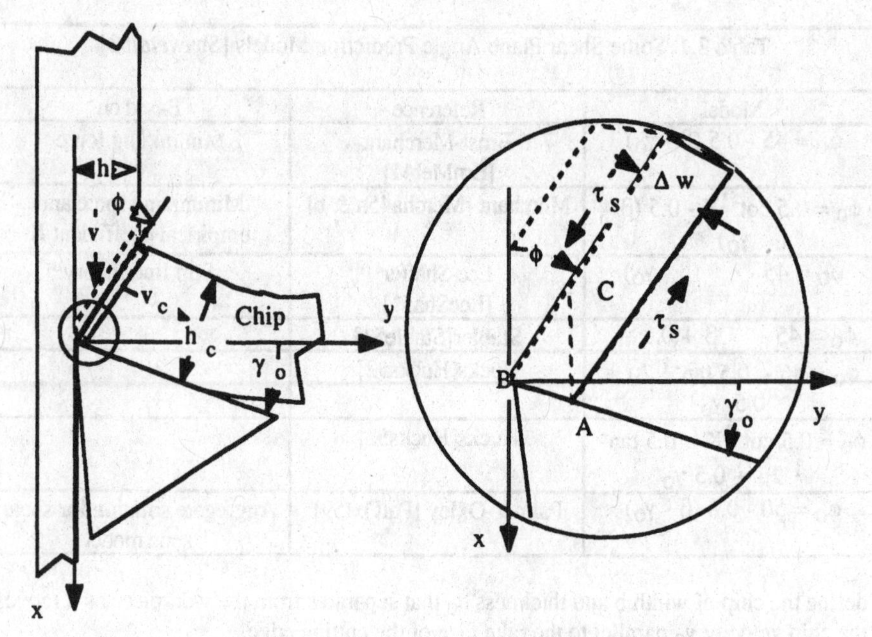

Figure 3.4a Macro geometry Figure 3.4b Cutting edge tip detail

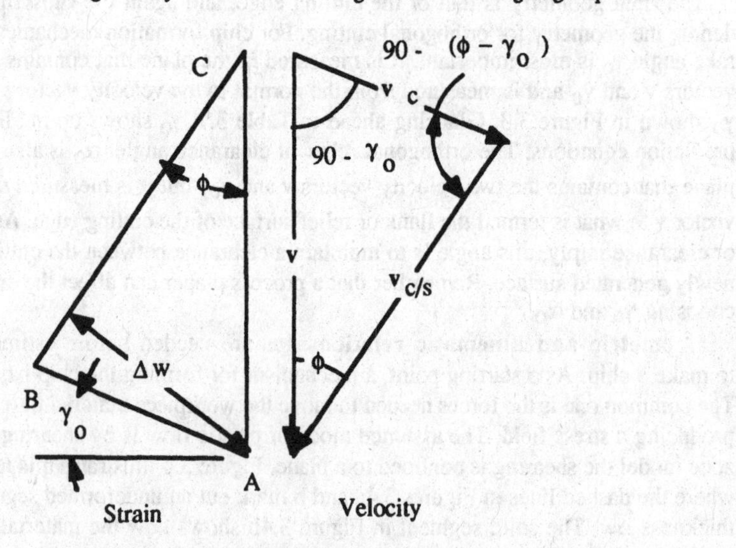

Figure 3.4c Similar triangles for strain and velocity

Figure 3.4 Strain and velocity relationships for Orthogonal Cutting.

The velocity triangle in Figure 3.4c is a graphical representation of Eq (3.1). The law of sines and cosines define the velocity magnitudes and geometric relationships as

$$\frac{v}{\cos(\phi_o - \gamma_o)} = \frac{v_c}{\sin(\phi_o)} = \frac{v_{c/s}}{\cos(\gamma_o)}. \tag{3.2a}$$

For estimating the shear strain and strain rate, triangle ABC, which is similar to the velocity triangle, can be used. The shear strain is

$$\gamma \quad = \frac{\overline{BC}}{\Delta w} = \frac{\Delta w \tan(\phi_O - \gamma_O) + \Delta w \cot(\phi_O)}{\Delta w} = \frac{\cos(\gamma_O)}{\sin(\phi_O)\cos(\phi_O - \gamma_O)}. \quad (3.3a)$$

The rate of change of segment $\overline{BC}$ is $v_{c/s}$ and the identities in Eq (3.2) allow the shear strain rate to be estimated in terms of the cutting speed

$$\dot{\gamma} \quad = \frac{d(\overline{BC})/dt}{\Delta w} = \frac{v_{c/s}}{\Delta w} = \frac{v \cos(\gamma_O)}{\Delta w \cos(\phi_O - \gamma_O)}. \quad (3.4a)$$

A final geometric calculation based on the continuity assumption is the relationship between the uncut chip thickness h and the chip thickness, h_c. The volumetric removal rate is the product of the cross section of the uncut chip and the cutting speed,

$$Z \quad = v \bullet b \bullet h. \quad (3.5)$$

Because b is a constant and the density doesn't change, the continuity equation is,

$$v \bullet b \bullet h \bullet \rho = v_c \bullet b \bullet h_c \bullet \rho = Z \bullet \rho.$$

The *cutting ratio* can be defined in several ways using this continuity equation and Eq (3.2a)

$$r \quad \equiv \frac{h}{h_c} = \frac{v_c}{v} = \frac{\sin(\phi_O)}{\cos(\phi_O - \gamma_O)} = \frac{\tan(\phi_O)}{\cos(\gamma_O) + \tan(\phi_O)\sin(\gamma_O)} < 1. \quad (3.6a)$$

The last form of the expression for r isn't needed now, but it is the basis for the experimental way to estimate ϕ_O given in Eq (3.6b).

Before estimating the forces in this thin zone model, consider two things about the development so far:

- Equations (3.2-6) and more to come all involve geometry: γ_O which can be specified and ϕ_O that needs to be calculated. Only some of the approaches to predicting ϕ_O will be covered here.
- The thin zone model used in this calculation assumes that $\Delta w \rightarrow 0$. But Eq (3.4a) points out if $\Delta w = 0$, there would be an infinite strain rate. While the strains are large and the strain rates are high, they are not infinitely large, and this is one of the reasons that shear *zone* models have been developed.

Estimating forces from mechanics follows the development of Ernst and Merchant [ErnMer41], where the chip is isolated as a free body. Forces act on the chip on two surfaces, the shear plane surface and the rake surface. Figure 3.5 shows uniform distributions of normal compressive stresses and shear stresses that are usually assumed on both surfaces. An external force applied on the shear plane is balanced by the the forces acting on the rake face. These external loads cause internal stresses in the chip, and these stresses cause the material to flow along the shear plane when the flow stress of the material τ_s, is reached. The contact stresses on the rake face are usually modeled assuming the traction or shear stress is proportional to the normal stress. This form of Coulomb's law is appropriate as long as the true contact area is less than the apparent contact area.

Estimating the forces on the two chip interfaces is is an elementary mechanics problem with this simple formulation. The forces parallel and normal to the shear plane can define a force vector $\mathbf{F_s}$ as the integral of the product of the stress and the area,

$$\mathbf{F_s} = \begin{bmatrix} \text{Parallel Shear Component} \\ \text{Compressive Component} \end{bmatrix} = \begin{bmatrix} \int^{A_s} \tau \, dA \\ \int^{A_s} \sigma \, dA \end{bmatrix} \text{(any stress distribution)} \qquad (3.7a)$$

$$= \begin{bmatrix} F_s \\ F_c \end{bmatrix} = \begin{bmatrix} \dfrac{\tau_s \, b \, h}{\sin(\phi_o)} \\ F_c \end{bmatrix} \text{(uniform stress distribution as in Figure 3.5).} \qquad (3.7b)$$

Usually F_s is assumed to be known completely once τ_s is specified, with only F_c's direction - normal to F_s - known.

The force vector on the rake face $\mathbf{F_r}$, can be written in a similar way,

$$\mathbf{F_r} = \begin{bmatrix} \text{Normal Rake Component} \\ \text{Friction Component} \end{bmatrix} = \begin{bmatrix} \int^{A_c} \sigma \, dA_{true} \\ \int^{A_c} \tau \, dA_{true} \end{bmatrix} \text{(any stress distribution)} \quad (3.8a)$$

$$= \begin{bmatrix} F_n \\ F_f \end{bmatrix} = \begin{bmatrix} F_n \\ \mu \, F_n \end{bmatrix} \text{(for Coulomb friction where } A_c = l_c \, b > A_{c,true}\text{).} \qquad (3.8b)$$

Note that at this contact interface, the true contact area should be used, as well as the correct stress distribution. When the nominal contact area A_c is greater than the true contact area, the Coulomb friction law defines μ, the friction coefficient, and β, the friction angle, in terms of the force components in Eq (3.8b):

$$\mu = \frac{F_f}{F_n} = \tan(\beta). \qquad (3.9a)$$

Equations (3.7) and (3.8) resolve the forces on the chip into orthogonal components relative to the shear plane and rake face respectively. Treating the chip as a two force member, the solution is determined from the vector equation $\mathbf{F_s} + \mathbf{F_r} = 0$.

Using the reference coordinates shown in Figure 3.6, the force vector $\mathbf{F_s}$ in the shear plane coordinates can be rotated counter clockwise (a positive rotation using the right hand rule) by ϕ_o. This defines the resultant force vector $\mathbf{R}$ in the x-y plane, which has components with special names, i.e.,

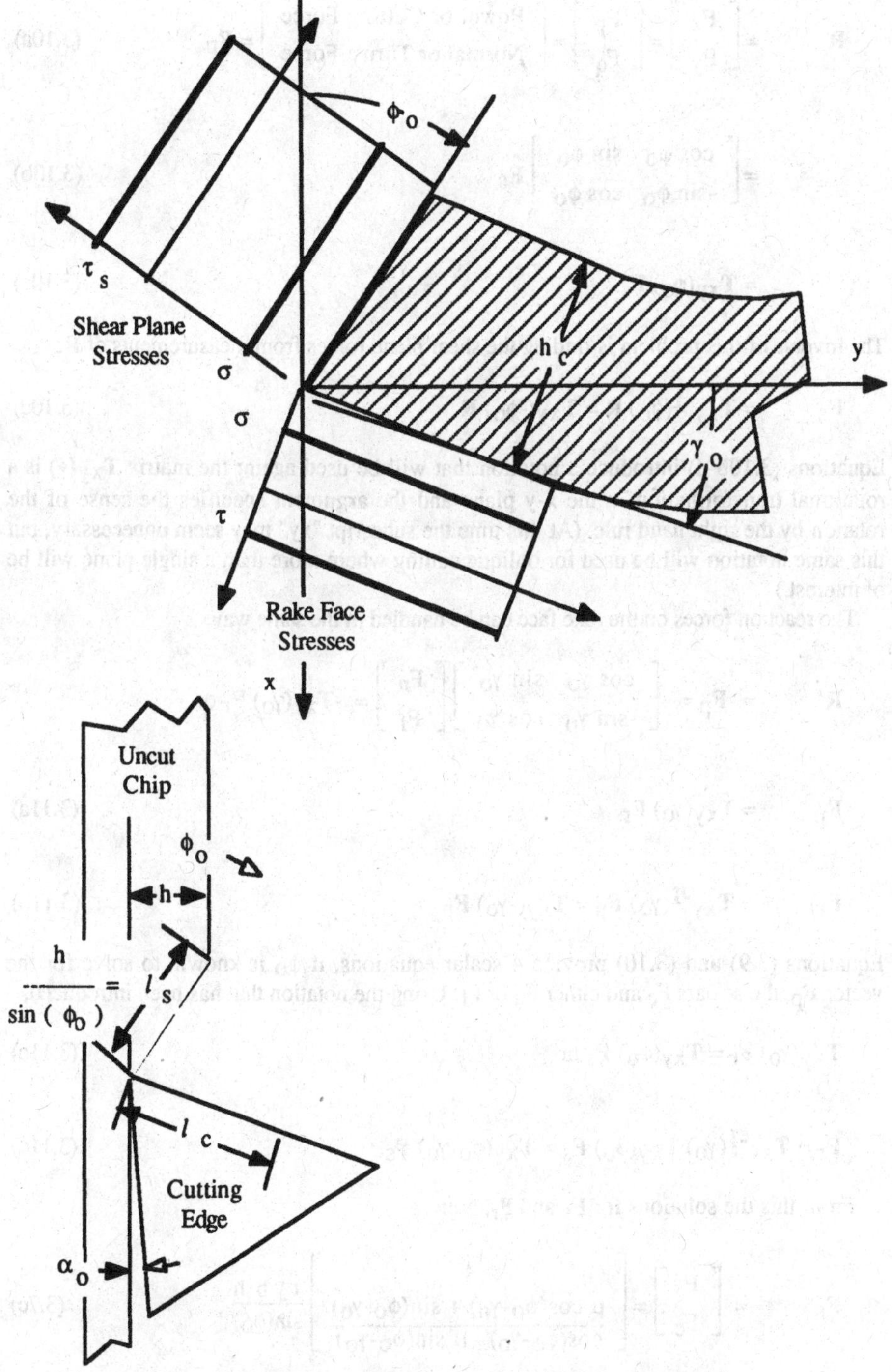

Figure 3.5 Uniform stress distribution with thin zone model.

$$\mathbf{R} = \begin{bmatrix} F_x \\ F_y \end{bmatrix} = \begin{bmatrix} F_p \\ F_q \end{bmatrix} = \begin{bmatrix} \text{Power or Cutting Force} \\ \text{Normal or Thrust Force} \end{bmatrix} = \mathbf{F_p} \tag{3.10a}$$

$$= \begin{bmatrix} \cos\phi_o & \sin\phi_o \\ -\sin\phi_o & \cos\phi_o \end{bmatrix} \mathbf{F_s} \tag{3.10b}$$

$$= \mathbf{T_{xy}}(\phi_o)\, \mathbf{F_s}. \tag{3.10c}$$

The inverse of this problem is finding the shear plane forces from measurements of **R**:

$$\mathbf{F_s} = \mathbf{T_{xy}}^{-1}(\phi_o)\, \mathbf{R} = \mathbf{T_{xy}}(-\phi_o)\, \mathbf{R}. \tag{3.10d}$$

Equations (3.10b-d) introduce a notation that will be used again: the matrix $\mathbf{T_{xy}}(\bullet)$ is a rotational transformation in the x-y plane and the argument specifies the sense of the rotation by the right hand rule. (At this time the subscript "xy" may seem unnecessary, but this same notation will be used for oblique cutting where more than a single plane will be of interest.)

The reaction forces on the rake face can be handled in the same way,

$$\mathbf{R'} = -\mathbf{F_p} = -\begin{bmatrix} \cos\gamma_o & \sin\gamma_o \\ -\sin\gamma_o & \cos\gamma_o \end{bmatrix} \begin{bmatrix} F_n \\ F_f \end{bmatrix} = -\mathbf{T_{xy}}(\gamma_o)\, \mathbf{F_r} \text{ or}$$

$$\mathbf{F_p} = \mathbf{T_{xy}}(\gamma_o)\, \mathbf{F_r} \tag{3.11a}$$

$$\mathbf{F_r} = \mathbf{T_{xy}}^{-1}(\gamma_o)\, \mathbf{F_p} = \mathbf{T_{xy}}(-\gamma_o)\, \mathbf{F_p} \tag{3.11b}$$

Equations (3.9) and (3.10) provide 4 scalar equations, if ϕ_o is known, to solve for the vector $\mathbf{F_p}$, the scalars F_c and either F_n or F_f. Using the notation that has been introduced,

$$\mathbf{T_{xy}}(\gamma_o)\, \mathbf{F_r} = \mathbf{T_{xy}}(\phi_o)\, \mathbf{F_s} \text{ or} \tag{3.11c}$$

$$\mathbf{F_r} = \mathbf{T_{xy}}^{-1}(\gamma_o)\, \mathbf{T_{xy}}(\phi_o)\, \mathbf{F_s} = \mathbf{T_{xy}}(\phi_o - \gamma_o)\, \mathbf{F_s}. \tag{3.11d}$$

From this the solutions for $\mathbf{F_s}$ and $\mathbf{F_r}$, are:

$$\mathbf{F_s} = \begin{bmatrix} F_s \\ F_c \end{bmatrix} = \begin{bmatrix} 1 \\ \dfrac{\mu \cos(\phi_o - \gamma_o) + \sin(\phi_o - \gamma_o)}{\cos(\phi_o - \gamma_o) - \mu \sin(\phi_o - \gamma_o)} \end{bmatrix} \frac{\tau_s\, b\, h}{\sin(\phi_o)}, \tag{3.7c}$$

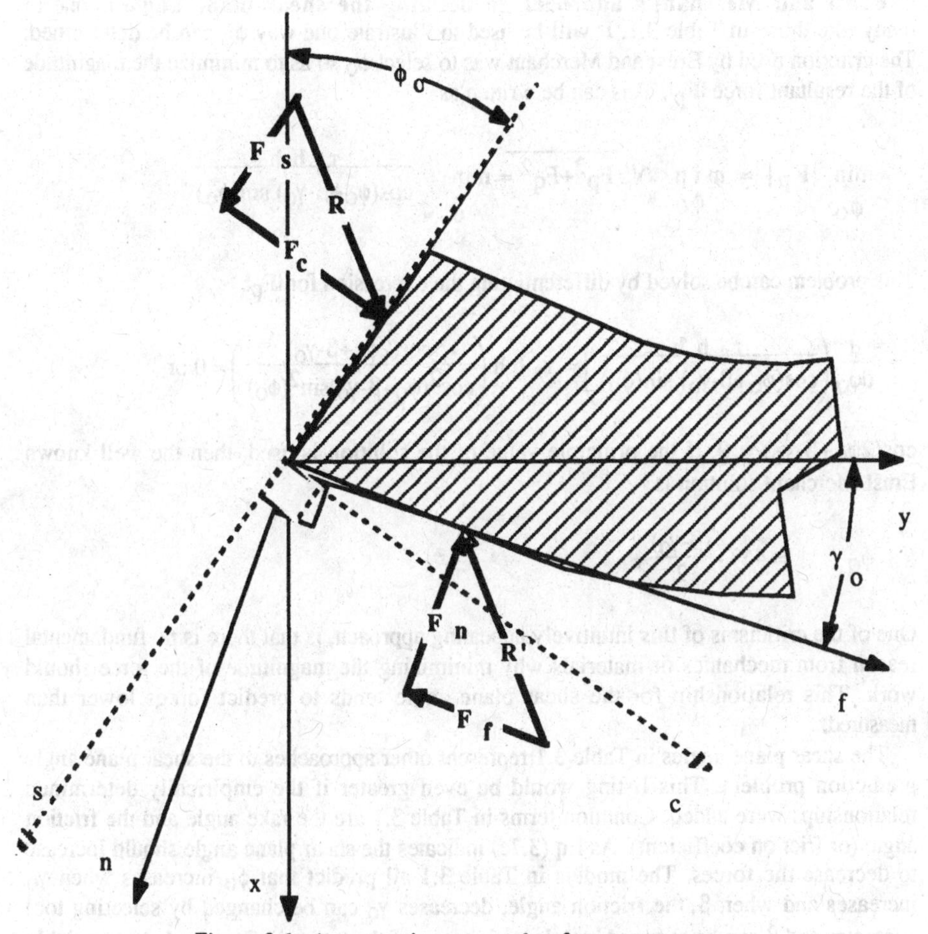

Figure 3.6 Axes rotations to resolve force components.

$$\mathbf{F_r} = \begin{bmatrix} F_n \\ F_f \end{bmatrix} = \begin{bmatrix} \dfrac{1}{\cos(\phi_o-\gamma_o) - \mu \sin(\phi_o-\gamma_o)} \\ \dfrac{\mu}{\cos(\phi_o-\gamma_o) - \mu \sin(\phi_o-\gamma_o)} \end{bmatrix} \frac{\tau_s \; b \; h}{\sin(\phi_o)}, \tag{3.8c}$$

and substituting Eq. (3.7c) into Eq (3.10c) and replacing μ by Eq (3.9) gives

$$\mathbf{F_p} = \begin{bmatrix} F_p \\ F_q \end{bmatrix} = \begin{bmatrix} \dfrac{\cos(\beta-\gamma_o)}{\cos(\phi_o+\beta-\gamma_o)} \\ \dfrac{\sin(\beta-\gamma_o)}{\cos(\phi_o+\beta-\gamma_o)} \end{bmatrix} \frac{\tau_s \; b \; h}{\sin(\phi_o)}. \tag{3.10e}$$

All these are different ways of resolving the same force vectors. The ϕ_o, the shear plane angle is still unknown.

Ernst and Merchant's approach to defining the shear plane angle is one of many like those in Table 3.1. It will be used to illustrate one way ϕ_o can be determined. The criterion used by Ernst and Merchant was to select ϕ_o so as to minimize the magnitude of the resultant force $|\mathbf{F_p}|$. This can be written as

$$\min_{\phi_o} |\mathbf{F}_p| = \min_{\phi_o} \sqrt{F_p^2+F_q^2} = \min_{\phi_o} \frac{\tau_s\, b\, h}{\cos(\phi_o+\beta-\gamma_o)\sin(\phi_o)}.$$

This problem can be solved by differentiating the expression for $|\mathbf{F_p}|$,

$$\frac{d}{d\phi_o}\left(\frac{\tau_s\, b\, h}{\cos(\phi_o+\beta-\gamma_o)\sin(\phi_o)}\right) = \tau_s\, b\, h\left(\frac{\cos(2\phi_o+\beta-\gamma_o)}{\cos^2(\phi_o+\beta-\gamma_o)\sin^2(\phi_o)}\right) = 0 \text{ or}$$

$\cos(2\phi_o+\beta-\gamma_o) = 0$. If the principle value of the solution is used, then the well known Ernst-Merchant solution is:

$$\phi_o = 45° - \frac{(\beta-\gamma_o)}{2}.$$

One of the criticisms of this intuitively appealing approach, is that there is no fundamental reason from mechanics or materials why minimizing the magnitude of the force should work. This relationship for the shear plane angle tends to predict forces lower than measured.

The shear plane angles in Table 3.1 represent other approaches to the shear plane angle prediction problem. This listing would be even greater if the empirically determined relationships were added. Common terms in Table 3.1 are the rake angle and the friction angle (or friction coefficient). As Eq (3.7c) indicates the shear plane angle should increase to decrease the forces. The models in Table 3.1 all predict that ϕ_o increases when γ_o increases and when β, the friction angle, decreases γ_o can be changed by selecting tool geometry and β can be changed by lubrication, one of the topics in Chapter 4. How quickly γ_o and β change ϕ_o is what differentiates the various relationships in Table 3.1.

Experimental and practical calculations can be made using the force prediction relationships and the models for strain. If $\mathbf{F_p}$ and the velocity vector $\mathbf{v} = [v\ 0]^T$ are known, then the power to make the cut is

$$P = \mathbf{v}^T \mathbf{F_p} \tag{3.12a}$$

$$= v\, F_p \text{ (Because } v_y = 0). \tag{3.12b}$$

This is why F_p is called the power component of force. Equation (3.12a) is valid for cases where there is motion in the y or z directions. However, in most of the practical machining processes the velocity components in the y or z directions are so small (1 or 2 percent of the total) that their contributions to P are neglected.

From the short discussion of predicting shear plane angles, it is apparent that there is often a need to compare predicted values with experimental values for ϕ_o. There are a number of ways to do this, ranging from high speed videos that "watch" the chip formation to quick stop devices that try to instantaneously "stop" the formation of a chip. In both cases the angle can be measured directly from a view of the chip cross section. A simpler approach uses a rearranged version of Eq (3.6a) to estimate ϕ_o from the cutting ratio r and the known orthogonal rake angle γ_o

$$\phi_o = \tan^{-1}\left(\frac{r\cos(\gamma_o)}{1 - r\sin(\gamma_o)}\right). \tag{3.6b}$$

Finding the best way to measure r is the difficult part with this method; should it be based on thickness or scribing a reference length on the chip. Once a good technique is developed, methods in Chapter 2 can be used to design an experiment to evaluate a theoretical model or develop an empirical shear plane angle prediction model.

In line with these experimental techniques, often the model for orthogonal cutting is used to "back out" some of the parameters that have been assumed to be known, for example μ or τ_s in Eq (3.7c), Eq (3.8c) or Eq (3.10e). By designing experiments that measure $\mathbf{F_p}$ these quantities can be determined. To estimate the friction coefficient in orthogonal machining, for example,

$$\mu = \frac{\sin(\gamma_o)\, F_p + \cos(\gamma_o)\, F_q}{\cos(\gamma_o)\, F_p - \sin(\gamma_o)\, F_q}. \tag{3.11e}$$

This suggests that the easiest experiment for measuring μ would be if $\gamma_o = 0$.

Equations (3.7c) and (3.10d) suggest that once ϕ_o is determined, then

$$\tau_s = \sin(\phi_o)\left(\frac{\cos(\phi_o)\, F_p - \sin(\phi_o)\, F_q}{b\ h}\right). \tag{3.10f}$$

In Example 2.6 this technique was used to provide the experimental data for τ_s. Power law models for τ_s that are a function of the average uncut chip thickness

$$\tau_s = \tau_0\left(\frac{h_{avg}}{h_R}\right)^{a} \tag{2.14}$$

are often used. The exponent is in the range $-1 < a < 0$, indicating that the shear strength increases as the chips get thinner. Several reasons have been proposed for this behavior. One is that when the shear strain rates for machining estimated using Eq (3.4a) are very high, then the stresses are usually larger. Another explanation is while the analysis assumes all the plastic deformation is ahead of the cutting edge, plastic deformation work hardens the newly generated workpiece surface too. For most practical machining processes, this work hardened region forms the free surface of the chip that is cut on a subsequent pass. This means that for thin chips, there is a greater amount of work hardened material removed, hence the power law dependence on chip thickness in Eq (2.14). Another explanation is the size effect [Shaw84]. Thin chips have a higher strength than thicker chips because the number of defects is less, the same mechanism that gives high strength to metal fibers.

The *specific cutting energy* is an empirically determined characteristic of machining. Later in Chapter 5 it will be used as an alternative to mechanics for estimating forces and power. The specific cutting energy is defined as the energy to cut a unit volume of material, or, the ratio of the power in Eq (3.12) to the volumetric removal rate of Eq (3.5a)

$$E_c = \left(\frac{\text{Cutting Energy}}{\text{Unit Volume of Material}}\right) = \frac{P}{Z}. \quad (3.13)$$

E_c is mainly material dependent, but if the power law expression for τ_s is allowed for, the specific cutting energy for orthogonal cutting is

$$E_c = \frac{\cos(\beta-\gamma_o)\,\tau_0\left(\frac{h}{h_R}\right)^a}{\sin(\phi_o)\cos(\phi_o+\beta-\gamma_o)}.$$

As can be seen from this expression, while the shear strength of the material being cut is a dominant characteristic, the friction angle and shear plane angle indicates a function of the rake interface conditions γ_o and h, represent how tool geometry and operating conditions affect E_c.

3.2.2 *Oblique Machining*

In oblique machining, the single cutting edge and the velocity vector **v** forms an oblique angle called the *inclination angle* λ. To analyze this problem, assume that the coordinates in Figure 3.1 will be used to define v, h and b. However, to look along the cutting edge and get a true view of the angle the edge makes with the y axis requires rotating the y-z plane by λ. This is a perspective similar to the one in Section 3.2.1 for orthogonal machining, but with modifications of course. Once the mechanics have been analyzed in the plane normal to the cutting edge, a rotation by λ will be applied to give a three dimensional solution to the oblique machining problem.

Geometric and kinematic relationships for oblique machining are similar to those for orthogonal cutting, but in addition to the inclination angle λ, another important new angle is needed: the *chip flow angle* η_c. The chip flow angle is measured in the plane defined by the rake face of the cutting edge, and it is measured from a normal to the cutting edge. The chip flow angle represents the direction of the velocity vector v_c on the face of the cutting edge. For orthogonal cutting $\eta_c = 0$, but in oblique machining the chip changes direction as it flows up the face of the cutting edge. In fact, this is one of the advantages of oblique machining over orthogonal machining, choosing λ affects η_c so it is possible to direct the chip flow and keep it from scratching the newly generated surface. Because friction always opposes the chip flow, in oblique machining v_c is parallel to the friction force vector. Several expressions for predicting η_c have been developed and Table 3.2 lists some of the more commonly used ones. All are functions of λ, with some including v, γ_n, β_n or ϕ_n. Figure 3.7 compares the predictions of η_c made with the models in Table 3.2 over a typical range of inclination angles. The predictions tend to be similar, with the magnitudes of η_c less than λ. Because of the uncertainty in knowing the friction or shear plane angles, simple models for η_c, like Stabler's or Russel and Browne's, are reasonable for making oblique machining calculations.

There are several ways to select a reference for measuring the geometry of the cutting edge. Because the rake angle is the most important for determining the shear plane orientation, most of the systems use the term *rake* in their definition. For example, the *velocity or true rake* is measured in a plane normal to the generated surface and parallel to v

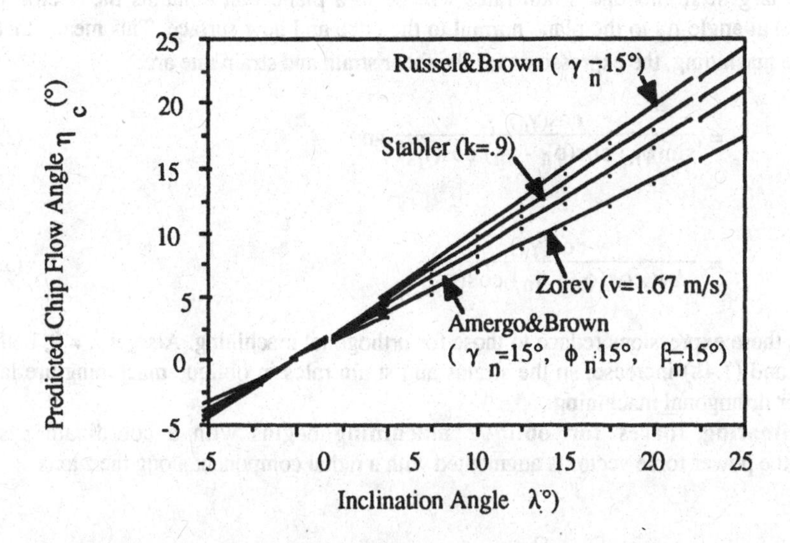

Figure 3.7 Comparing Chip flow prediction methods in Table 3.2.

[Kronen54]. This is an easy angle to visualize, but it depends on the angle of the cutting wedge and the inclination angle. The *effective rake* is measured in a plane that contains both v and v_c vectors [Stable51]. Table 3.2 indicates some of the problems with this definition, namely the different ways used to predict the chip flow direction as well as dependence on the wedge and inclination angles.

The *normal, oblique or primary rake* γ_n, will be the way the cutting edge is defined in this book. The normal rake is measured in a plane that is normal to both the inclined cutting edge and to the newly generated surface. This is a reference that can be used to grind or specify a cutting edge. The velocity and effective rake systems don't allow this because they depend on application specific geometry like η_c or λ. It also turns out that this reference frame results in expressions for oblique machining that are analogous of those developed for orthogonal machining when the shear plane angle is measured in the same plane as γ_n. The resulting angle is called the *normal shear plane angle* ϕ_n, where the subscript *n* stands for *normal*, as the *o* stood for *orthogonal* in Section 3.2.1. For example, Eq (3.1) describing the relationship between the velocity vectors is still valid. However the inclination and chip flow angles affect the magnitudes in a more complicated way

$$\frac{v\cos(\lambda)}{\cos(\phi_n - \gamma_n)} = \frac{v_c\cos(\eta_c)}{\sin(\phi_n)} = \frac{v_{c/s}\cos(\eta_s)}{\cos(\gamma_n)}, \text{ where} \tag{3.2b}$$

$$\eta_s = \tan^{-1}\left(\frac{\tan(\lambda)\cos(\phi_n - \gamma_n) - \tan(\eta_c)\cos(\phi_n)}{\cos(\gamma_n)}\right) \tag{3.14}$$

is referred to as the *shear flow direction*. All the terms in the numerators of Eq (3.2b) are projections of the velocity vectors v, v_c and $v_{c/s}$ on the plane normal to the cutting edge. All of the chip flow angle expressions in Table 3.2 are such that if $\lambda = 0$ then $\eta_c = \eta_s = 0$, so Eq (3.2b) for oblique machining reduces to Eq (3.2a) for orthogonal machining.

The largest strains and strain rates will be in a plane that contains the vector $v_{c/s}$, inclined at angle η_s to the plane normal to the edge and new surface. This means that for oblique machining, the expressions for the shear strain and strain rate are:

$$\gamma = \frac{\cos(\gamma_n)}{\sin(\phi_n)\cos(\phi_n - \gamma_n)\cos(\eta_s)} \text{ and} \tag{3.3b}$$

$$\dot{\gamma} = \frac{v\cos(\gamma_n)}{\Delta w \cos(\phi_n - \gamma_n)\cos(\eta_s)}. \tag{3.4b}$$

Again, these expressions reduce to those for orthogonal machining. Also, if $\lambda \neq 0$, both Eq (3.3b) and (3.4b) increase, so the strains and strain rates in oblique machining are larger than for orthogonal machining.

Estimating forces for oblique machining begins with a coordinate system where the power force vector is augmented with a radial component along the z axis

$$F_P = \begin{bmatrix} F_x \\ F_y \\ F_z \end{bmatrix} = \begin{bmatrix} F_p \\ F_q \\ F_r \end{bmatrix} = \begin{bmatrix} \text{Power or Cutting Force} \\ \text{Normal or Thrust Force} \\ \text{Radial Force} \end{bmatrix}. \tag{3.10g}$$

The y-z plane is rotated by λ about the x axis and forms a primed set of coordinates [ArmBro69]. The z' axis coincides with the cutting edge and the x'-y' plane is normal to the edge and the new surface so that

$$F_P' = \begin{bmatrix} F_x' \\ F_y' \\ F_z' \end{bmatrix} = \begin{bmatrix} 1 & 0 & 0 \\ 0 & \cos\lambda & \sin\lambda \\ 0 & -\sin\lambda & \cos\lambda \end{bmatrix} F_P = T_{yz}(\lambda)\, F_P. \tag{3.10h}$$

The problem is solved in this x'-y'-z' coordinate system where all the quantities with the "n" subscript are defined. This means determining the shear plane and rake face forces so that F_P' can be found. The final step is solving $F_P = F_P'\, T_{yz}(-\lambda)$.

Models for the shear plane and rake face are analogous to those for orthogonal machining. On the shear plane, the magnitude of the shear force is assumed to be given by the product of τ_s and the area of the shear plane. The width of the shear plane is $\overline{AB}$ in Figure 3.8, so b_c for oblique cutting is greater than for the orthogonal machining case. The direction of shearing η_s predicted by Eq (3.14) usually is not equal to the chip flow angle η_c. This means that there can be a shear force component along the cutting edge, as well as normal to it. The force vector in the shear plane, and the rotation to put it into "primed" coordinates are

$$\mathbf{F_s} = \begin{bmatrix} F_{s\_1} \\ F_c \\ F_{s\_2} \end{bmatrix} = \begin{bmatrix} \dfrac{\tau_s\, b\, h \cos(\eta_s)}{\sin(\phi_n)\cos(\lambda)} \\ F_c \\ \dfrac{\tau_s\, b\, h \sin(\eta_s)}{\sin(\phi_n)\cos(\lambda)} \end{bmatrix} = \begin{bmatrix} \text{Shear Normal to edge} \\ \text{Compressive Component} \\ \text{Shear Parallel to edge} \end{bmatrix}$$

$$= \begin{bmatrix} \cos\phi_n & -\sin\phi_n & 0 \\ \sin\phi_n & \cos\phi_n & 0 \\ 0 & 0 & 1 \end{bmatrix} \mathbf{F_p}' = \mathbf{T}_{x'y'}(-\phi_n)\,\mathbf{F_p}' . \qquad (3.10i)$$

The Coulomb friction model on the the rake face assumes that the direction of the friction force opposes $\mathbf{v_c}$. This means the rake face force vector has friction force components normal and parallel to the cutting edge that are balanced by $\mathbf{F_p}'$ as indicated by

$$\mathbf{F_r} = \begin{bmatrix} F_n \\ F_{f\_1} \\ F_{f\_2} \end{bmatrix} \begin{bmatrix} F_n \\ \mu F_n \cos(\eta_c) \\ \mu F_n \sin(\eta_c) \end{bmatrix} = \begin{bmatrix} \text{Normal Rake Component} \\ \text{Friction Normal to Edge} \\ \text{Friction Parallel to Edge} \end{bmatrix} \qquad (3.8d)$$

$$= \begin{bmatrix} \cos\gamma_n & -\sin\gamma_n & 0 \\ \sin\gamma_n & \cos\gamma_n & 0 \\ 0 & 0 & 1 \end{bmatrix} \mathbf{F_p}' = \mathbf{T}_{x'y'}(-\gamma_n)\,\mathbf{F_p}' . \qquad (3.11f)$$

The friction coefficient in Eq (3.8d) and the *normal friction angle* β_n that appears in Table 3.2 are related to μ and β in Eq (3.9a) by the chip flow angle η_c

$$\mu = \frac{\sqrt{F^2_{f\_1}+F^2_{f\_2}}}{F_n} = \frac{\tan(\beta_n)}{\cos(\eta_c)} = \tan(\beta). \qquad (3.9b)$$

This means that the same friction coefficient determined by, for example, the experimental methods in Eq (3.11e), can be used for oblique machining.

The solution for $\mathbf{F_p}$ can be outlined, but only the result will be given. The unknowns are: F_n on the rake plane, F_c on the shear plane, and η_s which is why a check of the solution in Eq (3.14). The static equilibrium equations in the x'-y'-z' plane are $\mathbf{T}_{x'y'}(\phi_n)\mathbf{F_s}=\mathbf{T}_{x'y'}(\gamma_n)\mathbf{F_r}$ or $\mathbf{F_s}=\mathbf{T}_{x'y'}(\gamma_n-\phi_n)\mathbf{F_r}$. Knowing $\mathbf{F_s}$ the transformations to $\mathbf{F_p}'$ and $\mathbf{F_p}$ are applied: $\mathbf{F_p}'=\mathbf{T}_{x'y'}(\phi_n)\mathbf{F_s}$ and $\mathbf{F_p}=\mathbf{T}_{yz}(-\lambda)\,\mathbf{F_p}' =\mathbf{T}_{yz}(-\lambda)\mathbf{T}_{x'y'}(\phi_n)\mathbf{F_s}$. The calculations are long and tedious, but straight forward; ideally suited for a spread sheet or computer program.

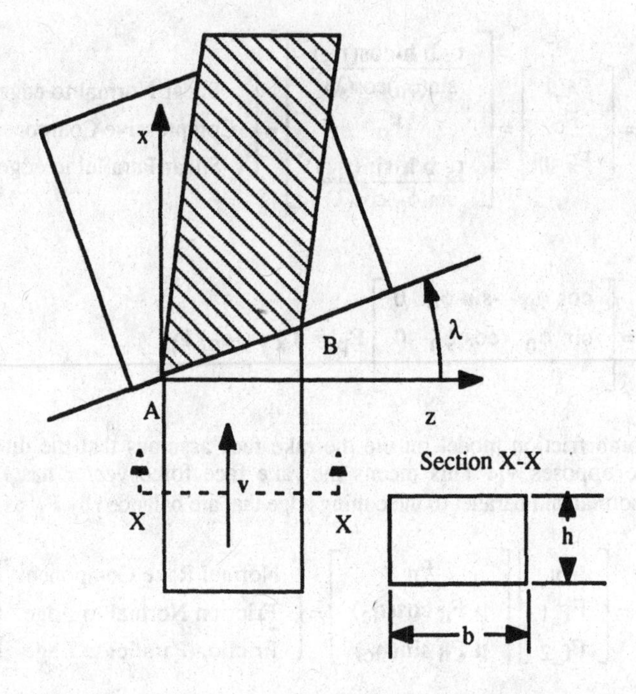

3.8a. View in plane parallel to new surface (true view of **v**).

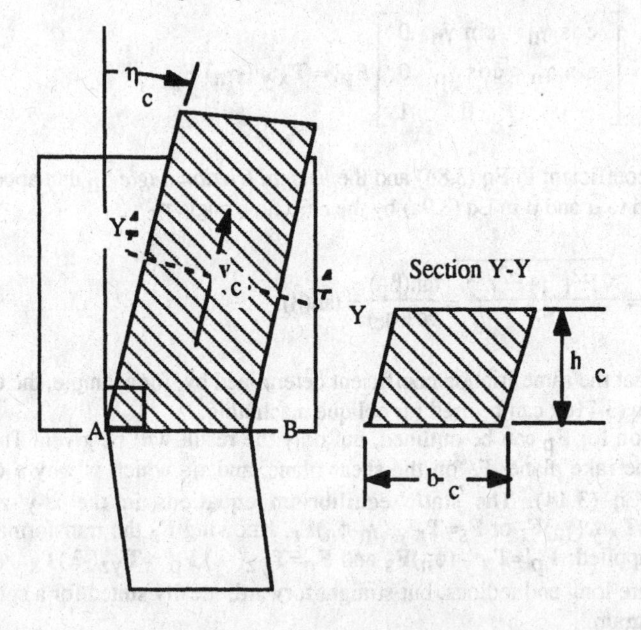

3.8b. View in plane parallel to rake face (true view of $\mathbf{v}_c$).

Figure 3.8 Chip Cross Sections for Oblique Machining.

The resulting expression for the power force vector is,

$$F_p = \begin{bmatrix} \dfrac{\cos(\beta_n-\gamma_n)\cos(\eta_s)\cos(\lambda)+\cos(\phi_n+\beta_n-\gamma_n)\sin(\eta_s)\sin(\lambda)}{\cos(\phi_n+\beta_n-\gamma_n)} \\ \dfrac{\sin(\beta_n-\gamma_n)\cos(\eta_s)}{\cos(\phi_n+\beta_n-\gamma_n)} \\ \dfrac{\cos(\beta_n-\gamma_n)\cos(\eta_s)\sin(\lambda)-\cos(\phi_n+\beta_n-\gamma_n)\sin(\eta_s)\cos(\lambda)}{\cos(\phi_n+\beta_n-\gamma_n)} \end{bmatrix} \frac{\tau_s\, b\, h}{\sin(\phi_n)\cos(\lambda)}. \quad (3.10i)$$

This expression is more complicated than Eq (3.10e) for orthogonal cutting, but so is the problem of oblique machining. Note that in this form, when the inclination angle $\lambda = 0°$, Eq (3.10i) reduces to the orthogonal machining case, Eq (3.10e). This means that basic concepts from orthogonal machining still hold, viz., low friction, large rakes and lower strength properties reduce forces. Also, the power at the cutter is $P = v^T F_p$ as given by Eq (3.12a). However, there is a bit more latitude and control in planning the cut by including the inclination angle, because some control over the chip flow direction η_c can be exercised. Example 3.1 is a sample calculation for both oblique and orthogonal machining that shows some of the differences and similarities in these calculations.

Before proceeding with writing the force vector for the rake face of the cutting edge, a word about finding ϕ_n. Experimentally it can be estimated from the cutting ratio $r = (h/h_c)$. Shear angle prediction equations like those in Table 3.1 for orthogonal machining are rare; two of them are noted in Table 3.2 that are compatible with specific chip flow rules. The relationships are all similar in form to those for orthogonal machining so that using shear plane prediction models for orthogonal machining is a reasonable way to estimate ϕ_n when all else fails.

Table 3.2 Some Chip Flow Angle Prediction Models

Model	Reference	Based on
$\eta_c = \lambda$ $\{\phi_n = 45 - 0.5(\beta_n - \gamma_n) - 0.5\,\beta_n\}$	Stabler [Stable51]	Intuition and observation
$\eta_c = k\,\lambda\ (0.9 < k \leq 1)$	Stabler [Stable51]	Experiments to confirm intuition
$\eta_c = \tan^{-1}(\tan(\lambda)\cos(\gamma_n))$ see 4.44 in A&B $\{\phi_n = 45 - 0.5(\beta_n - \gamma_n) - 0.5\,\beta_n\}$	Russel and Brown [RusBro66]	Empirical
$\eta_c = \dfrac{\lambda}{\left(\dfrac{v}{1\ \text{m/min}}\right)^{0.08}}$	Zorev [Zorev66]	Empirical to account for effect of cutting speed.
$\eta_c = \tan^{-1}\left(\tan(\lambda)\left(\dfrac{\cos(\gamma_n)+\sin(\gamma_n)}{\tan(\phi_n+\beta_n)}\right)\right)$	Armerego and Brown [ArmBro69]	As part of a shear plane model with friction.

Example 3.1 Force Calculations With Orthogonal and Oblique Models

This example compares the forces estimated with both an orthogonal and oblique machining model. This example illustrates that with a mechanics model. It is possible to get a feeling for how different variables that we have control over when planning a machining operation affect the process.

For a student taking a course, trying a range of solutions can be long and tedious, but real process analysis involves trying different alternatives. As a result, use calculation tools that are easy for you. Most of the examples and solutions in this text are either done on a spread sheet (EXCEL) or with simple computer codes (written in C).

For both parts of this example, assume the following are known:

- $\phi_{o/n} = 45 - (\gamma_{o/n} - \beta_{o/n})$, the Lee-Shafer model for shear plane angle from Table 3.1,
- $\eta_c = \tan^{-1}(\tan(\lambda)\cos(\gamma_n))$, Russel and Brown's rule for chip flow direction from Table 3.2,
- $\mu = 0.8$,
- $\tau_s = 338\ (\text{MPa}) \left(\frac{h}{1\ \text{mm}}\right)^{-0.11}$, the shear stress from data in [Cre*etal*57] for annealed 1113 steel turned without cutting fluid and
- $h = 0.75$ mm, $b = 5$ mm, and $v = 0.5$ m/s.

Orthogonal Machining: determine P, the force components F_s and F_f, and $|\mathbf{F_p}|$ for orthogonal rake angles from $-5° \le \gamma_o \le 14°$. The steps to the solution are straight forward applications of the equations in section 3.2.1. The calculations are given in Table 3.3 and the results in this table are plotted against γ_o in Figure 3.9a. One of the main points is that negative rake angles lead to large force and power requirements, primarily because of the small shear plane angles. Both the table and the figure indicate that forces go up quickly with negative rakes. The effect of large negative rake angles will be referred to again in Chapter 6 to explain the specific energy requirements in grinding. Also, for positive rakes there is a point where increasing the rake angle has little effect on reducing the forces.

Oblique Machining: determine P, the $\mathbf{F_p}$ and $|\mathbf{F_p}|$ for a normal rake angle $\gamma_n = 14°$ and for inclination angles in the range $0° \le \lambda \le 20°$. The components of the force vector $\mathbf{F_p}$ are found from Eq (3.10i). In addition to the inclination and chip flow angles, the shear flow angle in Eq (3.14) is used in the calculation of the normal friction angle β_n. Then the calculations, while long, are straight forward. Table 3.4 shows the numerical results and Figure 3.9b the plotted results of these computations. The first thing to notice is that the forces and power when $\lambda = 0$ in Table 3.4 are exactly the same as the last row in Table 3.3 when $\gamma_o = 14°$ as it should be! As the inclination angle increases, the radial or z-directional force increases, while the magnitudes of both F_p and F_q drop a little. Because this model assumes that there is no velocity component in the y or z-directions, the decrease in F_p causes a slight decrease in the power.

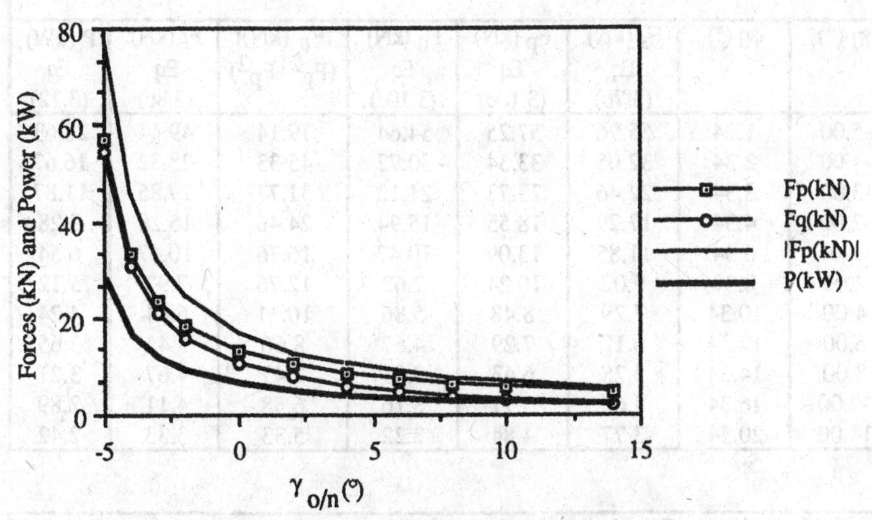

Figure 3.9a Forces with Orthogonal Machining Model.

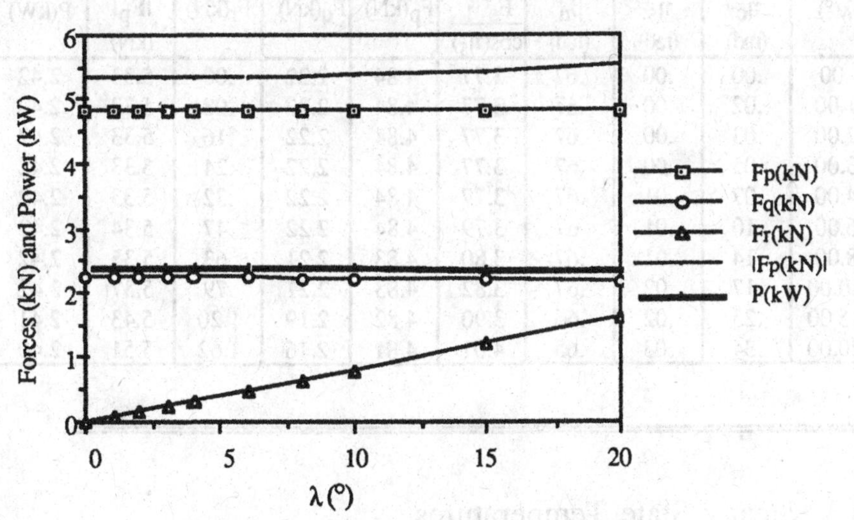

Figure 3.9b Forces with Oblique Cutting Model.

Figure 3.9 Forced Calculations for Example 3.1.

Table 3.3 Force Calculations for Orthogonal Machining, v = 0.5 m/s

γ_0 (°)	ϕ_0 (°)	F_s (kN) Eq (3.7b)	F_p (kN) Eq (3.10e)	F_q (kN) Eq (3.10e)	$\lvert F_p$ (kN)$\rvert$ $(F_p^2+F_p^2)$	F_f (kN) Eq (3.8c)	P (kW) Eq (3.12)
-5.00	1.34	55.96	57.25	54.64	79.14	49.44	28.63
-4.00	2.34	32.05	33.34	30.72	45.33	28.32	16.67
-3.00	3.34	22.46	23.73	21.12	31.77	19.85	11.87
-2.00	4.34	17.29	18.55	15.94	24.46	15.28	9.28
.00	6.34	11.85	13.09	10.47	16.76	10.47	6.54
2.00	8.34	9.02	10.24	7.62	12.76	7.97	5.12
4.00	10.34	7.29	8.48	5.86	10.31	6.44	4.24
6.00	12.34	6.12	7.29	4.67	8.66	5.41	3.65
8.00	14.34	5.28	6.43	3.81	7.47	4.67	3.21
10.00	16.34	4.65	5.77	3.16	6.58	4.11	2.89
14.00	20.34	3.77	4.84	2.22	5.33	3.33	2.42

Table 3.4 Force Calculations for Oblique Machining, v = 0.5 m/s and γ_n = 14°

λ(°)	η_c (rad)	η_s (rad)	β_n (rad)	$\frac{F_s\ 1}{\cos(\eta_s)}$	F_p(kN)	F_q(kN)	F_r(kN)	$\lvert F_p \rvert$ (kN)	P(kW)
.00	.00	.00	.67	3.77	4.84	2.22	.00	5.33	2.42
1.00	.02	.00	.67	3.77	4.84	2.22	.08	5.33	2.42
2.00	.03	.00	.67	3.77	4.84	2.22	.16	5.33	2.42
3.00	.05	.00	.67	3.77	4.84	2.22	.24	5.33	2.42
4.00	.07	.01	.67	3.77	4.84	2.22	.32	5.33	2.42
6.00	.10	.01	.67	3.79	4.84	2.22	.47	5.34	2.42
8.00	.14	.01	.67	3.80	4.83	2.21	.63	5.35	2.42
10.00	.17	.02	.67	3.82	4.83	2.21	.79	5.37	2.42
15.00	.25	.02	.66	3.90	4.82	2.19	1.20	5.43	2.41
20.00	.34	.03	.65	4.01	4.81	2.16	1.62	5.51	2.40

3.3 Steady State Temperatures

Elevated temperatures at the workpiece can affect the functional performance of a part due to residual stresses or thermal distortion. Elevated temperature at the cutting edge is the primary factor that accelerates cutting edge wear. As the calculations here will show, v is the main operating variable that affects temperature. This is why the empirical relationship called the Taylor's tool life equation depends most on speed as was illustrated by the factorial design screening experiments in Example 2.5.

This section deals with average steady state temperature calculations. The models for the average steady state calculations are relatively simple, traceable to Jaeger's work on interface temperatures for sliders like the one in Figure 3.10 [Jaeger42]. In dimensionlessform the solutions in Eq (3.15) are for the temperature rise along the slider length, the average temperature rise and maximum temperatures.

$$\Theta(l) = \frac{\Delta\theta(l)}{\left(\frac{q}{\rho\, c_p\, v}\right)} = \frac{2}{\sqrt{\pi}}\sqrt{\frac{l\ v}{\kappa}} \quad (3.15a)$$

$$\bar{\Theta} = \frac{\Delta\bar{\theta}}{\left(\frac{q}{\rho\, c_p\, v}\right)} = \frac{4}{3\sqrt{\pi}}\sqrt{\frac{l_{cc}\ v}{\kappa}} = \frac{4}{3\sqrt{\pi}}\sqrt{Pe} \quad (3.15b)$$

$$\Theta_{max} = \frac{\Delta\theta_{max}}{\left(\frac{q}{\rho\, c_p\, v}\right)} = \frac{2}{\sqrt{\pi}}\sqrt{\frac{l_c\ v}{\kappa}} = \frac{2}{\sqrt{\pi}}\sqrt{Pe} \quad (3.15c)$$

Applied to machining and grinding, these expressions are used to predict temperature rises on the rake face of a cutting edge or along the contact length in grinding. With these expressions determining q, the heat flux at the slider interface, is the key machining technology problem.

Two applications of these equations will be covered: estimating chip and rake face temperatures in orthogonal machining, and lumped parameter calculation of grinding temperatures.

3.3.1 Estimating Chip and Rake Temperatures in Orthogonal Machining

Heat sources in orthogonal machining are identified by the circled regions in Figure 3.11. In order of importance these sources are:

- The *shear zone* is where much of the mechanical energy supplied to the cutting process is converted into heat. This energy has to be allocated between that needed to cause plastic flow (a small amount) and that converted into heat (a large amount). This major heat source causes the average temperature of the chip to rise by $(\bar{\theta}_c - \theta_0)$ as it passes through the shear plane.
- The second heat source is the *rake face*, where friction heat is generated as the chip moves along the cutting edge at velocity v_c. This interface is analogous to the slider in Figure 3.10, so Eq (3.15b) is used to predict the average temperature rise of $(\bar{\theta}_r - \bar{\theta}_c)$ at the rake interface. This temperature can affect the hardness, toughness and wear rates of the cutting edge.
- The *clearance face* is the third heat source in order of importance. Heat is produced by rubbing or friction between the cutting edge and the newly generated surface, similar to the way heat is generated at the rake face. For the idealized picture in Figure 3.11, the contact area is zero because of the sharp cutting edge. As a result, this heat source is usually neglected in temperature calculations for sharp tools. However the clearance or

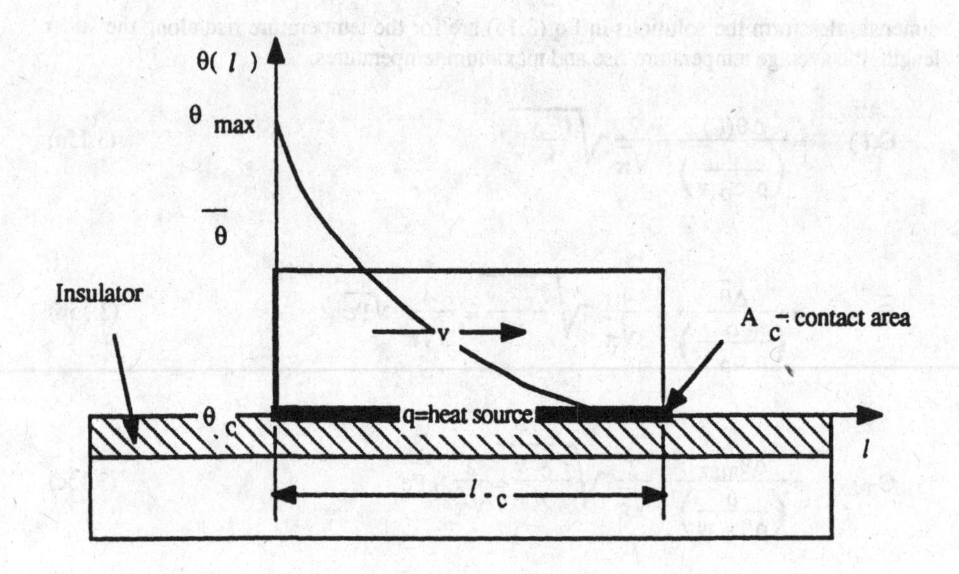

Figure 3.10 Jaeger's model for temperatures for a moving heat source.

flank face becomes a significant heat source when the contact area between the tool and workpiece increases, due to, for example tool wear.

Heat flows in orthogonal machining are identified by the arrows in Figure 3.11.

- While the heat sources remain stationary relative to the dashed boundaries of the thermal system, the *chip mass* forms the largest and most important heat sink. Heat is removed by mass transport through the system boundary and away from the cutting edge.
- Conduction across the interface between the chip and *cutting edge* is another heat flow.

  Calculations will show that $(\bar{\theta}_r - \theta_0) > (\bar{\theta}_c - \theta_0)$, so that on the average this interface is one way heat is removed from the cutting edge. In the lumped parameter temperature calculations, this flow will assumed to be minor, and this assumption will be checked by the Peclet number calculation in Eq (3.1).
- The third heat flow, in order of importance, is heat transfer into the *workpiece*. Some of this may be conduction through the uncut chip back into the workpiece and some may be conduction from the edge into the workpiece. For sharp cutting edges, conduction through the cutting edge can be neglected because the contact area is zero. Like the clearance face as a heat source, as the contact area increases due to wear, this becomes more important as a heat sink.

Identification of the heat sources and flows and making assumptions about their relative importance essentially solves the problem. Figure 3.12 is an idealization of the problem, indicating the energy flows that are used to make the average chip and rake face temperature calculations. Lumped parameter calculations will be used that assume no heat is generated or transfered between the cutting edge and the newly generated work surface, i.e., the third heat source and heat flow are neglected. Also, while the rake face serves as a heat source, it is assumed there is no heat transfer by conduction between the chip and cutting edge. A

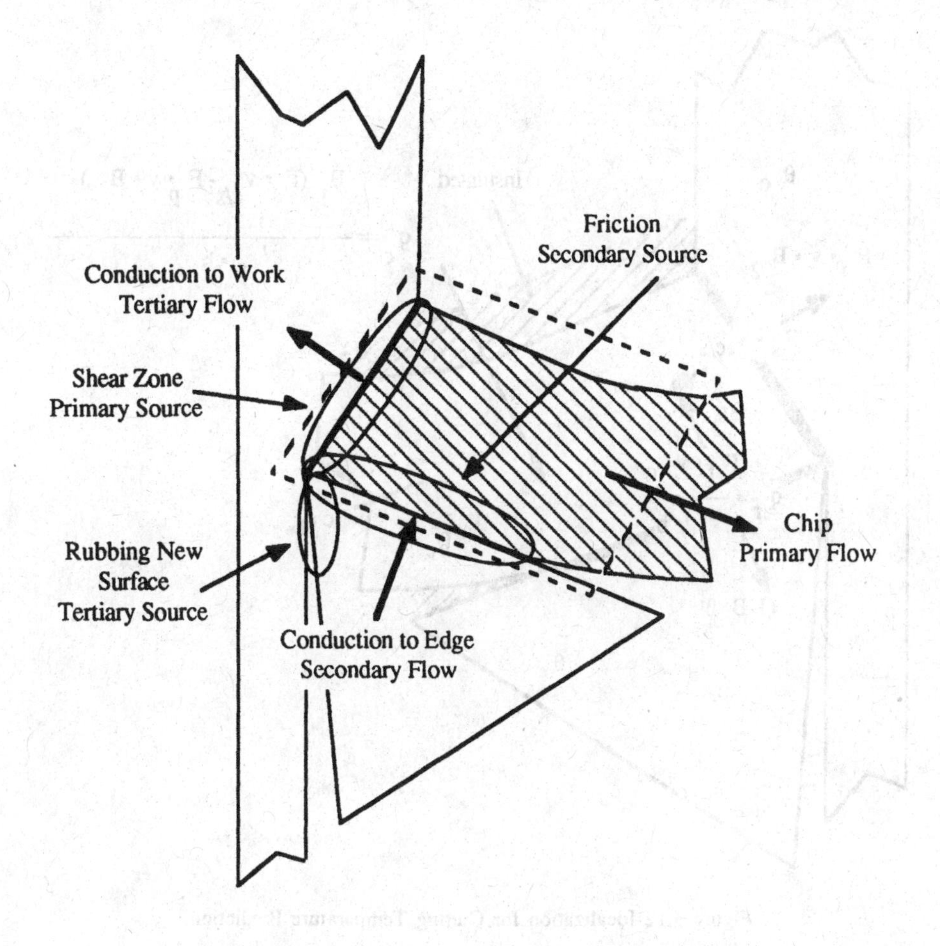

Figure 3.11 Heat sources and sinks in machining.

heuristic way to justify this is that the time the chip is in contact with the rake face is very small. For example, if $v_c = 1$ m/s and the contact length $l_c = 2$ mm, the contact interval is 2 ms. A more quantitative way to check whether the lumped analysis is valid is with the Peclet number. The Peclet number for the chip should be greater than 20

$$Pe_c = \frac{l_c \bullet v_c}{\kappa_c} = \frac{l_c \bullet v_c \bullet c_c \bullet \rho_c}{k_c} > 20 \qquad (3.16a)$$

where l_c is the contact length and as before v_c is the chip velocity. Other quantities to define the Peclet number for the chip are: k_c the thermal conductivity, c_c the heat capacity and ρ_c the density used to compute κ_c the thermal diffusivity. If the lumped parameter analysis is not valid, as it would be machining poor conductors at low speeds, numerical solutions like those in [Tay*etal*74] may be needed.

Estimating the average chip temperature $\bar{\theta}_c$, assumes the heat loss to the rake face of the tool is negligible, and the other sections are insulated. With this assumption, from an energy balance for the control volume, the heat flux through the

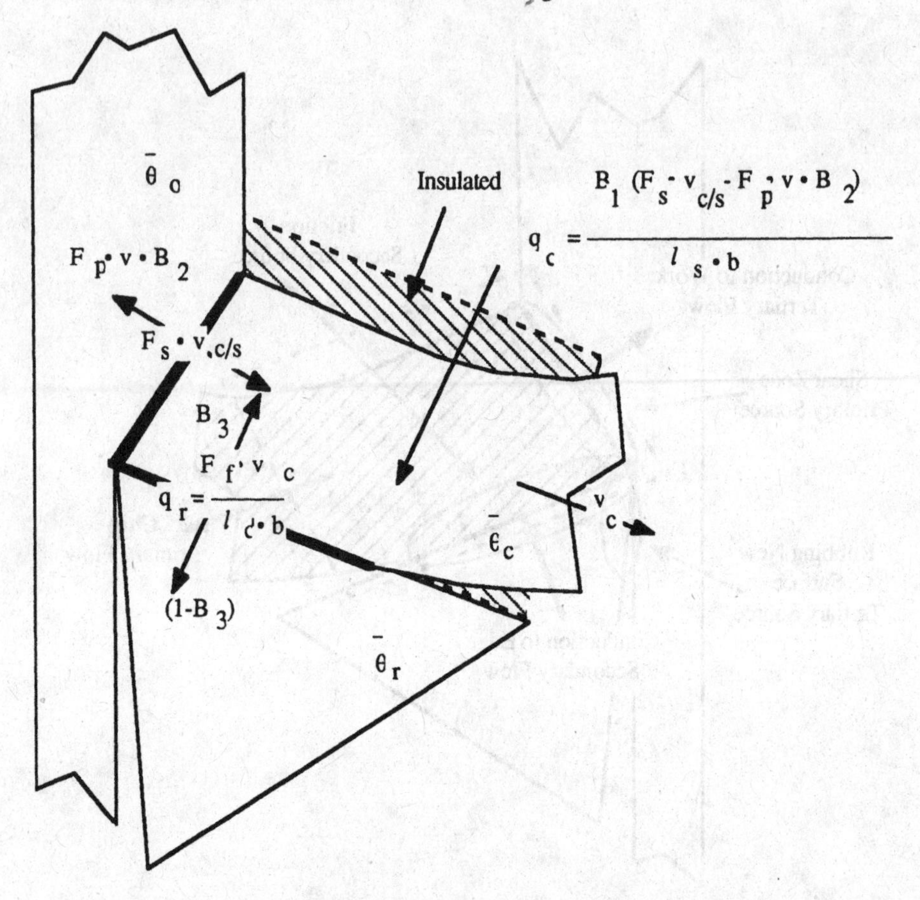

Figure 3.12 Idealization for Cutting Temperature Prediction.

uncut chip cross section (h•b) into the chip is,

$$q_c = (\bar{\theta}_c - \bar{\theta}_0)\, c_c \bullet \rho_c \bullet v \tag{3.17a}$$

where c_c is the specific heat of the chip, ρ_c is the chip density, and v is the speed as before. The problem is determining q_c, accounting for the fact that of the total rate at which mechanical energy goes into the process - the power in Eq (3.12) - some is conducted through the uncut chip into the workpiece (the tertiary heat flow) and most of what's left raises the temperature of the chip. However, a small amount of work must go into plastically deforming the chip. Calculating this energy partition is difficult in machining, but one way is in terms of the forces and velocities estimated in Section 3.2. Then,

$$q_c = \frac{B_1 \bullet [F_s \bullet v_{c/s} - F_p \bullet v \bullet B_2]}{b \bullet l_s} = \frac{B_1 \bullet [F_p \bullet v \bullet (1-B_2) - F_f \bullet v_c]}{b \bullet l_s} \tag{3.17b}$$

where B_1 is the fraction of energy at the shear zone that goes into raising the temperature of the chip, typically in the range $0.85 \leq B_1 \leq 0.95$. B_2 accounts for the division of shear zone heat between the chip and workpiece. It is the fraction of the power to make a chip that is conducted into the workpiece, found to be in the range $0.05 \le B_2 \le 0.15$. The quantity in brackets in Eq (3.17b) is the energy crossing the shear zone, and B_1 indicates how much of this energy raises the average chip temperature. After substitution the average chip temperature rise is:

$$(\bar{\theta}_c-\bar{\theta}_0) = \frac{B_1\bullet[F_s\bullet v_{c/s} - F_p\bullet v\bullet B_2]}{c_c\bullet\rho_c\bullet v\bullet h\bullet b} = \frac{B_1\bullet[F_p\bullet v\bullet(1-B_2) - F_f\bullet v_c]}{c_c\bullet\rho_c\bullet v\bullet h\bullet b} = \frac{q_c}{c_c\bullet\rho_c\bullet v}. \quad (3.17c)$$

Estimating the average rake interface temperature $\bar{\theta}_r$, is a bit more complicated than the average chip temperature. The calculations that follow are from Trigger and Chao [TriCho51], but Loewen and Shaw [LoeSha54] and Boothroyd [BooKni89] give solutions along similar lines. The average temperature of the interface rises from $\bar{\theta}_c$ to $\bar{\theta}_r$ because the friction heat of the chip sliding up the rake face ($F_f\bullet v_c$) acts as a heat source that can be modeled using Jaeger's model, Eq (3.15). This heat is transfered through the rake interface $A_r = b\bullet l_c$.

For the lumped analysis, the total rake face heat flux from this source is

$$q_r = \frac{F_f\bullet v_c}{A_r} = \frac{F_f\bullet v_c}{b\bullet l_c}. \quad (3.18a)$$

Part of this flux goes into the chip (B_3) and the rest goes into the cutting edge ($1-B_3$). This partition of the heat energy is found by solving for B_3 in

$$(\bar{\theta}_c-\bar{\theta}_0) + \frac{4}{3\sqrt{\pi}}\left(\frac{B_3\bullet q_r}{c_c\bullet\rho_c\bullet v_c}\right)\sqrt{Pe_c} = \frac{4(1-B_3)\bullet b\bullet q_r}{\pi\bullet k_e}\bullet SF(b/l_c) = (\bar{\theta}_r-\bar{\theta}_0) \quad (3.18b)$$

where k_e is the thermal conductivity of the cutting edge and the last term is a shape factor that depends on the ratio of the width to the length of the rake contact area (b/l_c). This function is

$$SF(b/l_c) = \frac{1}{3}\left(\frac{1}{(b/l_c)}\right)^2 + \frac{1}{6}(b/l_c)\left\{\left[\left(\frac{1}{(b/l_c)}\right)^2+1\right]^{3/2} - 1\right\}$$

$$+\frac{1}{2}\left(\frac{1}{(b/l_c)}\right)\left\{\sinh^{-1}(b/l_c) - \left[\left(\frac{1}{(b/l_c)}\right)^2+1\right]^{1/2}\right\}. \quad (3.18c)$$

Figure 3.13 is a plot of $SF(b/l_c)$ for $(b/l_c) < 5$. Once the shape factor is computed, the solution for B_3 from Eq (3.18b) is

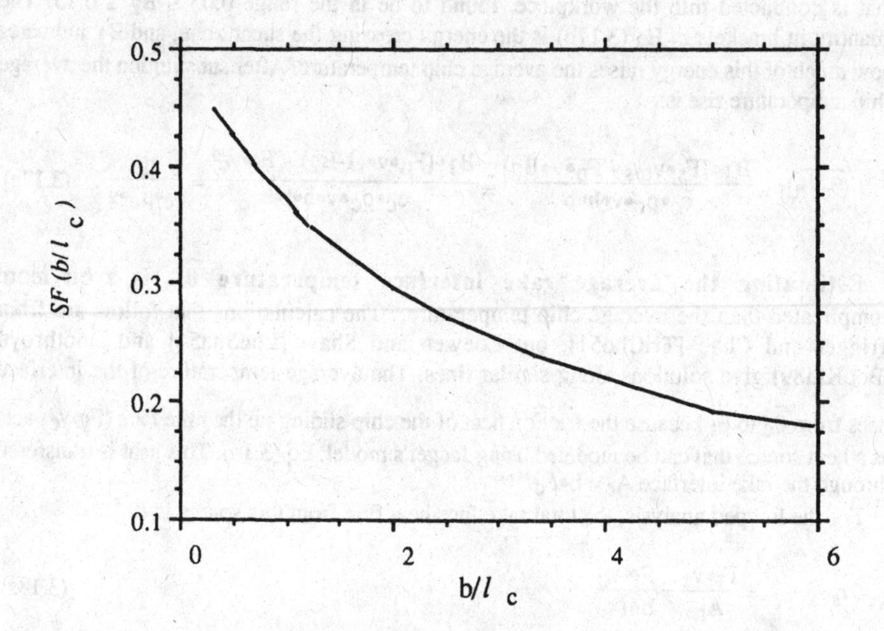

Figure 3.13 Shape Factor.

$$B_3 = \frac{\dfrac{4 \bullet b \bullet q_r \bullet SF(b/l_c)}{\pi \bullet k_e} - (\bar{\theta}_c - \bar{\theta}_0)}{\dfrac{4 \bullet b \bullet q_r \bullet SF(b/l_c)}{\pi \bullet k_e} + \dfrac{4}{3\sqrt{\pi}}\left(\dfrac{q_r}{c_c \bullet \rho_c \bullet v_c}\right)\sqrt{Pe_c}} . \tag{3.18d}$$

Once this fraction of the heat generated at the rake face and conducted into the chip is known, the average rake face temperature is easily estimated from Eq (3.15b) or Eq (3.18b)

$$\bar{\theta}_r = \bar{\theta}_c + \frac{4}{3\sqrt{\pi}}\left(\frac{B_3 \bullet q_r}{c_c \bullet \rho_c \bullet v_c}\right)\sqrt{Pe_c} = \bar{\theta}_0 + \frac{4(1-B_3) \bullet b \bullet q_r}{\pi \bullet k_e} \bullet SF(b/l_c). \tag{3.15d}$$

High temperatures on the rake face of a cutting edge have a negative effect on the life of a cutting edge. While thermal properties of the chip material and the cutting edge are important in determining the magnitude of $\bar{\theta}_r$, Eqs (3.18a) and (3.15d) indicate that v, which is proportional to v_c, is the operating variable that has the greatest effect on rake face temperature.

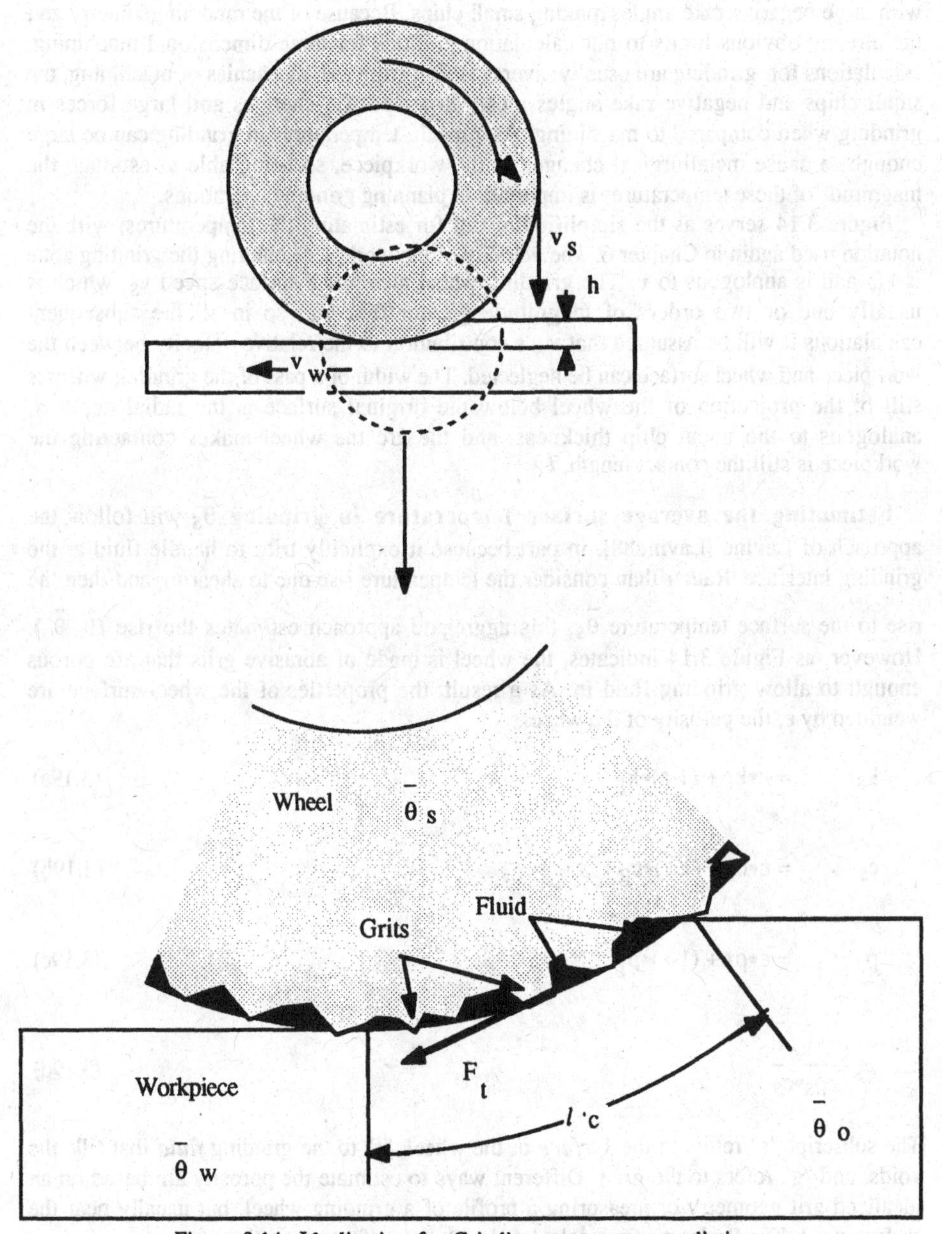

Figure 3.14 Idealization for Grinding temperature prediction.

3.3.2 *Estimating Interface and Workpiece Temperatures in Grinding*

The basic thermal model for temperatures in grinding is similar to the one in the previous section for orthogonal machining. While grinding will be covered in more detail in Chapter 6, it can be thought of as three dimensional machining with many random cutting edges with large negative rake angles making small chips. Because of the random geometry and the already obvious limits to our calculation abilities for three dimensional machining, calculations for grinding are usually averages. But from the mechanics of machining, the small chips and negative rake angles mean large specific energies and large forces in grinding when compared to machining. As a result, temperatures in grinding can be large enough to cause metallurgical changes in the workpiece, so being able to estimate the magnitude of these temperatures is important in planning grinding operations.

Figure 3.14 serves as the simplified model for estimating the temperatures, with the notation used again in Chapter 6. The velocity of the workpiece entering the grinding zone is v_W and is analogous to v. The grinding wheel also has a surface speed v_S, which is usually one or two orders of magnitude greater that v_W, so in all the subsequent calculations it will be assumed that v_W's contribution to the relative velocity between the workpiece and wheel surface can be neglected. The width of a pass of the grinding wheel is still b; the projection of the wheel below the original surface is the radial depth a, analogous to the uncut chip thickness, and the arc the wheel makes contacting the workpiece is still the contact length, l_c.

Estimating the average surface temperature in grinding $\bar{\theta}_s$ will follow the approach of Lavine [Lavine88], in part because it explicitly tries to handle fluid at the grinding interface. Rather than consider the temperature rise due to shearing and then the rise to the surface temperature $\bar{\theta}_S$, this aggregate approach estimates the rise $(\bar{\theta}_S-\bar{\theta}_O)$. However, as Figure 3.14 indicates, the wheel is made of abrasive grits that are porous enough to allow grinding fluid in. As a result, the properties of the wheel surface are weighted by ε, the porosity of the wheel.

$$k_S = \varepsilon \bullet k_f + (1-\varepsilon) \bullet k_g \tag{3.19a}$$

$$c_S = \varepsilon \bullet c_f + (1-\varepsilon) \bullet c_g \tag{3.19b}$$

$$\rho_S = \varepsilon \bullet \rho_f + (1-\varepsilon) \bullet \rho_g \tag{3.19c}$$

$$\kappa_S = \frac{k_S}{\rho_S \bullet c_S} \tag{3.19d}$$

The subscript "s" refers to the *surface* of the wheel, "f" to the grinding *fluid* that fills the voids, and "g" refers to the *grits*. Different ways to estimate the porosity are based on an idealized grit geometry or measuring a profile of a grinding wheel, but usually near the surface $\varepsilon \approx 1$. The Peclet number, calculated with the surface properties,

$$Pe_S = \frac{l_c \bullet v_S}{\kappa_S} = \frac{l_c \bullet v_S \bullet c_S \bullet \rho_S}{k_S} > 20 \tag{3.17b}$$

should again be used to check if it is appropriate to neglect conduction through the grinding zone. The heat flux in the grinding zone is

$$q_g = \frac{F_t \cdot v_s}{l_c \cdot b} \tag{3.20}$$

Instead of the empirical coefficients partition the energy, *all* the energy is assumed to be converted into heat. An energy balance can be used to estimate the fraction of energy into the workpiece so that the average workpiece surface temperature is

$$\bar{\theta}_w = \bar{\theta}_o + \frac{4}{3\sqrt{\pi}}\left(\frac{q_g}{c_w \cdot \rho_w \cdot v_s}\right)\sqrt{Pe_w}\,\frac{\sqrt{c_w \cdot \rho_w \cdot k_w}}{\sqrt{c_s \cdot \rho_s \cdot k_s}+\sqrt{c_w \cdot \rho_w \cdot k_w}} \tag{3.15e}$$

The average wheel surface temperature can be estimated in a similar way, i.e.,

$$\bar{\theta}_s = \bar{\theta}_o + \frac{4}{3\sqrt{\pi}}\left(\frac{q_g}{c_s \cdot \rho_s \cdot v_s}\right)\sqrt{Pe_s}\,\frac{\sqrt{c_s \cdot \rho_s \cdot k_s}}{\sqrt{c_s \cdot \rho_s \cdot k_s}+\sqrt{c_w \cdot \rho_w \cdot k_w}} \tag{3.15f}$$

When this is compared with the calculation of the average rake temperature for machining, it can be seen that B_3 in Eq (3.15d) is analogous to the last terms in Eq (3.15e and f). The thermal properties of the surface of the wheel and the workpiece material will play a significant role in determining how the heat generated in grinding is distributed.

These calculations are approximate. However, the approximation in the modeling is usually better than the approximation in finding the correct data to apply in the calculations. Finding data on the thermal properties of, for example, a new abrasive may be very difficult. Often times in these instances engineering judgement is needed to decide how closely a new abrasive's properties are to an existing abrasive. If grinding appears to be a key processing step, the other alternative is an efficient experimental program to find out these properties for yourself.

Table 3.5 Thermal Properties for Selected Workpiece, Cutting Edge, Abrasive or Cutting or Grinding Fluids

Material	Thermal Conductivity - k W/m-°K	Density - ρ kg/m³	Heat Capacity - c_p J/kg-°K
Water	0.613	997	4179
Kerosene	0.15	820	2000
Aluminum Oxide	6.74 - 46	3800 - 3970	765 - 770
CBN	20 - 130	3450 - 3490	810
Plain Carbon Steel	38.9 - 62.8	7800 - 7854	434- 574
Titanium	28.7	4540	527
Aluminum Alloys	223	2700 - 2100	920 - 962

Example 3.2 Temperature Calculations in Orthogonal Cutting and Grinding.

This example is aimed at comparing the temperatures in machining and grinding a 1045 steel with the following properties: ρ_{Fe} = 7800 (kg/m^3), c_{Fe} = 0.574 (J/g-K) and k_{Fe} = 38.9 (W/m-K). Also, the ambient temperature is 30°C.

In an orthogonal cutting experiment with this steel, a cutting speed of 1.52 m/s, a chip width of 3.02 mm and the uncut chip thickness led to l_s = 0.25 mm. A high speed steel (HSS) cutting edge was used, and the conduction coefficient for this edge material is k_{HSS} = 41.8 (W/m-K). Experimental measurements, rather than calculations were used to determine the forces, chip velocity and contact length. These values are: F_p = 1680 N, F_f = 970 N, v_c = 0.57 m/s and l_c = 1.2 mm. Values of B_1 = 0.875 and B_2 = 0.10 were available for this combination of workpiece and cutting edge.

Estimate the average chip and rake face temperature.

The first step is to check if the lumped parameter analysis is appropriate, which involves computing the Peclet number in Eq (3.16a)

$$\kappa_{Fe} = \frac{(38.9\ \text{W/m-°K})\ (10^6\ \text{mm}^2/\text{m}^2)}{(7800\ \text{kg/m}^3)\ (10^3\ \text{g/kg})\ (0.574\ \text{J/g-K})} = 8.69\ (\text{mm}^2/\text{s})\ \text{or}$$

$$Pe_{Fe} = \frac{(1.2\ \text{mm})\ (0.57\ \text{m/s})\ (10^3\ \text{mm/m})}{(8.69\ \text{mm}^2/\text{s})} = 78.73 > 20$$

so the lumped parameter analysis is valid.

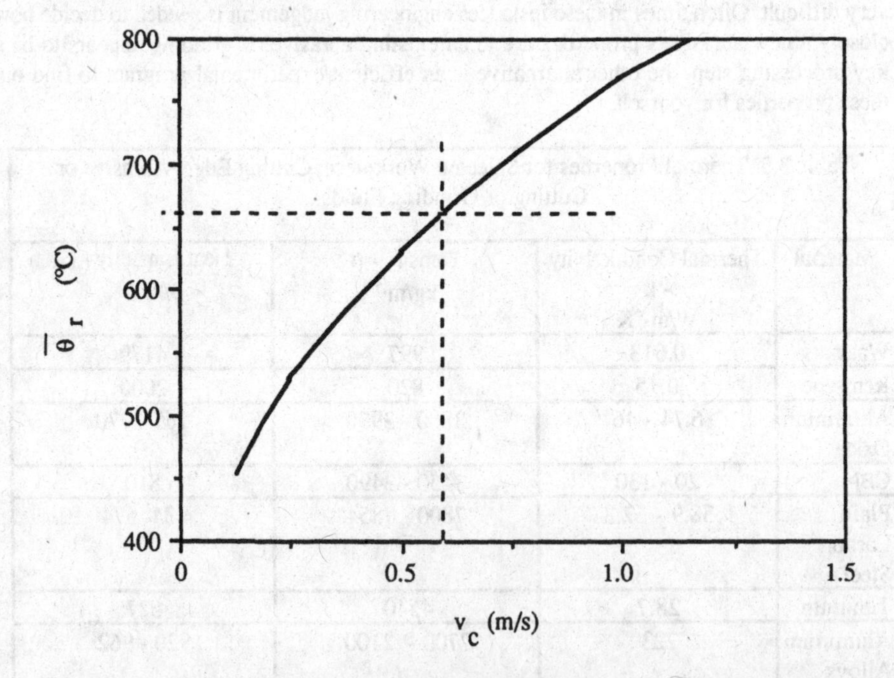

Figure 3.15 Estimated rake face temperature as a function of v_c for Example 3.2.

The average chip temperature is found based on the energy flux that raises the chip temperature. All the data to calculate q_c in Eq (3.17a) is given

$$q_c = \frac{0.875 \bullet [(1680\ N)(1.52\ m/s)(1\text{-}0.10)\ -\ (970\ N)(0.57\ m/s)]}{(3.02\ mm)\ (0.25\ mm)} = 2022.75\ (W/mm^2)$$

so the average chip temperature is

$$\bar{\theta}_c = 30°C + \frac{(2022.75\ W/mm^2)\ (10^6\ m^2/\ mm^2)}{(0.574\ J/g\text{-}°K)\ (7800\ kg/m^3)\ (10^3 g/kg)\ (1.52\ m/s)} = 327.23\ °C$$

Estimating the average rake face temperature requires estimating the rake face heat flux, the energy partition B_3, and then using the Jaeger relationship in Eq (3.15d). The energy flux is given by Eq (3.18a)

$$q_r = \frac{(970\ N)\ (0.57\ m/s)}{(3.02\ mm)\ (1.2\ mm)} = 152.57\ (W/mm^2)$$

The shape factor, by evaluating Eq (3.18c) is $SF(b/l_c{=}2.52) = 0.27$, which can be checked by referring to Figure 3.13. Then B_3 is

$$B_3 = \frac{\dfrac{4(3.03mm)(152.57W/mm^2)(0.27)}{\pi(41.8W/m\text{-}°K)(10^{-3}m/mm)} - (327.23°C - 30°C)}{\dfrac{4(3.03mm)(152.57W/mm^2)(0.27)}{\pi(41.8W/m\text{-}°K)(10^{-3}\ m/mm)} + \dfrac{4}{3\sqrt{\pi}}\left(\dfrac{(152.57W/mm^2)\sqrt{78.73}}{(0.574J/g\text{-}°K)(7800kg/m^3)(0.57m/s)}\right)}$$

$$= \frac{(3796.30 - 297.23)°C}{(3796.30 + 399.02)°C} = 0.83$$

or 83 percent of the heat generated at the contact interface goes into and away with the chip. The average rake face temperature, from the Jaeger solution in Eq (3.15d) is

$$\bar{\theta}_r = 327.23°C + \frac{4}{3\sqrt{\pi}}\left(\frac{(0.83)(152.57W/mm^2)\sqrt{78.73}}{(0.574J/g\text{-}°K)(7800kg/m^3)(0.57m/s)}\right) = 660.03\ °C$$

Figure 3.15 illustrates how this temperaure is affected by the chip velocity. To put these temperatures in perspective, steel tempering, which increases toughness while decreasing hardness, is done at temperatures above 300 °C. The eutectoid temperature, where the body centered cubic structure changes to face centered cubic, is about 730 °C.

In a plunge grinding experiment with the same 1045 steel and an AlO_2 abrasive wheel 15 mm wide. For this type of plunge grinding, the wheel width is the same as the grinding width, i.e., b = 15 mm. The workpiece speed is 0.50 m/s, the radial depth is 0.05 mm, and for the wheel diameter used $v_s = 20$ m/s and the contact length $l_c = 3.20$ mm. A tangential grinding force 550 N was measured. Grinding was done with a very light oil with thermal properties close to kerosene and a wheel where the porosity was 0.80. Thermal properties of the AlO_2 abrasive and the grinding fluid are tabulated in Table 3.6:

Table 3.6 Thermal Properties of Grinding Fluid and Abrasive for Example 3.2		
Property	Abrasive AlO_2	Fluid Kerosene
Density (kg/m^3)	3800	820
Heat Capacity (J/kg-°K)	770	200
Thermal Conductivity (W/m-°K)	6.74	0.15

Estimate the average workpiece temperature and the fraction of energy that goes into the workpiece and into the wheel.

As Eq (3.15e) indicates, finding $\bar{\theta}_{Fe}$ requires finding the energy partition, so that will be the starting point.

As a check of whether the analysis is valid, the Peclet number for the workpiece can be calculated using the same thermal properties as the machining example, so

$$Pe_{Fe} = \frac{(3.2\ \text{mm})\ (20\ \text{m/s})\ (10^3\ \text{mm/m})}{(8.69\ \text{mm}^2/\text{s})} = 7369.68 > 20.$$

Because of the much higher velocity with grinding, the Peclet number is much greater than for the machining example and indicates there is almost no conduction between the work and the wheel.

The composite properties of the wheel surface are needed to estimate the energy partition. From Eq (3.19) these properties are:

$$\rho_s = 0.80(820.00) + (1\text{-}0.80)(3800.00) = 1416.00\ (\text{kg/m3})$$

$$c_s = 0.80(2.00) + (1\text{-}0.80)(0.77) = 1.75\ (\text{J/g-K})$$

$$k_s = 0.80(0.15) + (1\text{-}0.80)(6.74) = 3.47\ (\text{W/m-K})$$

With these composite properties, the fraction of the energy going into the steel workpiece is

$$B_{Fe} = \frac{\sqrt{(0.574\text{J/g-K})(7800\text{kg/m}^3)(38.9\text{W/m-°K})}}{\sqrt{(1.75\text{J/g-°K})(1416\text{kg/m}^3)(3.47\text{W/m-°K})}+\sqrt{(0.574\text{J/g-°K})(7800\text{kg/m}^3)(38.9\text{W/m-°K})}}$$

$$= 0.82$$

$$B_s = \frac{\sqrt{(1.75\text{J/g-°K})(1416\text{kg/m}^3)(3.47\text{W/m-°K})}}{\sqrt{(1.75\text{J/g-°K})(1416\text{kg/m}^3)(3.47\text{W/m-°K})}+\sqrt{(0.574\text{J/g-°K})(7800\text{kg/m}^3)(38.9\text{W/m-°K})}}$$

$$= 0.18$$

Or over eighty percent of the heat goes into the workpiece. The heat flux is

$$q_g = \frac{(550\ \text{N})(20\ \text{m/s})}{(3.2\ \text{mm})(15\ \text{mm})} = 229.05\ \text{W/mm}^2.$$

This leads to a predicted workpiece temperature of

$$\bar{\theta}_{Fe} = 30°\text{C} + \frac{4}{3\sqrt{\pi}}\left(\frac{(0.82)(229.05\text{W/mm}^2)\sqrt{7369.68}}{(0.574\text{J/g-°K})(7800\text{kg/m}^3)(20\ \text{m/s})}\right) = 41.60°\text{C}$$

Fortunately, this temperature is much less than the temperature predicted for the rake face in machining. This means that in all probability, the workpiece will not be affected by the temperatures in grinding.

While this book, as well as most others treat the calculation of temperatures separately from the problem of estimating forces, in fact the two are coupled together. The deformation in the shear plan and the rubbing on the rake face of the cutting are assumed to be the source of heat in the machining process. The elevated temperatures usually reduce the stress required to cause plastic flow, which should reduce the forces. This should cause the intensity of the heat source to be reduced, so the cycle repeats, converging to both steady state temperatures and forces. While analytical solutions for time dependency and temperature distributions exist for certain special cases, these problems are most easily handled using numerical solutions.

3.4 References

[ArmBro69] Armarego, E.J.A, and R.H. Brown, (1969), *The Machining of Metals*, Prentice Hall, Inc., Englewood Cliffs, NJ.

[BooKni89] Boothroyd, Geoffrey, and Winston A. Knight, (1989), *Fundamentals of Machining and Machine Tools, Second Edition*, Marcel Dekker, Inc., New York and Basel.

[CarStr88] Carroll, J.T., and J.S. Strenkowski, (1988), "Finite Element Models of Orthogonal Cutting With Application to Single Point Diamond Turning, " *International Journal of Mechanical Science*, Vol. 30, No. 12., pp. 899-920.

[Cre*etal*57] Creveling, J.H., T.F. Jordan and E.G. Thomsen, (1957), "Some Studies of Angle Relationships in Metal Cutting," *ASME Transactions*, Vol. 79, pp.127-138.

[ErnMer41] Ernst, H. and M.E. Merchant, (1941), "Chip Formation, Friction, and High Quality Machined Surfaces," Surface Treatment of Metals, *Transactions of the American Society of Metals*, Vol. 29, pp. 299-378.

[Hucks52] Hucks, H., (1952), "Plastizitatsmechanische Theorie der Spanbildung," *Werkstatt und Betrieb*, No. 1.

[Jaeger42] Jaeger, J.C., (1942), "Moving Sources of Heat and the Temperatures of Sliding Contacts," *Proceedings of the Royal Society of New South Wales*, Vol. 76, pp.203-224.

[Kronen54] Kronenberg, M., (1954) *Grundzuge der Zerspanungslehre* (2nd ed) Berlin: Springer-Verlag.

[Lavine88] Lavine, Adrienne S., (1988), "A Simple Model for Convective Cooling During the Grinding Process," *ASME Transactions, Journal of Engineering for Industry*, Vol. 109, pp.1-6.

[Lee84] Lee, D., (1984), "The Nature of Chip Formation in Orthogonal Machining," *ASME Transactions Journal of Materials and Technology*, Vol. 106, pp. 9-15.

[LeeSha51] Lee, E.H., and B.W. Shaffer, (1951), "The Theory of Plasticity Applied to a Problem of Machining," *ASME Transactions, Journal of Applied Mechanics*, Vol. 73, pp.405-413.

[LoeSha54] Loewen, E.G., Shaw, M.C., (1954) "On the Analysis of Cutting Tool Temperatures," *ASME Transactions*, Vol. 76.

[Mercha45a] Merchant, M.E., (1945a), "Mechanics of the Metal Cutting Process," *Journal of Applied Physics, 16:5*, pp. 267-275.

[Mercha45b] Merchant, M.E., (1945b), "Mechanics of the Metal Cutting Process," *Journal of Applied Physics, 16:6* , pp. 318-324.

[Oxley63] Oxley, P.B.L., (1963), "Mechanics of Metal Cutting for a Material With Variable Flow Stress," *ASME Transactions, Journal of Engineering for Industry*, Series B, Vol. 85, pp. 339-345.

[Oxley89] Oxley, P.B.L., (1989), *Mechanics of Machining: An Analytical Approach to Assessing Machinability*, Ellis Horwood Limited, Chichester, England.

[PalOxl59] Palmer, W.B., and P.L.B. Oxley, "Mechanics of Orthogonal Machining," *Proceeding of the Institution of Mechanical Engineers*, Vol. 173, pp. 623.

[RusBro66] Russell, J.K., and R.H. Brown, (1966) "The Measurement of Chip Flow direction," *International Journal of Machine Tool Design and Research*, Vol. 6, pp. 129.

[Sha*etal*53] Shaw, M.C., N.H. Cook and I. Finnie, (1953), "Shear Angle Relationships in Metal Cutting," *ASME Transactions*, No. 2, Vol. 75.

[Shaw84] Shaw, Milton C., (1984), *Metal Cutting Principles*, Clarendon Press, Oxford, England.

[Stable51] Stabler, G,V., (1951), "The Fundamental Geometry of Cutting Tools," *Proceedings of the Institution of Mechanical Engineers*, Vol. 165, p. 14.

[Stephe91a] Stephenson, D.A., (1991a), "Assessment of Steady State Metal Cutting Temperature Models Based on Simultaneous Infrared and Thermocouple Data," *ASME Transaction, Journal of Engineering for Industry*, Vol. 113, No. 2. May, pp. 121-128.

[Tay*etal*74] Tay, A.O., Stevenson, M.G. and Davis, G., (1974) "Using the Finite Element Method to Determine Temperature Distributions in Orthogonal Machining", *Proceedings of Institution of Mechanical Engineers*, Vol. 188, pp. 627-638.

[TriCho51] Trigger, K.J., and Chao, B.T., (1951) "An Analytical Evaluation of Metal Cutting Temperatures," *ASME Transactions*, Vol. 73, pp. 57-68.

[Trent77] Trent, Edward Moor, (1977), *Metal Cutting*, Butterworths, London and Boston.

[Zorev66] Zorev, N.N., (1966), *Metal Cutting Mechanics*, translated from Russian by H.S.H. Massey and edited by Milton C. Shaw, Pergamon Press, Ltd., London.

3.5 Problems

3.1 Because the shear angle plays a major role in determining forces in metal cutting, a number of different models to predict ϕ_o were listed in Table 3.1. Compute the shear angle ϕ_o, predicted by the three models in the table below, for the indicated values of rake angle (γ_o) and friction coefficient (μ).

Table 3.7 Table of Shear Angles for Different Shear Angle Models

	$\mu = 0.50$			$\mu = 0.25$	$\mu = 1.0$
Models	$\gamma_o = -10°$	$\gamma_o = 0°$	$\gamma_o = 10°$	$\gamma_o = 0$	
Merchant					
Lee-Shaffer					
Palmer-Oxley					

3.2 Suppose that a new material has to be machined under conditions close to orthogonal cutting and you need to make some preliminary calculations to see if it is possible. Since this is a new material, no power law expressions for force and power are available, so a mechanics approach has to be used. The following decisions have already been made: $\gamma_o = 10°$, b = 5 mm, h= 1 mm, conservative estimate of $\mu = 1.0$ will be used to estimate friction, and of the shear angle prediction equations in Table 3.1 pick the one that should give the highest forces. Use the empirically determined shear strength in Example 2.1.

a. Estimate the force vector $\mathbf{F_p}$

b. If 5 kW is available at the cutting edge, what is the maximum cutting speed that can be used to machine this material?

c. There is no machine tool in this shop that can handle a thrust force of 0.5 kN. Based on your calculations with the cutting conditions specified, can this material be machined "in house"? If the answer is yes, you are wrong; if your answer is no, select a new h and v and go through the calculations to show these values will work.

3.3 Since shear is the way we assume metal is cut, estimate the shear strain and shear strain rate based on the following data:

a. v = 2 m/s, $\gamma_o = 15°$, $\phi_o = 30°$, $\Delta w = 1.0$ mm

b. v = 2 m/s, $\gamma_o = -5°$, $\phi_o = 20°$, $\Delta w = 0.1$ mm

3.4 Assume that the empirical models for predicting the shear plane angle and the shear stress given below, and the other data listed are to be used in this orthogonal cutting problem: $\phi_o = 60° - 1.1\ (\beta - \gamma_o)$, v = 0.4 m/s, b = 6 mm, h = 2 mm, both the rake and clearance angles are 20° and $\tau_s = 360$ MPa $(h/1mm)^{-0.2}$. An oil based cutting fluid will be used on this relatively low cutting speed operation. An empirical relationship for the friction angle can be approximated as: $\beta = 45°(v/1.6)^{0.25}$

a. Use this data to estimate the shear plane force vector $\mathbf{F_s}$

b. Estimate $\mathbf{F_p}$(kN).

c. If the cutting speed v, were double from 0.4 m/s to 0.8 m/s, using the data and assumptions in this problem, the power required to make this cut, would (*not quite double, exactly double, more than double*).

3.5 Assume that the power component of force and the cutting speed are $F_p = 4.0$ kN and $v = 1$ m/s when the chip width and the uncut chip thickness are 5 mm and 1 mm, respectively. The measured friction force is $F_f = 3.2$ kN and $v_c = 0.5$ m/s. The chip contact length was found to be 1 mm, and the thermal properties for the workpiece material are: $c_c = 0.50$ kJ/kg-°C, $\rho_c = 7500$ kg/m^3, $k_c = 37.5$ W/m-°C. For this problem, the ambient temperature is $\bar{\theta}_0 = 20$ °C.

a. Estimate the chip temperature as it leaves the shear zone, where for this problem $B_1 = 0.75$ and $B_2 = 0.10$.

b. Compute the quantity $\frac{l_c \bullet v_c}{\kappa_c} = \frac{l_c \bullet v_c \bullet c_c \bullet \rho_c}{k_c}$ to see if it is valid to use lumped parameter methods to estimate the rake face temperature (It is, so make sure that you continue). Assume that the conduction coefficient for the cutting tool is 40 W/m-°C and compute the average rake face temperature.

3.6. This problem is aimed at investigating the effect of the cutting edge material, viz. high speed steel or tungsten carbide, on the average rake face temperature $\bar{\theta}_r$. Assume that the following remain constant when the cutting edges are changed: $F_p = 4.0$ kN and $v = 1$ m/s , $F_f = 3.2$ kN and $v_c = 0.5$ m/s, $b = 3$ mm, $h = 1$ mm, and $l_c = 1.5$ mm, $c_c = 0.50$ kJ/kg-°C, $\rho_c = 7500$ kg/m^3, $k_c = 37.5$ W/m-°C, $\bar{\theta}_0 = 20$ °C. Assume that the properties of high speed steel are: $c_{HSS} = 0.50$ kJ/kg-°C, $\rho_{HSS} = 7500$ kg/m^3, $k_{HSS} = 40.0$ W/m-°C,. The properties of tungsten carbide are: $c_{WC} = 1.00$ kJ/kg-°C, $\rho_{WC} = 20{,}000$ kg/m^3, $k_{WC} = 10.0$ W/m-°C. With these assumptions

a. Estimate the average rake face temperature if the cutting edge is high speed steel, and

b. Estimate the average rake face temperature if the cutting edge is tungsten carbide.

3.7 A simple orthogonal machining experiment was run in the laboratory using a cutting edge with $\alpha_o = 5°$ and $\gamma_o = 5°$ without any cutting fluid. The uncut chip thickness was 0.5 mm, and the measured chip thickness was 1.0 mm. Separate experiments determined that for this workpiece-tool-fluid combination the friction angle was 35°. Based on this experimental data compute the following:

a. The experimental shear plane angle (in degrees):

b. The errors (in degrees) between this measured angle and any three of the shear plane angles you like that are listed in Table 3.1.?

3.8 Assume that an orthogonal cutting experiment was done with the following conditions: the uncut chip thickness was 2 mm, the chip width was 10 mm, the chip thickness ratio was 0.25 and a zero degree rake angle cutting edge was used. The cutting test was run at a cutting speed of 2.5 m/s and the measured forces were: $F_p = 2.0$ kN and $F_q = 1.5$ kN.

a. Compute the experimental shear plane angle. This angle is closest to the which of these three theoretical models: Merchant, Palmer-Oxley, Lee-Shaffer.

b. Compute the experimental shear stress.

3.9 Assume that an orthogonal cutting experiment was run with the following conditions: no coolant, the uncut chip thickness was 0.2 mm, the chip width was 3.2 mm, the chip contact length l_c was 1.0 mm, and a zero degree rake angle cutting edge was used. The cutting test was run at a cutting speed of 0.7 m/s and the measured force vector was $F_p = [1.1 \text{ kN}, 0.6 \text{ kN}]^T$. Nobody remembered to collect chips so the the experimental shear plane angle could not be computed; as a result the Lee-Shaffer shear angle prediction equation should be used if the shear plane angle is needed.

a. Estimate the friction angle with this data.

b. Estimate the apparent normal stress σ_{rake} and shear stress τ_{rake} on the rake face of the cutting edge.

c. Estimate the apparent normal stress σ_s and shear stress τ_s on the shear plane of the chip:

d. Your values for τ_{rake} in 3.9b and τ_s in 3.9c should not be equal if you made you calculations correctly. Give at least one explanation (or rationalize if you prefer) for this difference.

3.10 In an orthogonal cutting experiment, the friction force is $F_f = 0.6$ kN and $v_c = 0.2$ m/s when b = 3.2 mm, h = 0.2 mm and v = 0.7 m/s; the chip contact length was found to be 1 mm. By measurement and calculation the net rate that energy goes into the *shear zone* is q = 0.64 kW, and of this energy 80 percent (A=0.80) goes to raising the temperature of the chip. The thermal properties for the workpiece material are: $c_c = 0.50$ kJ/kg-°C, $\rho_c = 7500$ kg/m^3, $k_c = 37.5$ W/m-°C and the conductivity of the tool material is 40 W/m-°C. For this problem, the ambient temperature is $\theta_0 = 20$ °C.

a. Estimate the average chip temperature $\bar{\theta}_c$.

b. For this problem the quantity $\frac{l_c \bullet v_c}{\kappa_c} = \frac{l_c \bullet v_c \bullet C_c \bullet \rho_c}{K_c} \geq 20$ so it is valid to use lumped parameter methods to estimate the rake face temperature, so find $\bar{\theta}_r$.

c. If the cutting speed and the chip contact length both doubled, with the other cutting conditions and material properties staying constant, based on the equations used to solve for the chip and rake temperatures: q would probably (*increase, decrease, remain the same*), $\bar{\theta}_c$ would probably (*increase, decrease, remain the same*), B_3 would probably (*increase, decrease, remain the same*), and $\bar{\theta}_r$ would probably (*increase, decrease, remain the same*).

4

Edge Materials, Wear and Fluids for Cutting and Grinding

By comparison with the mechanics and thermal models for machining in Chapter 3, this chapter will be less quantitative and more qualitative. The calculations in Chapter 3 treated the interfaces at the rake face of the cutting edge or wheel contact region in grinding as if everything was known in a very precise and quantitative way. In reality this is not the case. This chapter covers cutting edge materials and fluids applied at the interface, where they interact with the workpiece to affect both the friction, temperatures and wear in machining and grinding.

The coverage in this chapter, summarized in Figure 4.1, is aimed at providing guidelines for choosing edge materials and fluids. The tribological mechanisms at the interface lead to an understanding of the wear of cutting edges and the empirical tool life models used in subsequent chapters, particularly Chapter 8 on Machining Economics and Optimization. Knowledge from Chapter 3 on the relationships between the cutting conditions that define the cutting speed and chip cross section and the forces and temperatures are assumed as inputs to this selection process. While specification of the edge material is an output of this process, the workpiece material is an input that affects the selection of the edge material and the fluids for grinding and cutting. Constraints remain those throughout the various stages of process planning: capital equipment, geometric tolerances and quality specifications. The mechanisms to understand what is going on at the interface is an abbreviated discussion of an entire field called tribology, but it serves as a way of explaining how conditions at the interface can be improved. Another important mechanism for deciding on tool materials and fluids is experience. While relying on vendors to be completely objective in their recommendation is naive, they often have good advice that should be interpreted in the light of engineering principles.

4.1 Cutting Edge Materials

Before starting the discussion of cutting edge materials, a few "forward references", i.e., material that is important, but hasn't been covered yet. First, the emphasis in this chapter is on materials for cutting, rather than grinding. However, several of the materials that will be discussed are used for both cutting edges, e.g., certain carbides (WC as an edge material and SiC as an abrasive), ceramics (Al_2O_3), diamond, and man made hard materials like cubic boron nitride are discussed in this chapter and are listed as abrasives in Tables 6.1 and 6.2. The second forward reference is in Table 4.1 which recommends cutting edge materials based on the combination of workpiece material and machining process. Chapter 5 covers the analysis of cyclic forces and thermal loading in processes like milling that affect cutting edge performance.

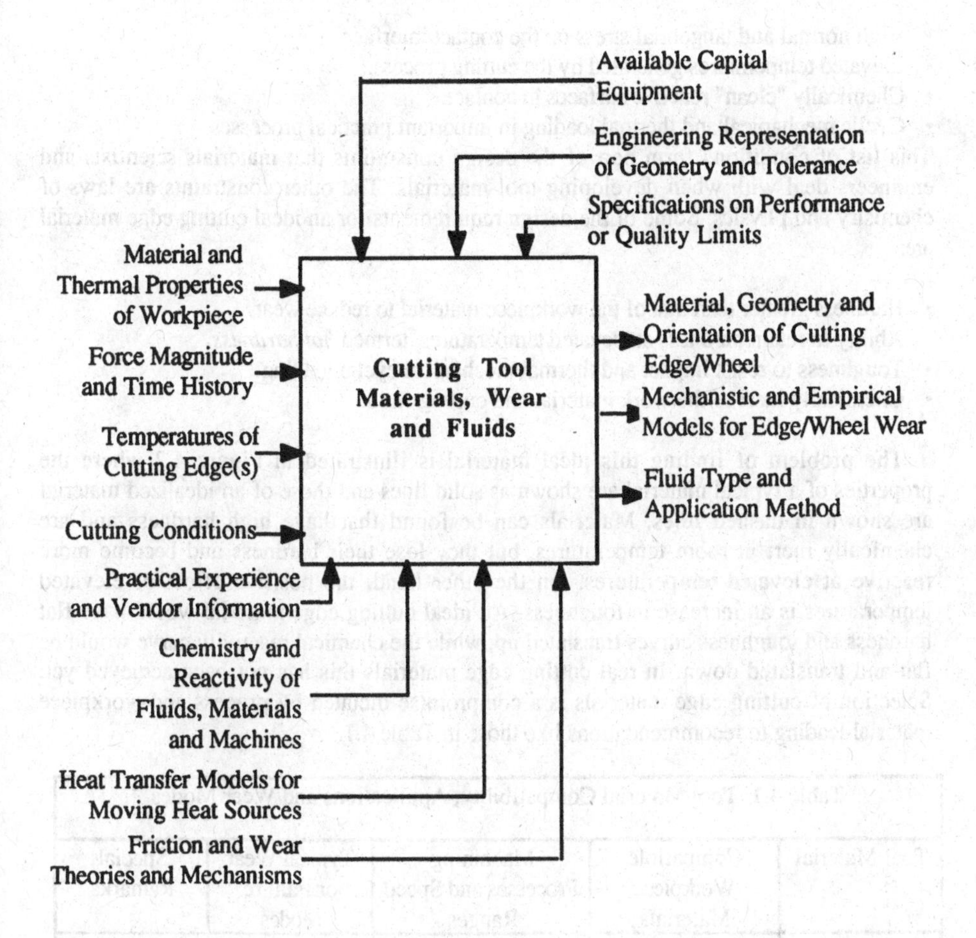

Figure 4.1 The Inputs, Constraints, Outputs and Mechanisms considered when analyzing and selecting a fluid for machining or grinding.

4.1.1 Ideal Cutting Edge Material

The environment that a cutting edge experiences is rather harsh. The important part of this region is the rake contact area A_r defined by the contact length l_c and the chip width b. In this region, high normal and tangential forces like those predicted in Section 3.1 generate high contact stresses on the face of the cutting edge. In Section 3.3, the tangential or friction force and chip velocity were used to calculate the heat flux at the rake face to predict the interface temperatures for both machining and grinding. When making a chip, a new surface is generated that is chemically ready to react with either a fluid or the edge material it contacts moving up the rake face of the cutting edge. In processes like milling that have multiple cutting edges that intermittently generate chips, the edge undergoes cyclic mechanical and thermal loading. The severity of the loading depends in part on the cutting conditions selected by the process planner. The adverse conditions that affect the cutting edge can be summarized as:

- High normal and tangential stress on the contact interface.
- Elevated temperatures generated by the cutting process.
- Chemically "clean" reactive surfaces in contact.
- Cyclic mechanical and thermal loading in important practical processes.

This list of conditions form part of the design constraints that materials scientists and engineers deal with when developing tool materials. The other constraints are laws of chemistry and physics. Some of the design requirements for an ideal cutting edge material are:

- Hardness greater than that of the workpiece material to reduce wear.
- Ability to retain hardness at elevated temperatures, termed *hot hardness*.
- Toughness to resist impact and thermal/mechanical cyclic loading.
- Chemically inert to the work material and cutting fluid.

The problem of finding this ideal material is illustrated in Figure 4.2 where the properties of a typical material are shown as solid lines and those of an idealized material are shown in dashed lines. Materials can be found that have high hardness and are chemically inert at room temperatures, but they lose their hardness and become more reactive at elevated temperatures. On the other hand, the positive effect of elevated temperatures is an increase in toughness. An ideal cutting edge material would have flat hardness and toughness curves translated up, while the chemical reactivity curve would be flat and translated down. In real cutting edge materials this has not been achieved yet. Selection of cutting edge materials is a compromise dictated by process and workpiece material leading to recommendations like those in Table 4.1.

Table 4.1 Tool Material Compatibility, Applications and Wear Modes

Tool Material	Compatible Workpiece Materials	Machining Processes and Speed Ranges	Typical Wear or Failure Modes	Special Remarks
High Carbon or Medium/Low Alloy Steels	Low strength and hardness materials, nonferrous alloys and plastics	Single-point turning, drilling and tapping (v < 0.5 m/s)	Buildup, plastic deformation, abrasive wear, microchipping	
High Speed Steel	All materials of low to medium strength and hardness	Single-point turning, drilling, reaming, tapping, broaching & both face & end milling (0.5 < v < 2.5 m/s)	Flank and crater wear	Used in almost every machining application
Cemented carbides	All materials of low to medium strength and hardness	Single-point turning, drilling, reaming, tapping, broaching and both face and end milling (0.5 < v < 2.5 m/s)	Flank and crater wear	At low speeds, chips cold weld to carbide and micro chip

Table 4.1 (Cont.) Tool Material Compatibility, Applications and Wear Modes				
Tool Material	Compatible Workpiece Materials	Machining Processes and Speed Ranges	Typical Wear or Failure Modes	Special Remarks
Coated tools	Cast iron, alloy and stainless steels, superalloys (Ti is a notable exception)	Single-point turning, (0.5 < v < 5 m/s)	Flank and crater wear	At low speeds, chips cold weld to carbide and micro chip
Ceramic	Cast iron, Ni-based superalloys, nonferrous alloys, plastics	Single-point turning, (v > 2.5 m/s)	Depth of cut notching, micro chipping, gross fracture	Low thermo-mechanical fatigue strength means no interrupted cuts
Diamond	Pure Cu and Al, Si-Al alloys, cemented carbides, rock, cement, plastics, glass-epoxy and fibrous composites, nonferrous alloys	Single-point turning,face milling (v > 2.5 m/s)	Chipping, oxidation, graphitization	Not for machining low carbon steels, Co, Ni, Ti and Zr
Cubic Boron Nitride	Hardened steel alloys and chilled cast iron, HSS, commercially pure Ni and Ni-based superalloys	Single-point turning,face milling (0.5 < v < 5 m/s)	Depth of cut notching, chipping, oxidation, graphitization	Can handle most of the materials that diamond cannot

4.1.2 Real Cutting Edge Materials

Developments in cutting tools have tried to reach the ideal of a hard, tough, chemically inert material that is unaffected by the temperatures generated at the machining interface. Since the calculations in Chapter 3 and practical experience indicate that v has the greatest effect on temperature, there has also been a link between the structural design of practical machine tools, their power and stiffness in particular, and tool materials that could be used at the higher speeds and temperatures. Another important development has been the concept of engineering or designing a material tailored to the application. Some of this evolution of real cutting tool materials is summarized below, and is based on Komanduri and Desai's

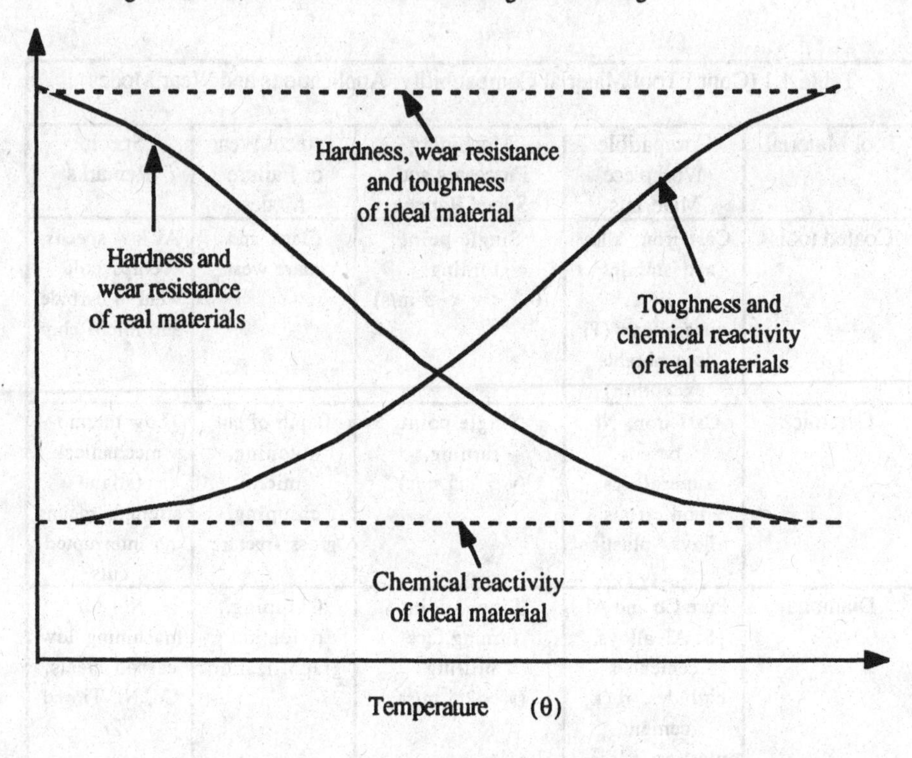

Figure 4.2 Real materials and an ideal tool material as a function of temperature.

report which gives an excellent discussion of tool materials, properties and selection principles [KomDes82].

High carbon or medium/low alloy steels were the first tool materials used in the late 1700's when the basic structure and design of machine tools we know today was introduced. It was understood that the hard phases caused by carbon in controlled amounts increased the strength of steel, producing hardnesses of about $H_{RA} \approx 80$. Improvements over high carbon steels are achieved by alloying with molybdenum (Mo) and chromium (Cr) for hardenability and tungsten (W) for wear. Low hot hardness and wear resistance at temperatures above 150-350 °C reduces the useful range of cutting speeds for these materials. These materials are also susceptible to thermal cracks and distortion. But because these materials are inexpensive, easily fabricated and heat treated and can be used for machining low strength materials, high carbon and alloy steels are still fairly common for these applications and for "hobbyist's" tools.

High speed steel (HSS) was invented by F.W. Taylor and White in the late 19th century. With good wear resistance, but more importantly because it is the toughest tool material, HSS is still used in many milling cutters, drills and broaches. Even with the term *high* in the name, room temperature hardness of $82 < H_{RA} < 87$ drops off quickly, at speeds above 0.5 m/s where temperatures are above 550 °C. At these temperatures, HSS tends to react chemically with the "new" surfaces on a chip. With relatively large amounts of Mo/W, Cr, cobalt (Co) and vanadium (V) for hardenability, uniform hardness can be achieved without the rapid cooling rates that can cause quench cracks.

Cast cobalt, also known by the trade name Stellite, was developed about the same time as HSS. It has about the same room temperature hardness as HSS, but retains hardness at high temperatures, permitting higher speeds (≈ 25% greater than for HSS). Cast cobalt tools are "hard as cast;" this makes them difficult to fabricate. Also, large amounts of Co, W and Cr, all strategic materials, have caused this tool material to be phased out and replaced with others, particularly the cemented carbides that use less Co and W but can be used at higher speeds than Stellite.

Cemented carbides, around since the 1920's, are probably the most common high production tool materials with hardness in the range $89 < H_{RA} < 93$. Fabricated by powder metallurgy technology where fine particles of hard carbides are sintered in a metal binder, the cemented carbides produce superior hardness at elevated temperatures and are chemically more stable than HSS. As a powder metallurgy product, it is possible to tailor the carbide to the material and application which gives rise to a number of *carbide grades*. Toughness and ductility may be low, but this depends a lot on the grade and density of the carbide. Also, the powder metallurgy process produces inserts of various geometries that either require a special tool holder (more about that in Chapter 5) or the insert can be brazed onto a tougher, less expensive shank. Carbides are usually tungsten carbide (WC) applicable for speeds on the order of 2.5 m/s, but titanium carbide (TiC) gives higher speed (≈ 5.0 m/s) at higher cost.

Coated tools are "engineered materials" introduced in the 1960's. The idea behind them is simple: "Engineer" a material that has the toughness and ductility on the inside and a hard, low friction, chemically stable surface on the outside. The way this has been accomplishes is by coating materials like a tough carbide grade, a HSS milling cutter or drill with a material like TiC, titanium nitride (TiN) or a ceramic like alumina (Al_2O_3). The technology problem is how to put the coating on: chemical vapor deposition at high temperatures (≈ 1000 °C), physical vapor deposition or ion sputtering. The performance of these tools has been excellent.

Ceramics were available since 1905, about the same time as HSS, but it wasn't until the mid 1950's that ceramics became of commercial interest. Most often, ceramics are alumina based, but silicon based ones are also available. Of the cutting tool materials available at reasonable cost, ceramics have the best wear resistance, highest hardness at room ($91 < H_{RA} < 95$) and elevated temperatures, so they are capable of being applied at the highest speeds ($2.5 < v < 5$ m/s). However, what has held ceramics back has been low toughness and resistance to thermal and mechanical shock and catastrophic failure due to notching at the "Depth of Cut Line". The secret to success in using ceramics is choosing applications that avoid these short comings. Ceramics are used almost exclusively in turning where the cutting edge is in continuous contact and with either no cutting fluid or a flood to avoid thermal shocks.

Diamonds, to paraphrase a well known marketer of them, have been around since forever, and are the hardest material ($H_K \approx 8000$ kg/mm^2), natural or man-made. As a tool material it has three other ideal characteristics: low friction, high wear resistance in most cases, and maintains a sharp edge (determined by the cleavage planes). The negative aspects of using diamond as cutting edge or abrasive are: low toughness, very high cost due to materials (mono-crystal or polycrystal) and fabrication, and the fact it reacts chemically with some steels which causes a very expensive edge to break down and wear rapidly. It is this reaction with one of the most common engineering materials - steel - that is a real limitation of diamond as a cutting edge material. When the edge does not react with the workpiece material, for example in ultra-precise machining of complex geometries, diamond is unsurpassed.

Cubic boron nitride (CBN) is another engineered material, developed in the 1960's as a material with hardness ($H_K \approx 4700$ kg/mm^2) second only to diamond, but has one important advantage over diamond as a cutting edge or abrasive material; CBN can be used to machine ferrous materials. The crystal structure of CBN is similar to diamond, i.e., a meta-stable cubic structure or the more stable hexagonal crystal. The cost of CBN as a cutting edge or as an abrasive is high, almost that of diamond and one or two orders of magnitude greater than HSS. Because of the high hardness, the wear resistance is also high, so rather than gradual wear ending the useful life of a CBN edge, failure tends to be catastrophic after micro-chipping begins on the cutting edge or a depth of cut notch develops. Even with the high cost and risk of catastrophic failure, CBN or diamond if used under the correct conditions can be economical because of the higher production rates at $v > 2.5$ m/s and the greater number of units per cutting edge.

It is impossible to be up to date on tool materials. It is also difficult to cover all the specifications needed to select a cutting edge material, but Table 4.1 gives some guidelines. Other important practical information comes from knowledge of the different grades of tool materials, with some sources of this information listed in the references or from vendors.

4.2 Cutting and Grinding Fluids

Nearly every practical production machining or grinding process uses a fluid. Some instances where ceramic tools are used is an exception noted in Section 4.1. Reasons why cutting fluids are used in many cases include:

- Reducing the forces and power in machining and grinding by reducing friction at the contact interface, as predicted by the chip forming mechanics in Section 3.2.
- Increasing the tool life of cutting edges or increasing the time between dressing cycles in grinding, can be achieved reducing friction so less heat is generated. Also the fluid removes some of the heat generated in making a chip.
- Improving surface finish, mainly by reducing the chances of metal welding to the cutting edge, forming a built up edge (BUE) and then gouging the newly generated workpiece surface when the BUE breaks off.
- Reducing the thermal distortion and sub-surface damage of workpieces by convecting heat. Reducing thermal distortion is a macro effect that improves the dimensional accuracy that can be achieved in a given process. Subsurface damage to a workpiece is on a smaller and often times invisible scale. The level of the temperatures can cause metallurgical changes, and the temperature gradients can cause residual stresses that are detrimental to functional performance and safety.
- Assisting in chip removal by, in some cases, helping promote the formation of discontinuous or Type I chips, and more commonly by washing the chips away from the cutting region.

In most cases, a fluid is used for the first two or three of these reasons, with the others being important benefits. Two of the most important actions underlying all fluids for cutting and grinding, cooling and lubrication, will be covered in the next section, followed by practical guidelines for selecting an applying fluids.

4.2.1 Basic Actions of Lubrication and Cooling

While reduction in the shear strength τ_s of the workpiece material is suggested as a secondary action of a fluid for cutting or grinding, the two major actions of cutting fluids are cooling and lubrication.

The coolant action of a fluid is dependent on properties and conditions that enhance heat transfer between the cutting edge or workpiece and the fluid. This means that surface tension or how the fluid "wets" the surface will surface to enhance heat transfer. Properties of a fluid that are most important for coolant action are high heat transfer coefficients to get the heat into the fluid and a large heat capacity so that this heat energy can be carried away. For these reasons, water based emulsions with wetting agents and corrosion inhibitors are fluids that make excellent coolants. Once a coolant is selected based on its chemistry and heat transfer properties, the main process design considerations are designing nozzles to apply the fluid to the heat sources: the shear zone and the rake and clearance faces in cutting and the contact zone in grinding, and the volumetric flow rates, ranging from a mist to flood application of coolant.

The lubrication actions of a fluid are aimed at reducing the friction or tangential forces in cutting and grinding, and in that way reducing the need for coolant action. One way of looking at the lubricating action of fluids is to look at the extremes of lubrication, starting with no fluid present. Referring to Figure 4.3 and also going back to the way forces on the rake face were modeled by Eq (3.8), the macro way of looking at Coulomb friction used in our calculations and the micro view used to explain the actions of lubricants should come together. Using Eq (3.8a) to re-write Eq (3.9a)

$$\mu = \frac{F_f}{F_n} \quad \text{(macro definition)} \tag{3.9a}$$

$$= \frac{\int^{A_c} \tau_{min}\, dA_{true}}{\int^{A_c} \sigma\, dA_{true}} = \frac{\tau_{min} A_{true}}{\sigma A_{true}} \quad \text{(micro definition).}$$

These equations and Figure 4.3 together describe Coulomb's law, where the stresses σ and τ_{min} are assumed to be constant. The true area of contact A_{true} comprised of the contacting asperities on the scale of surface roughness or less, is assumed to be proportional to the normal load - just like in the Archard model for wear, c.f., Eq (4.1) - because the asperities flatten out. This proportionality continues until the true area and nominal area are equal, i.e., $A_{true} = A_c$; then F_f is no longer proportional to F_n. As long as $A_{true} < A_c$ and no fluid is used, the way to reduce μ is to shear a material with a low τ_{min}. This may be the workpiece material itself, compounds in the workpiece material that form a solid lubricant (sulfur or lead in free machining steels are an example) or compounds or oxides that form a solid film in the presence of a fluid; τ_{min} is the shear strength of whichever is the smallest. Remember that Coulomb's law assumes solid surfaces in contact.

On the other extreme is *hydrodynamic lubrication* like that idealized in Figure 4.4. Reynold's equations predict that when a fluid of viscosity ν is in a moving convergent channel, a pressure develops that causes separation of the surfaces so that there is no longer solid-solid contact. Shearing of the fluid replaces shearing the τ_{min} of a solid, so the friction drops. Common sense and solving Reynold's equations says that for actual machining or grinding processes, pure hydrodynamic lubrication doesn't occur. But when solving Reynold's equations for machining conditions, the film thickness are about 4 or 5 orders of magnitude less than the surface roughness, so there must be some hydrodynamic lubrication, with asperities crashing into each other. This is termed *boundary lubrication.*

Actual machining with a fluid that acts as a lubricant falls into this last category as shown in Figure 4.5, where the fluid may generate enough pressure to reduce some of the asperity contacts. The Stribeck curve in Figure 4.6 is another way of visualizing how the speed, viscosity and contact pressure affects the ratio of the friction and normal force. What moves machining to the left are relatively low velocities (when compared to bearings) and relatively high pressures because of the small contact area A_c. Taken together, for a fluid to act as a lubricant it should help form low strength compounds on the asperities that act as a solid lubricant, while at the same time have good wetting properties and high viscosity so the benefits of decreasing friction in the boundary lubrication can be achieved.

4.2.2 Considerations in Selecting Fluids

Selecting fluids for cutting and grinding still depends very much on experience and a qualitative understanding of the basic cooling and lubrication actions they provide. They do in fact affect what has been modeled by the single coefficient μ and temperatures at the cutting or grinding interface. The ideal fluid, which would act as both a fine coolant and lubricant, would have a high heat transfer coefficient, specific heat capacity, viscosity, and promote the formation of low strength solid lubricant, and low surface tension to promote wetting for both heat transfer and lubrication. As in most engineering, balance has to be maintained because no fluid possesses all these attributes, so guidelines and experience are the way to make an informed selection.

Technological considerations in selecting a fluid for cutting or grinding are:

- Low speed applications (v < 1 m/s) or at higher speeds where the fluid can be introduced to the contact region are where fluids act best as a lubricant, c.f., Figure 4.6. This means mineral or neat oils with extreme pressure additives that will form sulfides or chlorides as low strength layers.
- Normal speed applications($\geq$ 1 m/s) where it is usually difficult to introduce the fluid to reduce friction, the fluid will act primarily as a coolant, e.g., water based emulsions with additives to prevent rust.
- A solid lubricant in the workpiece - a "free machining" material - if the production volume, performance and material cost are acceptable. This is an alternative in small, machined and mass produced parts, e.g., those produced on an automatic screw machine.
- Application of the fluid to enhance the coolant and lubrication actions, e.g., volumes such as mist or flood, nozzle and jet design or through the tool or wheel. These are design decisions that usually are made for an application.

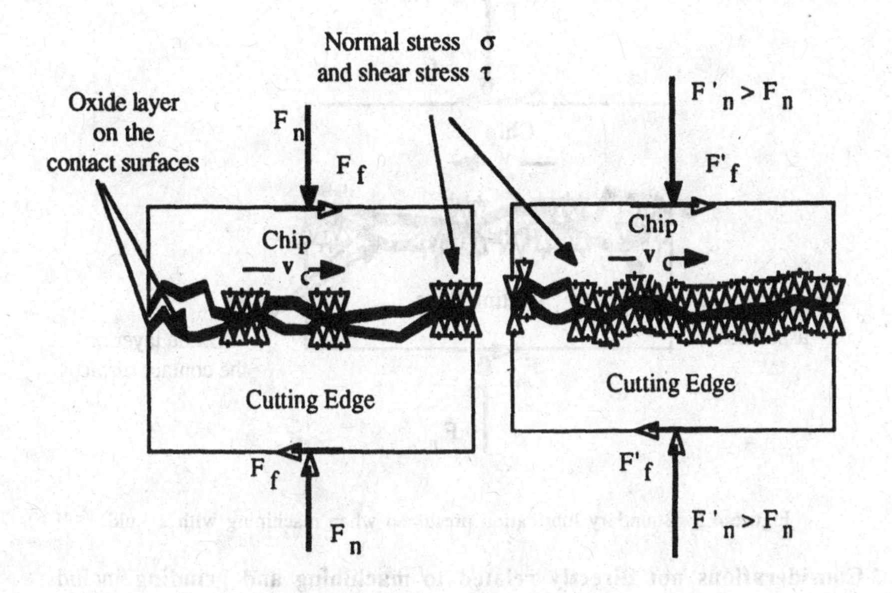

Figure 4.3 Contact conditions for Coulomb's law.

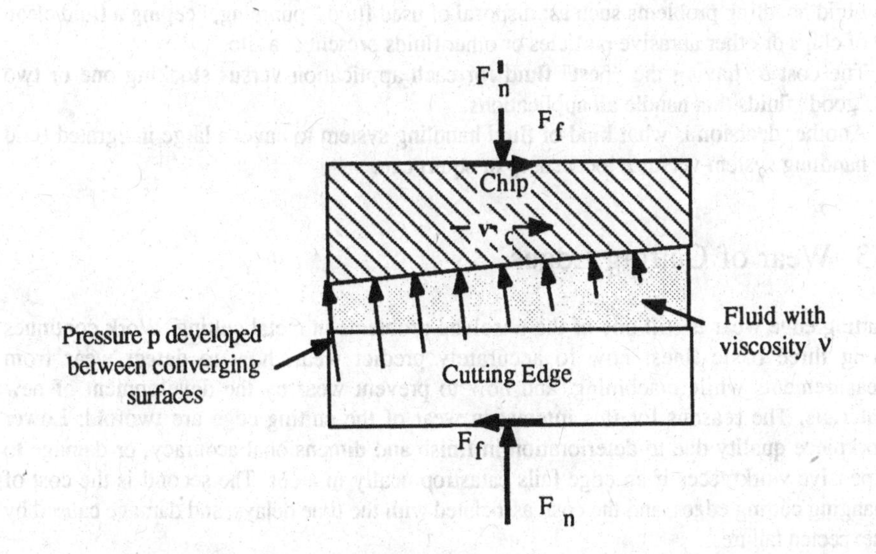

Figure 4.4 Hydrodynamic lubrication never achieved in machining.

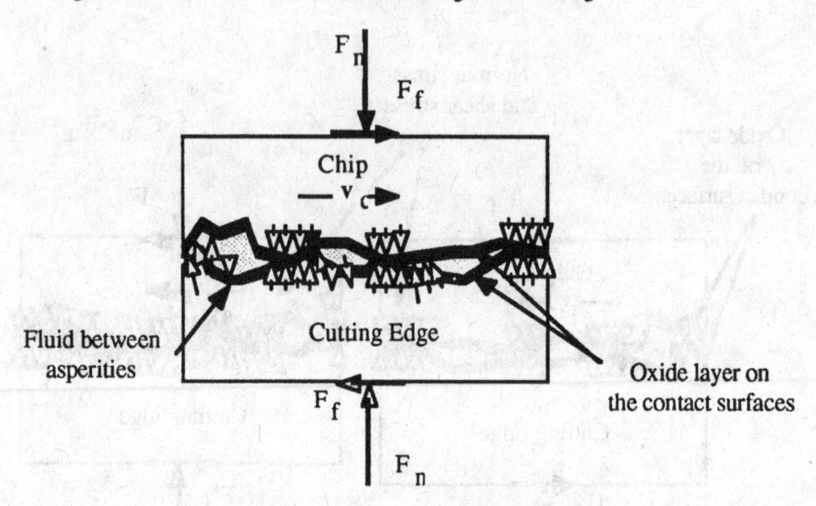

Figure 4.5 Boundary lubrication presumed when machining with a fluid.

Considerations not directly related to machining and grinding include:

- Physiological and health effects of fluids on people when used on a machine or in a sump system include: toxicity, odors, dermatitis, etc.
- Effects of a fluid on workpiece materials that cause corrosion, stains or reactions that reduce fatigue life.
- Effects of a fluid in contact with a machine tool over an extended time include: corrosion, and contamination of hydraulic or lubrication fluids for drives and bearings.
- Fluid handling problems such as: disposal of used fluids, pumping, keeping a fluid clean of chips or other abrasive particles or other fluids present at a site.
- The cost of having the "best" fluid for each application versus stocking one or two "good" fluids that handle all applications.
- Another decision is what kind of fluid handling system to have; a large integrated fluid handling system versus a local sump or squirt can.

4.3 Wear of Cutting Edges

Cutting edge wear is still one of the unsolved problems in metal cutting. Work continues along three basic lines: how to accurately predict wear, how to detect wear from measurements while machining, and how to prevent wear by the development of new materials. The reasons for this interest in wear of the cutting edge are twofold: Lower workpiece quality due to deterioration in finish and dimensional accuracy, or damage to expensive workpieces if an edge fails catastrophically in a cut. The second is the cost of changing cutting edges, and the cost associated with the time delays and damage caused by unexpected failures.

Sections 4.3.1 and 4.3.2 concentrate on this problem of wear; the first section on the different mechanisms of wear, and the second section on how to model wear empirically in machining.

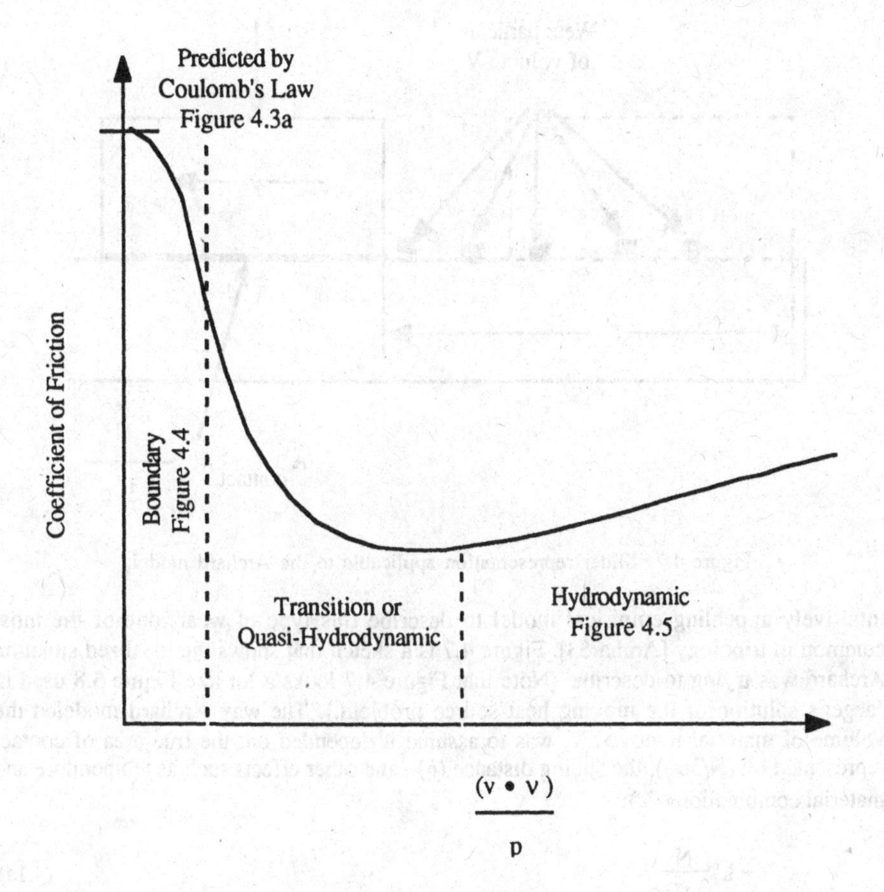

Figure 4.6 Stribeck curve to explain the lubrication action of cutting and grinding fluids.

4.3.1 Mechanisms

Adhesion is the localized welding of materials together and then, during sliding, the welds break transferring a wear particle by pulling it out of the cutting edge. This mechanism is enhanced by elevated temperatures, high pressures and "clean" reactive surfaces; all things occurring on the rake face. Adhesion explains wear due to the built up edge (c.f., Section 3.1). It does not explain the empirical observation of wear particles smaller than BUE debris, because adhesion means that one surface (the chip) gains the material lost by the other (the edge).

Abrasion can be thought of as small scale cutting done by hard particles (the same ones not explained by adhesion) as they move across a surface under load. It depends on the hardness and elastic properties (indentation strain), asperity geometry and relative motion between two surfaces. Abrasion is primarily a mechanical form of wear, Archard used an

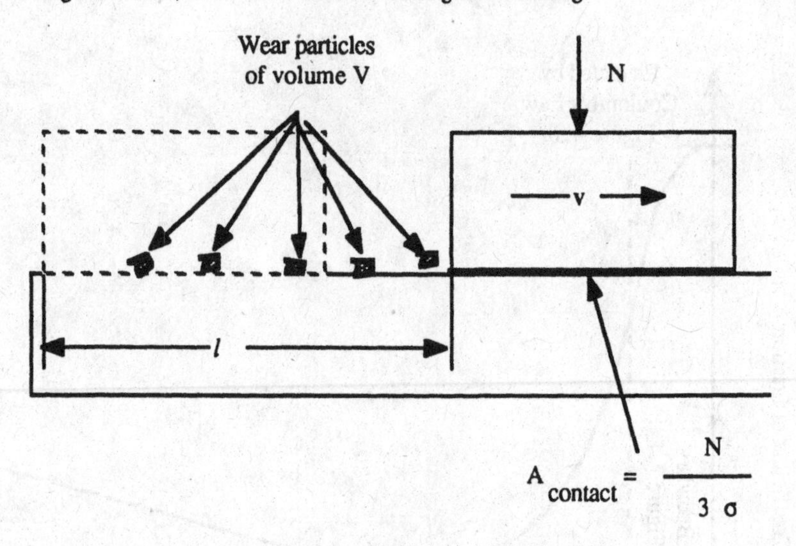

Figure 4.7 Slider representation applicable to the Archard model.

intuitively appealing empirical model to describe this type of wear, one of the most common in tribology [Archar53]. Figure 4.7 is a sketch that shows the idealized situation Archard was trying to describe. (Note that Figure 4.7 looks a lot like Figure 3.8 used in Jaeger's solution of the moving heat source problem.). The way Archard modeled the volume of material removed V, was to assume it depended on: the true area of contact represented by ($N/3\sigma_y$), the sliding distance (l), and other effects such as temperature and material combinations (k'):

$$V = k' \frac{N}{3\,\sigma_y}\, l \tag{4.1a}$$

Taking the time derivative of this expression the wear rate is

$$\dot{V} = k' \frac{N}{3\,\sigma_y}\, v. \tag{4.1b}$$

This model is appealing when describing things that affect the wear of a cutting edge; for example the normal load N, analogous to the normal rake force F_n, and velocity both increase the wear rate. The Archard abrasive wear model is trying to describe wear in the *softer* material so the σ_y in Eq (4.1) is for the softer material, but in Section 4.1.1 the first requirement of an ideal cutting edge material was being harder than the material machined. The form of the Archard model is probably correct to describe the wear of a cutting edge, but rather than all the wear occurring on the softer material, wear occurs at a much lower rate - fortunately - on the cutting edge.

Diffusion is the atomic transfer of material when asperities are in contact due to chemical affinity and difference in concentration. Diffusion wear is enhanced by elevated temperature, increased time of contact, and increased contact pressure which increases the asperity contact area (like $N/3\sigma_y$ in the Archard model). Theoretically this type of wear doesn't need relative motion between the chip and the cutting edge to occur. In machining, the chip velocity v_C is the heat source that raises the average rake face temperature to $\bar{\theta}_r$ as predicted in Section 3.3. The point of maximum temperature on the rake face is probably more relevant to determining the rate and location of diffusion wear, and that is the reason Trigger and Chao's [TriCho51] work on measuring the rake face temperature was so important. They identified the point of maximum temperature as $l_c/3$ up the rake face. This and a diffusion rate equation are the basis for models that have been developed to describe this type of wear in machining.

Fatigue is a mechanism caused primarily by the compression and tension cycling of the sub-surface inherent in some of the practical machining processes where the cutting edge makes intermittent contact with the workpiece. Face milling is an example of such a process. At the same time the cutting edge undergoes thermal cycling. This cycling causes fatigue and results in sub-surface separation or spalling out of the cutting edge material. Fatigue is usually presented as an alternative explanation to the adhesive wear theory which did not explain the empirical observation of small wear particles; this theory predicts they are spalled from the cutting edge. The other reason for considering the fatigue mechanisms aid in explaining catastrophic failure of edges that undergo cyclic thermal and mechanical loading.

None of these four wear mechanisms *completely* describes wear in machining or grinding. In most real machining cases more than one wear mechanism is involved, so efforts to describe the effects of this interaction have been the approach used in machining.

4.3.2 Tool Wear and Empirical Tool Life Models

Considering how machining is done and the brief review of wear mechanisms, conditions are such that wear of the cutting edge is inevitable. Even with the best design for applying a cutting or grinding fluid, the complete separation of the chip and cutting edge surfaces by hydrodynamic lubrication does not occur. There is little time for protective oxides to form on the surface of the cutting edge or the new surfaces of the chip as they slide up the rake face. (This short duration argument was used in the temperature calculations to say that no heat was conducted through the chip contact interface.) The large amount of plastic deformation in the shear zone severely work hardens the chip, making it more abrasive. Calculations and experimental measurements confirm the high temperatures and pressures at the sliding interface. The higher temperatures tend to decrease strength properties, as well as accelerate diffusion wear. The high contact pressures increase the true area of contact mechanically, again causing wear rates to increase as predicted by an Archard like expression. And finally, while velocity indirectly increases the wear rate because it increases the temperatures and decreases strength properties, velocity also directly increases the wear rate as in Eq (4.1b). It is clear that several wear mechanisms are active at once in machining, so rather than measure the progress of each mechanism, functional wear measurements are made.

Several of the common types of wear are identified on a cutting edge in Figure 4.8. On the clearance or relief surface, also called the flank of the cutting edge, a flat surface develops where the edge rubs the newly generated workpiece surface. This kind of wear, which is mainly due to abrasive wear like that idealized in Figure 4.7, is termed *flank wear*. It is measured by the width of the *flank wear land* VB. On the rake face where the chip flows over the cutting edge, a crater can develop, due to diffusion, abrasion and adhesion. Diffusion is likely to be the main mechanism, because the crater develops a distance up the face of the edge, just like the point of maximum temperature. This form of wear is referred

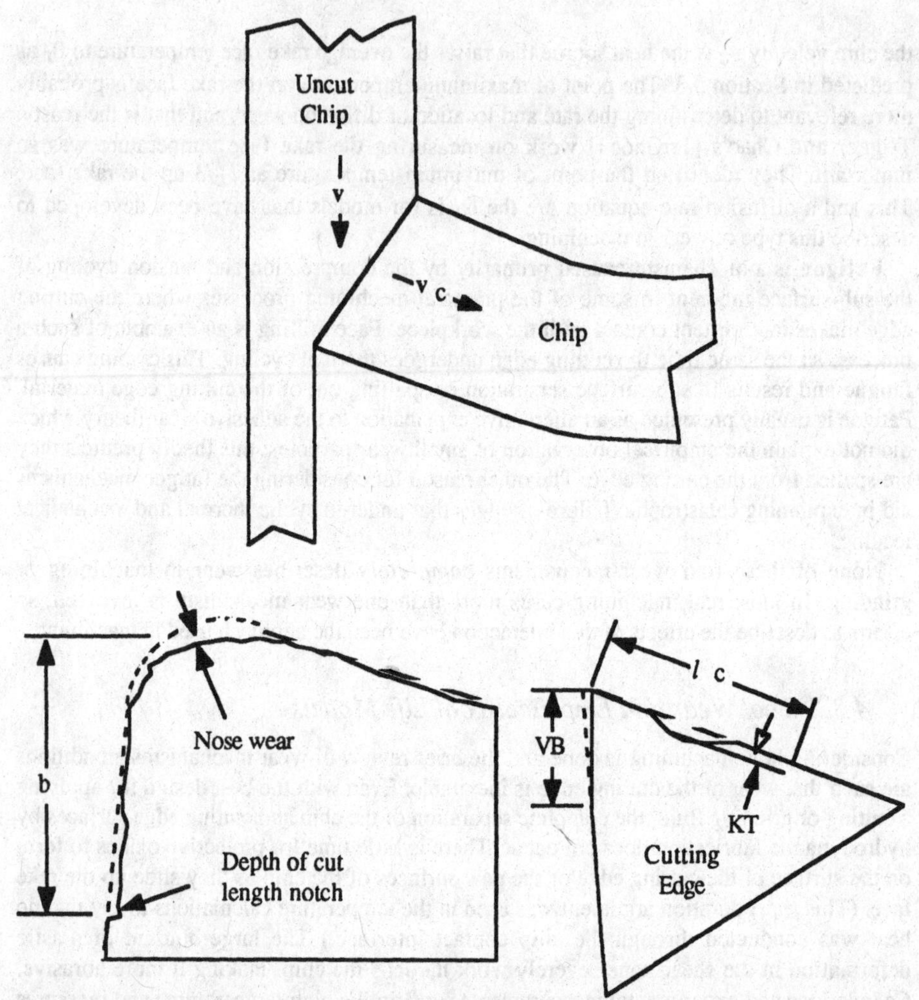

Figure 4.8 Top and edge view of a cutting edge showing nose, depth of cut notching, crater and flank wear.

to as *crater wear* and is measured by the crater depth, KT. *Nose wear* on the tip of the cutting edge is like flank wear in that it is driven by the same predominant mechanism, abrasion. *Depth of cut notching* occurs along the cutting edge because the entire edge is not cutting. Multiple wear mechanisms are involved and the danger is that the notch acts as a stress riser so that the edge may fail catastrophically.

As a criterion to measure the amount of wear that occurs in machining, flank wear is probably the most widely reported, followed by results for crater wear and nose wear. But while flank and crater wear are important, for the dimensional accuracy, finish and potential for damaging a workpiece, nose wear and notching are probably more significant.
The reason VB and KT are used most often is that they can be measured in a fairly objective way, while nose wear and notching are more difficult to quantify. Fortunately, the time history of each of these types of wear are similar. As a result, the discussion that follows, based on flank wear, could equally be applied to quantitative measurements of KT, nose wear or depth of cut notching.

Figure 4.9 is how the net effect of all the different wear mechanisms on, in this case, flank wear, occurs on a new cutting edge over time. A changing and initially rapid wear rate, usually attributed to break-in, begins the process. One way to reduce the variability of this break-in stage is to *hone* the edge. This type of edge preparation puts a controlled radius on the edge that minimizes the chances of chipping, a cause of erratic wear patterns. A premium is paid for this uniformity in initial performance of the cutting edges. After the initial break-in the wear rate reduces to a nearly constant value, dependent primarily on the chip velocity, if a wear model like Archard's in Eq (4.1b) applies. This is the region where most of the theories of wear and tool life are defined. The last region of the wear curve is characterized by a rapidly increasing wear rate that usually precedes catastrophic failure of the cutting edge. Tool wear predictions, either empirical or analytical, are aimed at predicting this point, because we do not want to operate in this region. Explanations for the increasing VB are usually related to the heat generated by the flank rubbing on the newly generated workpiece raising the temperature, starting a chain reaction of more rubbing and more heat.

Wear curves like Figure 4.9 are the starting point for the empirical models used to describe the effects of cutting conditions on what is termed *tool life*. Tool life is how long it takes in time, volume of material removed or, in production, the part count; to reach the end of the useful life of a cutting edge. While there are several ways to measure tool life, what can be more confusing is the criteria for measuring the end of the useful life of the edge. When the amount of either flank or crater wear is used, a quantitative criterion, denoted by a "*" can be used that depends on the application. For flank wear of high speed steel tools used for: turning and face milling - 1.5 mm, end milling and drilling - 0.5 mm, and 0.15 mm for reaming. For carbide tools where both crater wear and flank wear are important, usually the amount of wear used for VB* is less than for HSS, e.g., 0.8 mm for rough turning, 0.4 mm for finishing and 0.15 mm for reaming. Because measurements of the amount of wear are tedious, other less quantitative criteria for defining the end of the useful life of a cutting tool have been developed, e.g., visible changes in surface finish caused by nose wear, chipping or fracture along the depth of cut notch, or other indirect measurements like the forces, power vibration or acoustic emission measurements during machining.

Once a way of defining and measuring tool life has been established - here it is the amount of flank wear VB* - the tool life T is the total elapsed time in continuous cutting to reach this criterion. The procedures F.W. Taylor used to develop the empirical model that bears his name is a way to show this relationship between wear, tool life and cutting conditions. If experiments on new cutting edges of the same material are used at different velocities, with all other things being equal, the wear plots would look like Figure 4.10. From this the tool lives, T_1, T_2 and T_3 are obtained. While plotting T versus v shows there is an inverse relationship between T and v, a logarithmic transformation as in Figure 4.11 makes the power law relationship that is called Taylor's tool life equation clear:

$$\left(\frac{T}{T_R}\right)^n = \left(\frac{v_R}{v}\right) \tag{4.2a}$$

The terms in this expression are:

T_R is the reference tool life, a specified constant that depends on the units used, e.g., in the English system T_R = 1 min/edge and using SI units T_R = 60 s/edge.

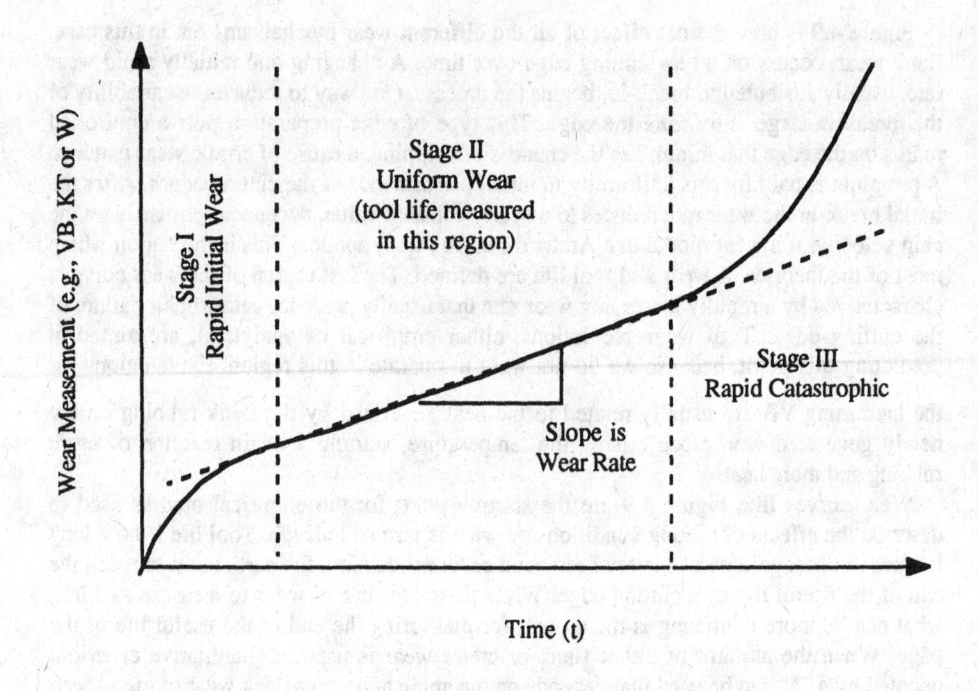

Figure 4.9 Flank wear curve.

v_R is an empirical function of cutting conditions, and both the workpiece and tool material. The larger v_R, the better because it can be interpreted as the cutting speed that gives a tool life of T_R.

n is the Taylor exponent, is considered a characteristic of the tool material with typical values listed in Table 4.2. Note that $0 < n < 1$, and as n gets smaller the tool life T of the cutting edge drops drastically for small increases in v.

Many times when English units are used, the Taylor equation is written as:

$$v\,T^n = C \tag{4.2b}$$

which is equivalent to Eq (4.2a) but not necessarily dimensionally correct. Another variation is the "generalized" tool life equation that includes other cutting conditions like the chip thickness (h) and width (b) in an empirical model for tool life:

$$\left(\frac{T}{T_R}\right) = \frac{K}{\left(\frac{v}{v'_R}\right)^{1/n}\left(\frac{h}{h_R}\right)^{1/n_1}\left(\frac{b}{b_R}\right)^{1/n_2}} \tag{4.2c}$$

where this time v'_R is a specified constant like h_R and b_R and the exponents n (the same as in Eq (4.2a), n_1 and n_2 as well as K are empirical constants determined from experiments where the thickness and width of the chip are changed. It has been found that $(1/n) > (1/n_1) > (1/n_2)$ is consistent with the conclusions made in Example 2.7 using a factorial design screening experiment. For this reason, the simpler form in Eq (4.2a) will

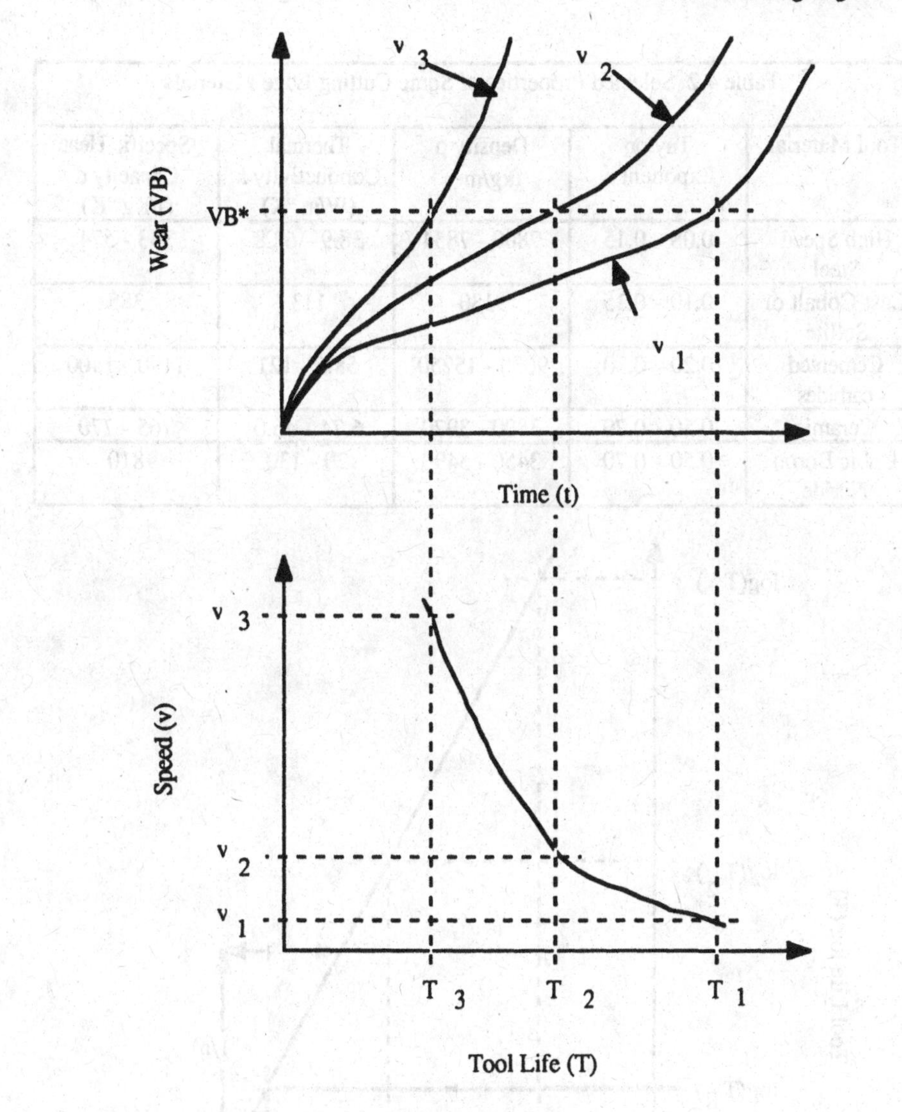

Figure 4.10 Graphical development to get tool life.

be used most of the time. But remember, by rearranging Eq (4.2c), the v_R in Eq (4.2a) is

$$v_R = v'_R \left(\frac{K}{\left(\frac{h}{h_R}\right)^{1/n_1} \left(\frac{b}{b_R}\right)^{1/n_2}} \right)^n ; \tag{4.2d}$$

one explanation of v_R's dependence on the cutting edge - workpiece material combinations and cutting conditions.

Table 4.2 Selected Properties of Some Cutting Edge Materials				
Tool Material	Taylor Exponent n	Density ρ (kg/m^3)	Thermal Conductivity k (W/m °K)	Specific Heat Capacity c (J/Kg °K)
High Speed Steel	0.08 - 0.15	7800 - 7854	38.9 - 62.8	343 - 574
Cast Cobalt or Sellite	0.10 - 0.15	9130	113	385
Cemented carbides	0.20 - 0.30	9030 - 15250	58.6 - 121	1190 - 1400
Ceramic	0.50 - 0.70	3800 - 3970	6.74 - 46.0	765 - 770
Cubic Boron Nitride	0.50 - 0.70	3450 - 3490	20 - 130	810

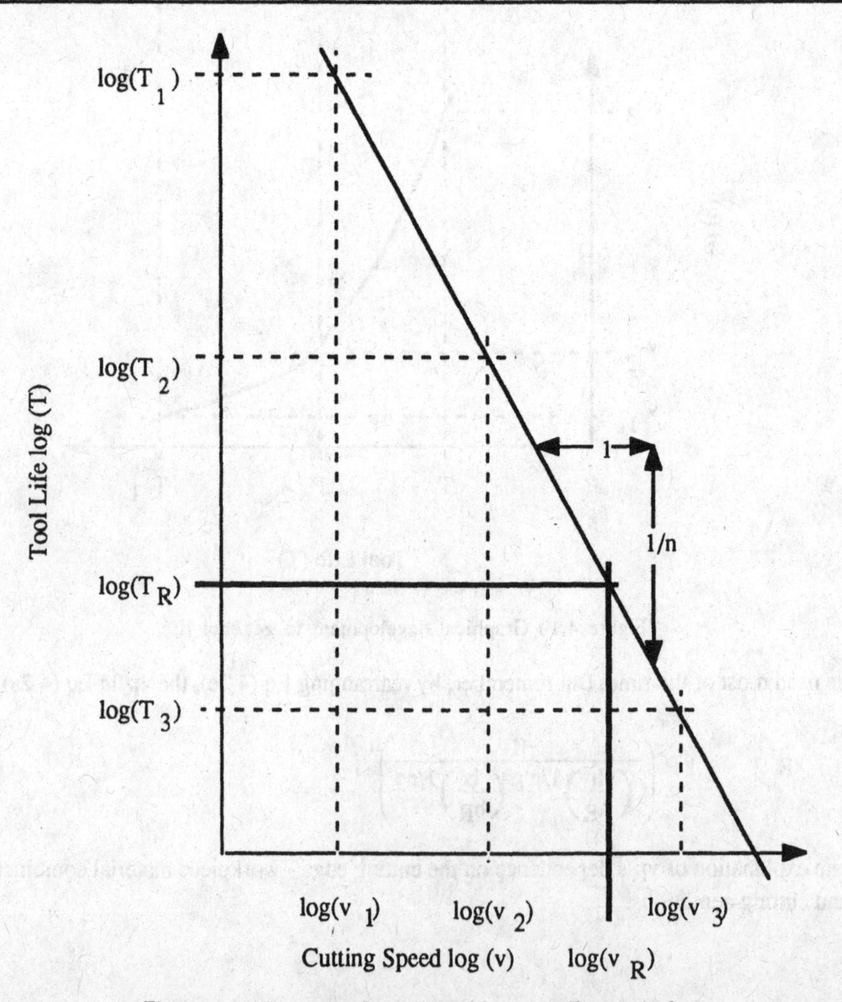

Figure 4.11 Log transformations of data in Figure 4.10.

While the Taylor equation will be used in subsequent chapters as if the material constants are readily available, this is not always the case. Many times the data is not available for a particular edge-workpiece material combination. Reasonable guesses are appropriate, but often experiments have to be used to get the data. This is where experimental design to lay out the test matrix can be used to economize on the number of tests, and the least squares method can be used to estimate the empirical coefficients n and v_R. After all Eq (4.2a) is the same form as the power law equation in Example 2.6.

4.4 References

[Archar53] Archard, J.F., (1953), "Contact and Rubbing of Flat Surfaces," *Journal of Applied Physics*, Vol. 24, pp. 981-988.

[BowTab71] Bowden, F.P., and D. Tabor, (1971 printing), *The Friction and Lubrication of Solids, Part 1*, Oxford University Press, London.

[KalJai80] Kalpakjian, Serope, and Sulekh Jain, Ed.s, (1980), *Metalworking Lubrication, Proceedings of the International Symposium on Metal Working Lubrication*, Century 2 Emerging Technologies Conferences, ASME Publication H00159, San Francisco, August 18-19.

[KinWhe66] King, Alan G., and W.M. Wheildon, (1966), *Ceramics in Machining Processes*, Academic Press, New York and London.

[Kir*etal*77] Kirk, J. A., Cardenas-Garcia, J.F., and Allison, C.R., (1977), "Evaluation of Grinding Lubricants - Simulation Testing and Grinding Performance," *Wear*, Vol.20, No.4, pp.333-339.

[KirCar77] Kirk, J.A., Cardenas-Garcia, J.F., (1977), "Evaluation of Grinding Lubricants-Simulation Testing and Grinding Performance," *Transactions of the American Society of Lubrication Engineers*, Vol. 20, pp.333-339.

[KobTho60] Kobayashi, S. and E.G. Thomsen, (1960), "The Role of Friction in Metal Cutting," *ASME Transactions, Journal of Engineering for Industry*, Series B, Vol. 82, pp. 324-332.

[KomDes82] Komanduri, R. and Desai, J.D., (1982), "Tool Materials for Machining", Technical Information Series #82CRD220, General Electric Company, Technical Information Exchange.

[TriCho51] Trigger, K.J., and Chao, B.T., (1951) "An Analytical Evaluation of Metal Cutting Temperatures," *ASME Transactions*, Vol. 73, pp. 57-68.

[Tri*etal*77] Tripathi, K.C., Nicol, A.W., Rowe,G.W., (1977), "Observation of Wheel-Metal-Coolant Interactions in Grinding," *Transactions of the American Society of Lubrication Engineers*, Vol. 20, pp.249-256.

4.5 Problems

4.1 Consult some references on tool materials to find:

a. What tool materials, at least theoretically, can machine primary martensite, which has a hardness of H_{RC} = 65, because they meet one of the basic requirements of a tool material?

b. Based on the mechanical properties data for different grades of carbides, what property seems to be the most important for differentiating carbide grades based on their application?

4.2 Tool life data for high speed steel was obtained using a turning setup, with a 2.5 mm depth of cut and a feed of 0.25 mm/rev. Three pairs of cutting speed and tool life data are:

Experimental Tool Life Test	
v(m/s)	T(s/edge)
1.27	48
1.13	228
1.02	846

a. Based on these data, use whatever method is correct and that you find easiest to estimate n and v_R in the Taylor equation $(T/T_R)^n=(v_R/v)$, where T_R= 60 s/edge.

b. Based on your results, what cutting speed would you expect to give a tool life of 30 s/edge.

c. Very often you will find the Taylor equation given in the form $vT^n = C$, this is particularly true when the English system is used. Then v and C are in units of ft/min, and T is in min/edge. Convert your results in 4.2.a to English units, i.e., what are C and n?

4.3 This problem is aimed at comparing the effects of possibly changing one of the cutting conditions, cutting speed, and the cutting tool material. Throughout, assume that except for cutting speed, all the other cutting conditions are kept constant and that the *machining time is inversely proportional to the cutting speed*, v.

a. Assume that the current cutting edge material is high speed steel (HSS) that has a tool life equation given by:

$$\left(\frac{T}{60\text{ s/edge}}\right) = \left(\frac{0.4\text{ m/s}}{v}\right)^{(1/0.1)}$$

and assume that the current cutting speed is 0.15 m/s. Estimate the tool life T (s/edge) for this HSS tool material at this cutting speed.

b. As a change in operating strategy, cutting speeds will be selected to give a tool life of 3.5 hours, so that cutting edges are changed about twice per shift. For this tool life and the tool life equation for HSS in 4.3.a, find the cutting speed (m/s) that will give a tool life of T= 3.5 h/edge, and compute the *ratio* of this new machining time to the machining time in 4.3.a.

c. Retain the operating strategy from 4.3.b, i.e., the cutting speed is selected so that the tool life is 3.5 hours. This time cemented carbides will be used as a tool material, and this grade of carbide has a Taylor equation

$$\left(\frac{T}{60\text{ s/edge}}\right) = \left(\frac{1.25\text{ m/s}}{v}\right)^{(1/0.25)}$$

With this new tool material, for T= 3.5 h/edge, compute the cutting speed (m/s) and the *ratio* of this new machining time to the machining time in 4.3.a

4.4 A salesman has a car load of cutting edges at a price too good to pass up. The claim is that this material has a Taylor tool life equation of $\left(\frac{T}{60 \text{ s/edge}}\right)^{0.45} = \left(\frac{2.75 \text{ m/s}}{v}\right)$. The deal sounds too good to be true, so you decide to get two of the edges out of the lot and run a simple tool life test at two cutting speeds, v = 1.00 m/s and 1.25 m/s. The resulting tool wear data is both plotted in Figure 4.12 and tabulated.

a. If the criterion for tool life is $VB^* = 0.75$ mm, use the experimental data to determine n and v_R.

b. Assume the values you found have an error of +/- 10%. Based on your analysis of this data, the samples you tested probably *(were, were not, cannot tell)* the same as the salesman was claiming and the cutting conditions you ran probably *(were, were not, cannot tell)* the same as would be expected based on the salesman's tool life equation.

c. For a cutting speed of 0.4 m/s, what is the tool life based on the salesman's tool life equation and the one you found based on experimental data.

Tabulated Version of the data in Figure 4.12

Time (s)	VB(v=1.00 m/s) (mm)	VB(v=1.25 m/s) (mm)
300	0.381	0.584
600	0.533	0.787
900	0.660	1.067
1200	0.813	1.321

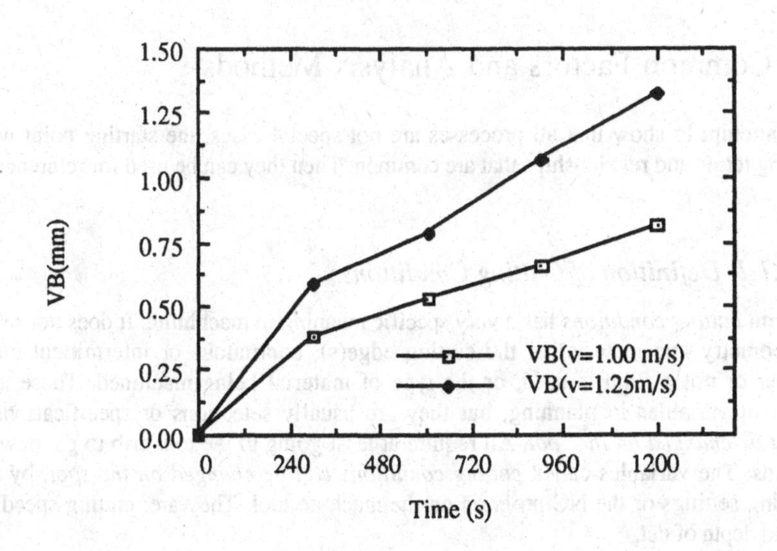

Figure 4.12 Wear data for Problem 4.3.

5
Practical Machining Processes

Finally after the background on mechanics, simple micro models of what happens at the contact interface in machining or grinding, and continuously being reminded about needing empirical materials data, it is time to consider actual production machining processes. The background and cautions are still valid, but most of the time this chapter will proceed as if most things are known. This chapter will consider some of the analysis methods that are common, i.e., how to estimate the theoretical surface roughness or use the specific energy method to estimate forces when mechanics will not do or is too complicated for simple planning of a process. It will also consider three common machining processes: turning, milling and drilling. Using the cutting edge geometry and number of edges as a classification scheme, this progression can be thought of as going from a case that is similar to the orthogonal and oblique models of Chapter 3, to multiple cutting edges that make intermittent cuts, to the complicated multiple edges of the drill.

In each case, as Figure 5.1 illustrates, the geometry and materials or the cutting edge and the workpiece will be assumed known, with the kinematics, mechanics and empirical relationships serving as the basis for estimating the practical output of finish, forces and power and estimated time to produce a workpiece. While each process is different, the basics are common and have been discussed in the previous chapters. This chapter and Chapter 6 on grinding are aimed at applying the basic principles to calculations on real processes. However, one important assumption remains, the practical process and workpieces are perfectly rigid and all inertial effects are neglected. Not until chapter 8 will this restriction be relaxed.

5.1 Common Factors and Analysis Methods

In an attempt to show that all processes are not special cases, the starting point will be defining terms and relationships that are common. Then they can be used for reference later on.

5.1.1 Definition of Cutting Conditions

The term *cutting conditions* has a very specific meaning in machining. It does not refer to: the geometry and material of the cutting edge(s), continuous or intermittent cutting, whether or not a fluid is used, or the type of material being machined. These are all important variables in planning, but they are usually selections or specifications that *cannot be changed on the spot*. All require at least going to the tool crib to get new tools or fluids. The variables called *cutting conditions can be changed on the spot*, by either changing settings or the NC program on the machine tool. They are: cutting speed, feed rate and depth of cut.

Cutting Speed (v) is the *largest* relative velocity between the cutting edge and workpiece, and it may be generated either by the workpiece moving (lathes or planing

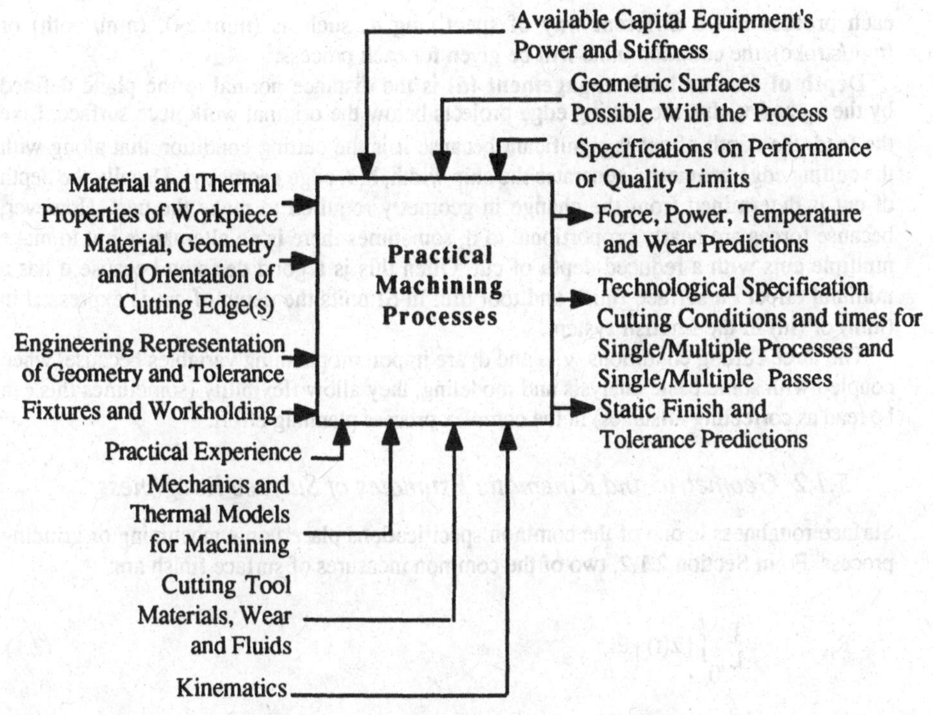

Figure 5.1 The Inputs, Constraints, Outputs and Mechanisms considered when analyzing and modeling a practical machining process.

mills) or by the cutting edges moving (milling or drilling). The cutting speed is the same as the scalar v component of the velocity vector in Section 3.1 for orthogonal and oblique cutting. Also, from consulting the equations in Chapter 3, it has the least effect on forces (but force components in this direction represent most of the power consumed at the cutting edge). From Example 2.7 and empirical modeling of tool life in Section 4.3, of the three cutting conditions v has the greatest effects on temperatures and tool life. At low v, because of the built up edge, surface finish is poor, but operating at high speed which is desirable anyway, or using a fluid that acts as a lubricant are ways to improve the surface finish when BUE occurs. The standard SI units for specifying v are (m/s), with (ft/min) the common English units.

Feed or feed engagement (*a*) specifies the relative lateral motion between the cutting edge and workpiece normal to v. There are two types of feed: the *relative feed, a* , that depends on the specific machining process and how the lateral motion is generated, and the *absolute feed, feed speed or feed rate* v_a. The feed rate is actually a second velocity component in the velocity vector $\mathbf{v} = [v, v_a, 0]^T$; in Chapter 3 it would be the velocity component in the F_q or y direction. The most important thing to remember about the relative feed *a*, is that it is the most important variable in the function to compute the uncut chip thickness i.e., h(*a*, edge geometry). Over the years a number of jargon terms have been associated with *a* like *chip load,* but it can always be determined based on the number of cutting edges, geometry and cutting speed. Because it determines h, the feed has an important effect on forces and power. As will be shown in the next section, it has an important geometric effect on surface finish too, but only a secondary effect on tool life. The common SI units for the feed rate are (mm/s); the English units are (in/min). Because each process has a different way of specifying a, such as (mm/rev), (mm/tooth) or (mm/stroke), the common units will be given for each process.

Depth of Cut or back engagement (d) is the distance normal to the plane defined by the v_a and v, that the cutting edge projects below the original workpiece surface. Like the feed, the depth of cut is significant because it is the cutting condition that along with the cutting edge geometry, generates the chip width, b(d, edge geometry). Usually the depth of cut is determined from the change in geometry required to make the part. However, because forces are nearly proportional to d, sometimes there is no alternative but to make multiple cuts with a reduced depth of cut. Often this is a good decision because d has a minimal effect on surface finish and tool life. In SI units the depth of cut is expressed in (mm) or (in) in the English system.

The three cutting conditions, v, a and d, are important planning variables because, when coupled with some basic analysis and modeling, they allow flexibility (sometimes this can be read as correcting mistakes) in the complex process planning effort.

5.1.2 Geometric and Kinematic Estimates of Surface Roughness

Surface roughness is one of the common specifications placed on a machining or grinding process. From Section 2.1.2, two of the common measures of surface finish are:

$$R_a = \frac{1}{L} \int_0^L | z(t) | \, dt, \tag{2.1}$$

the Arithmetic Average (AA) or Center Line Average (CLA) roughness and the maximum peak to valley height roughness

$$R_t = \max[z(0)....z(L)] - \min[z(0)....z(L)] \, . \tag{2.3}$$

Because both R_a and R_t measure geometric quantities, they can be estimated with the geometric and kinematic quantities that are common in the different machining and grinding processes. The roughness estimates that follow are only *theoretical lower limits* because they ignore such material related phenomena as built up edge or fracture and vibrations that usually increase the roughness of a part. They serve their purpose as ways to plan a machining operation.

Two simple geometries that show up on real cutting edges will be considered, and both will be presented as the profiles generated in a plane normal to the cutting speed component of v. Because the nomenclature for specifying the angles changes, generic angles on the cutting edges will be used, and referred to later by their specific jargon term. The first type of cutting edge geometry considered is like that shown in Figure 5.2a. This geometry generates a triangular cross section with base a, with one side inclined at angle ψ measured from the direction of a and with the other side inclined at Λ measured from the normal to a. From the geometry in Figure 5.2a and the definition of R_t, it is easiest to determine the maximum peak-to-valley roughness.

$$R_t(a, \psi, \Lambda) = \frac{a}{\cot(\psi) + \tan(\Lambda)} \, . \tag{5.1a}$$

The R_a roughness is referenced to the height center line ($R_t/2$), then the portions below the center line are rectified, resulting in two triangles of height ($R_t/2$) and base (a /2). As

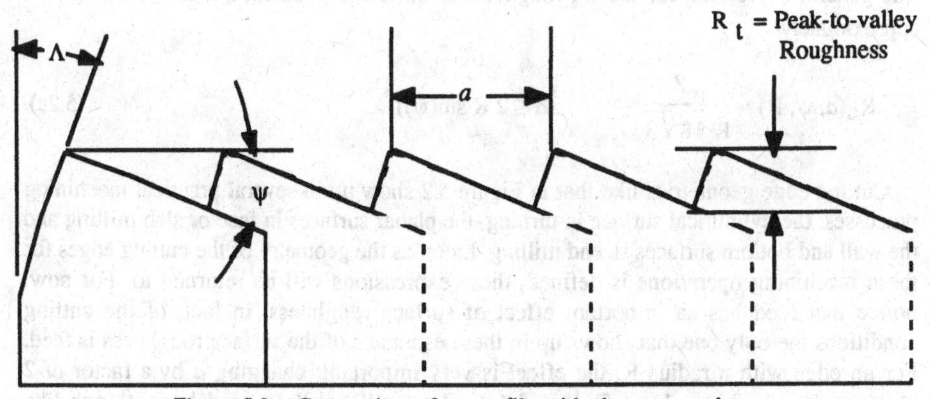

Figure 5.2a Geometric surface profile with sharp nose edge.

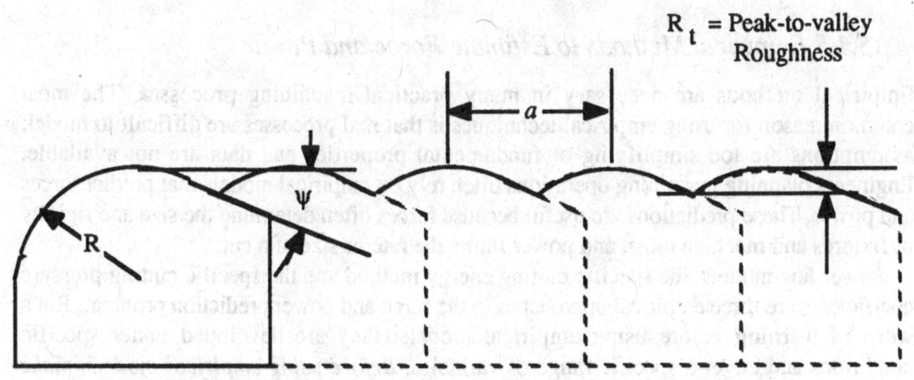

Figure 5.2b Geometric surface profile with finite nose radius.

Figure 5.2 Geometric surface profiles viewed in a plane normal to the cutting speed component of the velocity vector.

Eq (2.1) indicates, these triangular areas are divided by the a or

$$R_a(a, \psi, \Lambda) = \frac{a}{4(\cot(\psi)+\tan(\Lambda))}. \tag{5.1b}$$

A radius R, is more common on the nose or corner of a cutting edge$_t$. This produces a smoother finish as shown in Figure 5.2b. The R_t roughness in this case is

$$R_t(a, \psi, R) = [1-\cos(\psi)]\, R + a \sin(\psi)\cos(\psi) - \sqrt{2a\, R \sin^3(\psi) - a^2 \sin^4(\psi)} \tag{5.2a}$$

or if only the circular arc generates the profile, which occurs when $a \leq 2\, R \sin(\psi)$ for a finishing cut, Eq (5.2a) reduces to

$$R_t(a, \psi, R) \approx \frac{a^2}{R\ 8} \qquad (a \leq 2\, R \sin(\psi)). \tag{5.2b}$$

The general expression for the R_a roughness is quite complicated, but for finishing it is approximately

$$R_a(a, \psi, R) \approx \frac{a^2}{R\, 18\sqrt{3}} \qquad (a \le 2\, R \sin(\psi)). \tag{5.2c}$$

Cutting edge geometries like that in Figure 5.2 show up in several practical machining processes: the cylindrical surface in turning, the planar surfaces in face or slab milling and the wall and bottom surfaces in end milling. Later, as the geometry of the cutting edges for these machining operations is defined, these expressions will be returned to. For now, notice that feed has an important effect of surface roughness; in fact, of the cutting conditions the only one that shows up in these estimates of the surface roughness is feed. For an edge with a radius R, the effect is very important; changing a by a factor of 2 changes the roughness by a factor of 4 (and as will be discovered later, things like machining time, power or force will usually change only in proportion to a).

5.1.3 Empirical Methods to Estimate Force and Power

Empirical methods are necessary in many practical machining processes. The most common reason for using empirical techniques is that real processes are difficult to model; assumptions are too simplifying or fundamental properties and data are not available. Engineers planning machining operations often rely on empirical models that predict forces and power. These predictions are useful because forces often determine the size and rigidity of fixtures and machine tools, and power limits the rate or size of a cut.

Power law models, the specific cutting energy method and the specific cutting pressure coefficients are three empirical approaches to the force and power prediction problem. But a word of warning before using empirical models; they are developed under specific conditions and cover a specific range of variables. *Before using empirical models, make sure to read the footnotes and fine print.* It usually is not permissible to assume that an empirical model can be used in all applications, so use them with caution and know their limitations.

Table 5.1 lists selected power law models that have appeared in the literature over the years. Sources in the literature for these models include: [Boston41], [ASME52] [MachDa80]. If data for a specific application is not available, someone must decide how close the empirical model is to the problem at hand. For example, when planning a turning operation on a low carbon steel workpiece with a hardness of $H_B = 104$ with a high speed steel cutting edge that has a larger end relief angle than is listed in Table 5.1, can the empirical model in Table 5.1 still be used? The basic principles covered in Chapters 3 and 4 and practical knowledge about turning can help make the decision. It turns out that using the empirical model in Table 5.1 would be valid because the variation in hardness of four percent is normal and the relief angle has little effect on forces. If existing empirical models cannot be used - very possible when new materials have to be processed - then a combination of efficient experimental designs and empirical model fitting can be used in the development laboratory to build empirical models tailored to the application.

Specific cutting energy (often called unit horsepower in the English system of measurement) is another empirical technique for estimating the *average* power. Knowing the average power and the kinematics and mechanics of the specific process, average forces or torques can also be estimated. The basic idea behind this method was presented in Chapter 3. Orthogonal cutting experiments are run at standard cutting edge geometry,

Table 5.1 Selected Empirical Power Law Models for Estimating Forces, Torques and Specific Cutting Energy [Boston41]

Process	Notes	Power Law Models
Turning	Workpiece: Low carbon steel, H_B=100 Cutting edge: HSS, 8-14-6-6-6-0-1.2mm Cutting conditions: v constant =0.41 m/s, about the points *a* = 0.76 mm/rev and d=3.8 mm Without fluid	F_p=592(kN)(*a*/25.4mm/rev)$^{0.83}$(d/25.4mm) F_q=150(kN)(*a*/25.4mm/rev)$^{0.48}$(d/25.4mm)$^{1.45}$ F_r=4.10(kN)(*a*/25.4mm/rev)$^{0.56}$
	Workpiece: Low carbon steel, H_B=100 Cutting edge: HSS, 8-22-6-6-6-0-1.2mm Cutting conditions: v constant =0.41 m/s, about the points *a* = 0.76 mm/rev and d=3.8 mm Without fluid	F_p=456(kN)(*a*/25.4mm/rev)$^{0.8}$(d/25.4mm) F_q=56.0(kN)(*a*/25.4mm/rev)$^{0.42}$(d/25.4mm)$^{1.35}$ F_r=3.13(kN)(*a*/25.4mm/rev)$^{0.56}$(d/25.4mm)$^{0.13}$
	Workpiece: SAE 3135 steel, H_B=207 Cutting edge: HSS, 8-14-6-6-6-15-1.2mm Cutting conditions: v constant =0.25 m/s, about the points *a* = 0.76 mm/rev and d=3.8 mm Without fluid	F_p=1150(kN)(*a*/25.4mm/rev)$^{0.98}$(d/25.4mm) F_q=159(kN)(*a*/25.4mm/rev)$^{0.39}$(d/25.4mm)$^{1.55}$ F_r=32.5(kN)(*a*/25.4mm/rev)$^{0.74}$(d/25.4mm)$^{0.28}$
	Workpiece: SAE 3135 steel, H_B=207 Cutting edge: HSS, 8-22-6-6-6-15-1.2mm Cutting conditions: v constant =0.25 m/s, about the points *a* = 0.76 mm/rev and d=3.8 mm Without fluid	F_p=774(kN)(*a*/25.4mm/rev)$^{0.90}$(d/25.4mm) F_q=4.23(kN)(*a*/25.4mm/rev)$^{1.08}$ F_r=114(kN)(*a*/25.4mm/rev)$^{0.84}$(d/25.4mm)$^{0.58}$

cutting conditions and edge wear, and the power at the cutting edge is measured. From this data and the cutting conditions, the specific cutting energy can be computed from Eq (3.13)

$$E_c \equiv \left(\frac{\text{Cutting Energy}}{\text{Unit Volume of Material}}\right) = \frac{P}{Z} \tag{3.13}$$

Table 5.1b Selected Empirical Power Law Models for Estimating Forces, Torques and Specific Cutting Energy [Boston41]

Process	Notes	Power Law Models
Slab Milling	Workpiece: cast iron Cutting edge: HSS, 12 tooth, 75 mm x 75 mm, 25° helix, 15° rake Cutting conditions: about the points a = 0.25 mm/rev and d=3.175 mm Without fluid For up and down milling	E_c^{up}= 49.2(J/mm^3)(a/25.4mm/rev)$^{-.59}$ (d/25.4mm)$^{-.44}$ E_c^{dn}= 53.2(J/mm^3)(a/25.4mm/rev)$^{-.59}$ (d/25.4mm)$^{-.44}$
Drilling	Workpiece: cast iron, H_B=163 Cutting edge: HSS twist drills 12.5 <D< 37.5 mm with 31° helix, 121° point angle, 136° chisel edge and 5° relief angle. Cutting conditions: constant v=0.30 m/s, 0.23<a <0.38mm/rev with emulsion of soluble oil 1:16 water.	T=515(N-m)(a/25.4mm/rev)$^{0.60}$(D/25.4mm)^2 F_z=65.5(kN)(a/25.4mm/rev)$^{0.60}$(D/25.4mm)
	Workpiece: SAE 6150, H_B=187 Cutting edge: HSS twist drills 12.5 <D< 37.5 mm with 31° helix, 121° point angle, 136° chisel edge and 5° relief angle. Cutting conditions: constant v=0.30 m/s, 0.23<a <0.38mm/rev with emulsion of soluble oil 1:16 water.	T= 2490(N-m) (a/25.4mm/rev)$^{0.78}$(D/25.4mm)$^{1.8}$ F_z=238(kN)(a/25.4mm/rev)$^{0.7}$(D/25.4mm)

This is assumed to be a property of the material and contact conditions, and is how data like that in Table 5.2 is produced. To use this data, this experimental procedure is reversed. Treating E_c as a material property, the cutting conditions for a proposed machining application are used to compute Z, and then the average force or torque can be estimated. This assumes that corrections are made for deviations from the standardized test conditions, like the graphs in Figures 5.3. The forces or torques that contribute the most to the power, designated by the subscript "p", are estimated from

$$F_p = \left(\frac{Z \bullet E_c \bullet [K_c(h_{avg}) \bullet K_r(\gamma) \bullet K_w(VB)]}{v}\right) \text{ and} \tag{5.3a}$$

Table 5.2 Specific Cutting and Specific Grinding Energy

Material		Brinell Hardness	Cutting Energy (J/mm^3)	Grinding Energy (J/mm^3)	
Ferrous Materials				Small Swarf	Large Swarf
AISI Steels	1019	147-150	2.07		21.51
	1020	109-110	1.99	82.73	16.55
	1045	190-200	2.26		24.82
	1070	217	2.48		
	1095	182	2.35		
	1113	170	1.34		
	1340	192	2.56		
	2340	197	2.18		
	3115	131-150	1.88		14.89
	3130	169	2.10		
	3140	185	2.48		
	4135	220-250	1.77		28.13
	4135	330-350	2.37		33.09
	4340	210	2.56		
	4340	300-302	4.12		29.78
	4340	400-514	5.46		41.37
	52100	186	2.15		
	81B45	180	1.61		
Cast Irons	All	100-125	1.04		9.93
		150	1.39		13.24
		200-215	1.61	74.46	14.06
		250	1.96		33.09
Tool & Die Steels	Peerless 56	352	3.82		
	Super Tricent	352	3.55		
	A-6 Tool Steel	240		82.73	13.24
	A-6 Tool Steel	530		82.73	14.89
	T-15 HSS	700		99.28	20.68
Stainless Steels	302	160	3.08		
	303	162	2.21		
	304	162-182	2.92	99.28	18.20
	410	217	1.99		
	416	215	1.80		
	430	156	2.02		
	17-7PH	170	4.01		

Table 5.2 Specific Cutting and Specific Grinding Energy (Continued)

Material		Brinell Hardness	Cutting Energy (J/mm^3)	Grinding Energy (J/mm^3)	
Aerospace	A-286	321	3.85		
Alloys	AM-350	444	3.96		
	D6 AC	560	9.41		
	INCONEL 700	320	5.10		
	RENE 41	340-365	9.60	99.28	16.55
	UDIMET 500	360	4.94		
Non-Ferrous				Small Swarf	Large Swarf
Aluminum	2024-T4		0.87		
	6061-T6		0.76		
	"Soft"	80		41.37	12.41
	"Hard"	150		41.37	8.27
Brass	Free Machining		0.68	9.93	5.79
	Hard		1.50		18.20
	Red Brass 8/15		1.34		
Titanium	Hard Alloy		2.73		
	Soft Alloy		1.77		
	Alloy	295-300		66.19	19.03
	Cobalt (HS-25)		4.36		
	Columbium		3.82		
	Copper		1.23		28.96
	Magnesium		0.33	4.96	3.31
	Molybdenum (TZM)		6.82		
	Tantalum Alloy (90%)		8.18		
	Tungsten		6.82		
	Zinc		0.68		

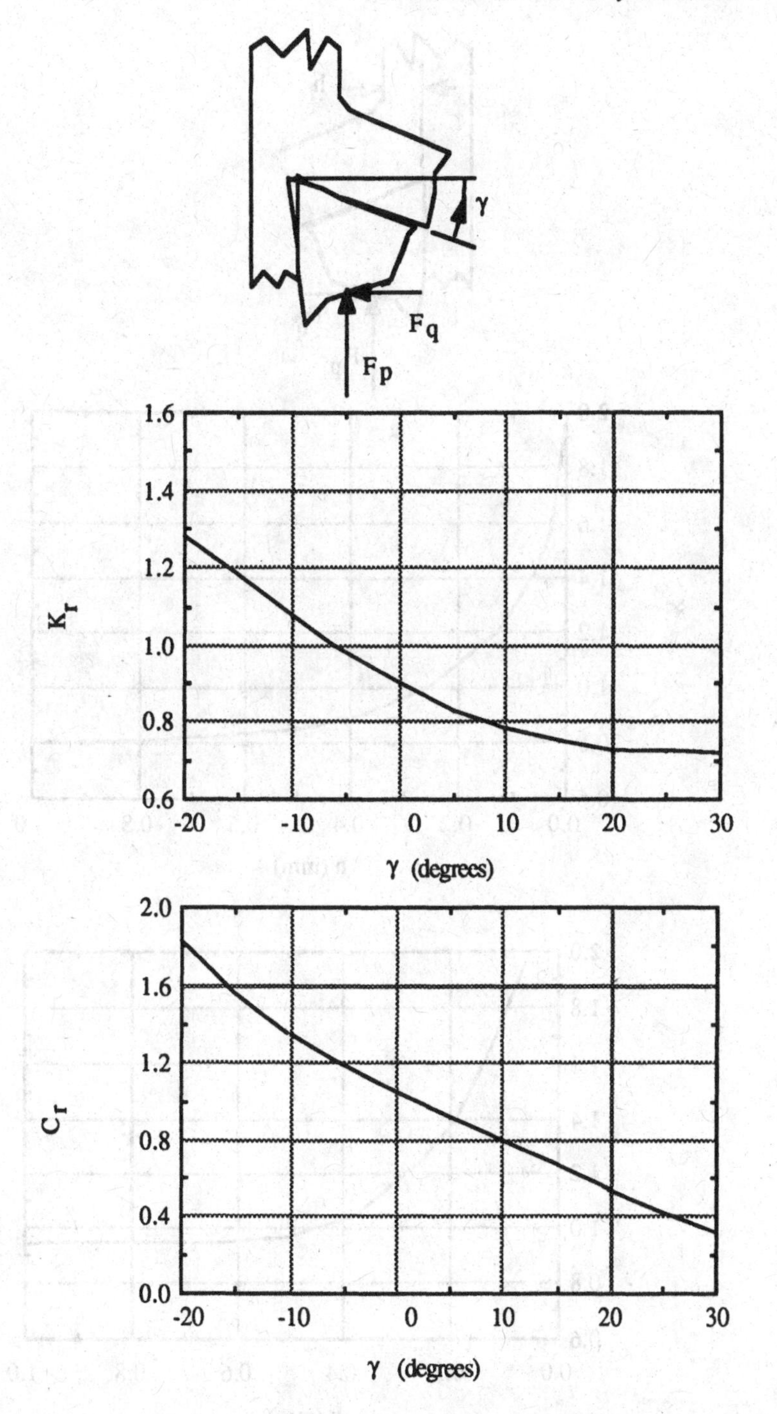

Figure 5.3a Charts for rake angle corrections using the specific cutting energy method.

Figure 5.3 Graphs to be used with the E_c values in Table 5.2.

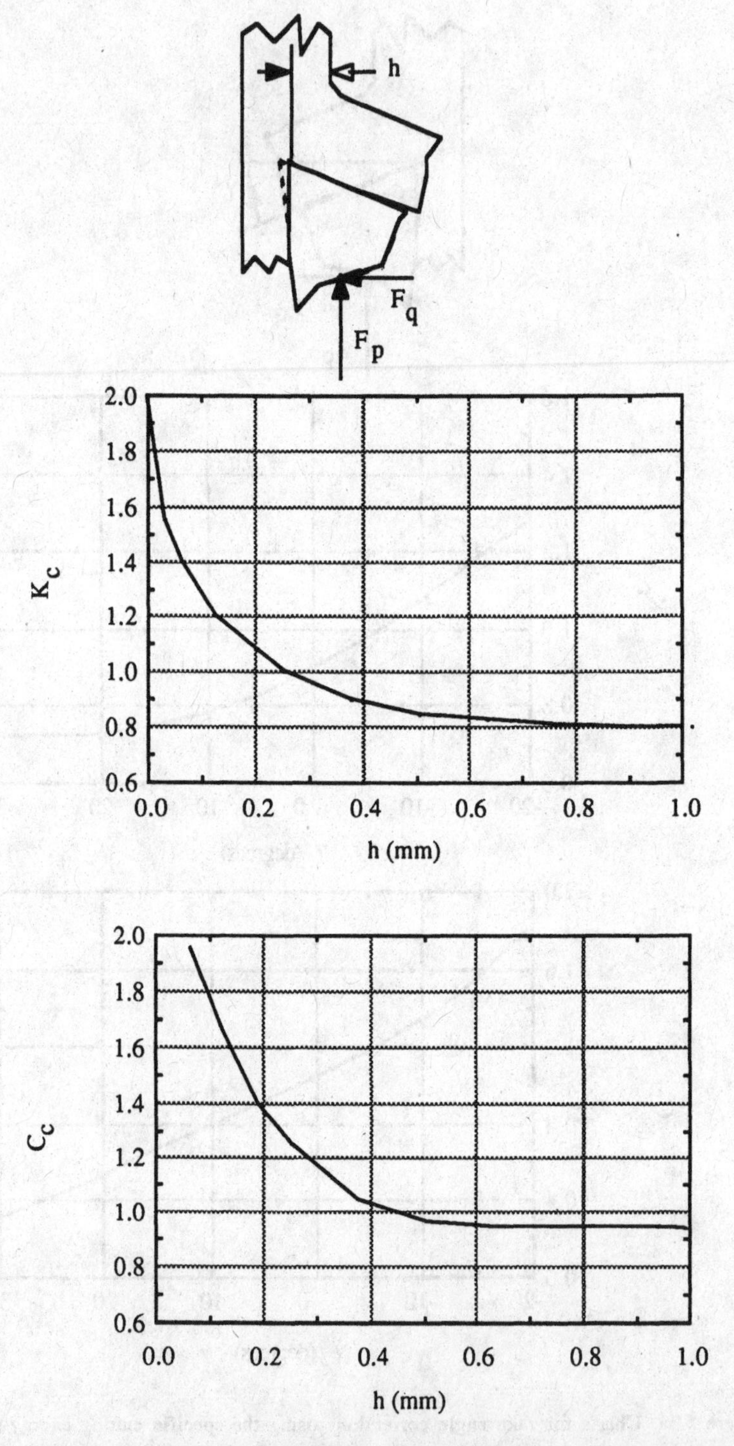

Figure 5.3b Charts for chip thickness corrections using the specific cutting energy method.

Figure 5.3 Graphs to be used with the E_c values in Table 5.2.

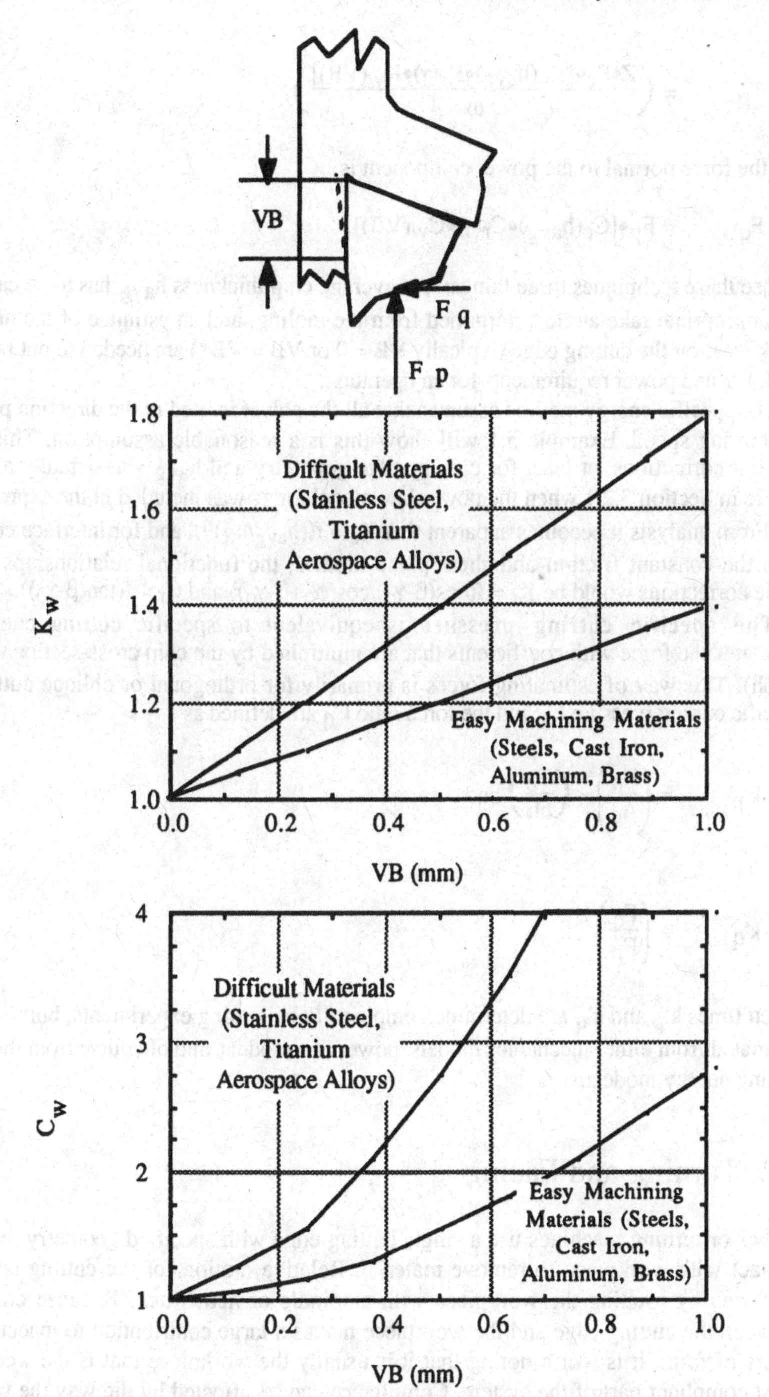

Figure 5.3c Charts for flank wear corrections using the specific cutting energy method.

Figure 5.3 Graphs to be used with the E_c values in Table 5.2.

$$T_p = \left(\frac{Z \bullet E_c \bullet [K_c(h_{avg}) \bullet K_r(\gamma) \bullet K_w(VB)]}{\omega}\right), \tag{5.3b}$$

and the force normal to the power component is

$$F_q = F_p \bullet [C_c(h_{avg}) \bullet C_r(\gamma) \bullet C_w(VB)]. \tag{5.3c}$$

To use these techniques three things: the average chip thickness h_{avg} has to be calculated, the appropriate rake angle determined from the tooling, and an estimate of the amount of flank wear on the cutting edge (typically VB = 0 or VB = VB*) are needed to put bounds on the force and power requirements for an operation.

The specific energy method assumes that all the power is used in the direction parallel to the cutting speed. Example 5.1 will show this is a reasonable assumption. This idea of making corrections, at least for cutting edge geometry and h_{avg} was actually alluded to before in Section 3.2.1 when the power law model for τ_s was included in the expression for E_c. From analysis it becomes apparent that $K_c = f((h_{avg}/h_R)^a)$, and for interface conditions with the constant friction and shear plane models, the functional relationships for rake angle corrections would be $K_r = f(\cos(\beta-\gamma_o)/\cos(\phi_o+\beta-\gamma_o))$ and $C_r = f(\tan(\beta-\gamma_o))$.

The specific cutting pressure is equivalent to specific cutting energy, but estimates the force with coefficients that are multiplied by the chip cross-sectional area A_c = (bh). This way of estimating forces is primarily for orthogonal or oblique cutting. The specific cutting pressure k'_p and the force ratio k'_q are defined as

$$k'_p \equiv \left(\frac{F_p}{A_c}\right) = \left(\frac{F_p}{bh}\right) \text{ and} \tag{5.4a}$$

$$k'_q \equiv \left(\frac{F_q}{F_p}\right). \tag{5.4b}$$

Often times k'_p and k'_q are determined empirically by cutting experiments, but they can be estimated from either mechanics models, power law models, and of course from the specific cutting energy models.

5.2 Turning and Facing

Lathes or turning machines use a single cutting edge with specified geometry in constant contact with workpiece to remove material. Relative motion for the cutting speed v, is produced by rotating the workpiece with a spindle or *headstock*. Because compliance between the cutting edge and the workpiece makes a large contribution to inaccuracy and errors of form, it is worth noting that it is usually the workpiece that is the weakest and most compliant part of the system. Compliance can be affected by the way the workpiece is held. Several common ways of holding a workpiece on a lathe are: *Chucking* by holding the workpiece in a 3 or 4 jaw chuck mounted on the headstock is usually the easiest and fastest setup, but because the workpiece is cantilevered this is also the most compliant setup. A more rigid setup is turning *between centers*, where the headstock end may use a chuck or *dog* to drive the workpiece. The *tailstock* end has either a bearing mounted *live center* that rotates or is a non-rotating *dead center*. The dead center tends to heat up due to friction, but is usually more rigid than a live center. To increase the stiffness of long slender workpieces, a special bearing termed a *steady rest* may be used to support the workpiece at its midpoint.

Because lathes rotate the workpiece as the cutting edge moves either parallel or normal to the axis of rotation, lathes are best at producing surfaces of revolution. Operations that produce these surfaces of revolution have specific names associated with them: The sides on an external cylinders are produced by *turning* when the feed motion is parallel to the axis of rotation. The ends of a cylinder are generated by *facing*, where the feed motion is radial and normal to the axis of rotation. Internal cylinders of moderate accuracy are produced by *drilling* and high accuracy internal cylinders are produced by *boring*; in both cases the feed motion is axial. Accurate *threads* are generated by the helical path produced when the axial motion and rotation of the workpiece are coordinated and the relative feed is large. Contoured shapes are generated by simultaneous axial and radial feed motions common on CNC machine tools, which makes these cuts a combination of turning and facing.

Figure 5.4 shows a schematic of an engine lathe, along with the machine tool coordinate system that will be used to describe the motion of the cutting edge. A right hand coordinate system represents the Z-axis along the spindle, the X-axis in the radial direction and the Y-axis determined by the right hand rule. These coordinates conform to the standards used to specify and program motions on a CNC machine tool.

5.2.1 Edge Geometry

Cutting edge geometry for turning depends on the tool or insert geometry and how the edge is mounted on the machine tool. For inserts the type of tool holder is important in defining the final edge geometry. A number of ways to specify cutting edge geometry are available, but the *tool-in-hand* specification shown in Figure 5.5 is used in most tool catalogs. The nomenclature for the cutting edges is like that in Chapter 3, i.e., γ designates rake angles and α relief angles. Instead of having one edge, there is a primary and secondary cutting edge blended with a radius to form the nose of the cutting tool. For turning, unless R is large or the depth of cut is small, most of the cutting is done by the primary or side cutting edge. As a result, γ_f is treated like γ_n, and κ_f plus the amount the shank of the tool is rotated when mounted on
the machine tool is treated like the inclination angle λ in Section 3.2. The tool signature is a shorthand for specifying the tool-in-hand geometry, and the order is *always* as shown in Figure 5.5. Carbide, ceramic or coated tool inserts need both the tool holder and insert geometry to determine the equivalent tool-in-hand edge geometry.

5.2.2 Kinematics

Terms for specifying cutting conditions for turning include the depth of cut, d, which is the amount the nose projects below the uncut workpiece surface. The relative feed in turning and facing is specified in terms of the axial or radial motion per revolution of the spindle a , with units of (mm/rev). The feed rate v_f is related to the rotational speed of the spindle N (rev/s) by

$$v_f = a \bullet N \tag{5.5}$$

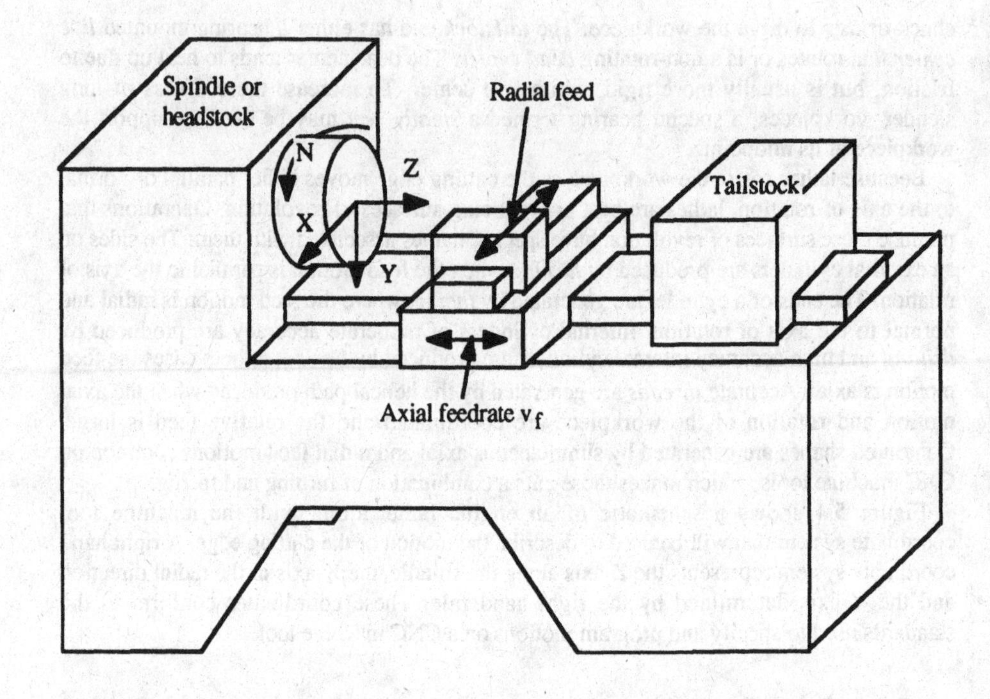

Figure 5.4 Lathe nomenclature.

in the radial or axial direction. The agreed upon convention is that cutting speed is always the maximum tangential velocity, so that the maximum diameter is used to determine the tangential velocity

$$v = \pi \bullet D \bullet N \qquad \text{(turning, boring, drilling)} . \qquad (5.6a)$$

For facing, unless the machine tool has a controller that increases the rotational speed to maintain constant tangential speed, the velocity decreases is given by:

$$v = \pi \bullet N \left(D - \int_0^t a \; ds \right) \qquad \text{(facing)}. \qquad (5.6b)$$

Taken together, Eq (5.5) and Eq. (5.6) point out that the velocity vector has two components,

$$\mathbf{v} = \begin{bmatrix} 0 \\ v \\ v_f \end{bmatrix} \text{(turning, boring, drilling) or} = \begin{bmatrix} v_f \\ v \\ 0 \end{bmatrix} \text{(facing).} \qquad (5.7)$$

Because the rake and clearance angles in Chapter 3 were defined normal and parallel to the velocity *vector*, this second component changes the geometry that the workpiece "sees". The velocity vector is inclined from the cutting speed component by $\tan^{-1}(v_f/v)$, reducing both the tool-in-hand rake and clearance angles to *effective rake and clearance angles*, γ_{fe} and α_{fe}. Other angles may be affected in a similar way.

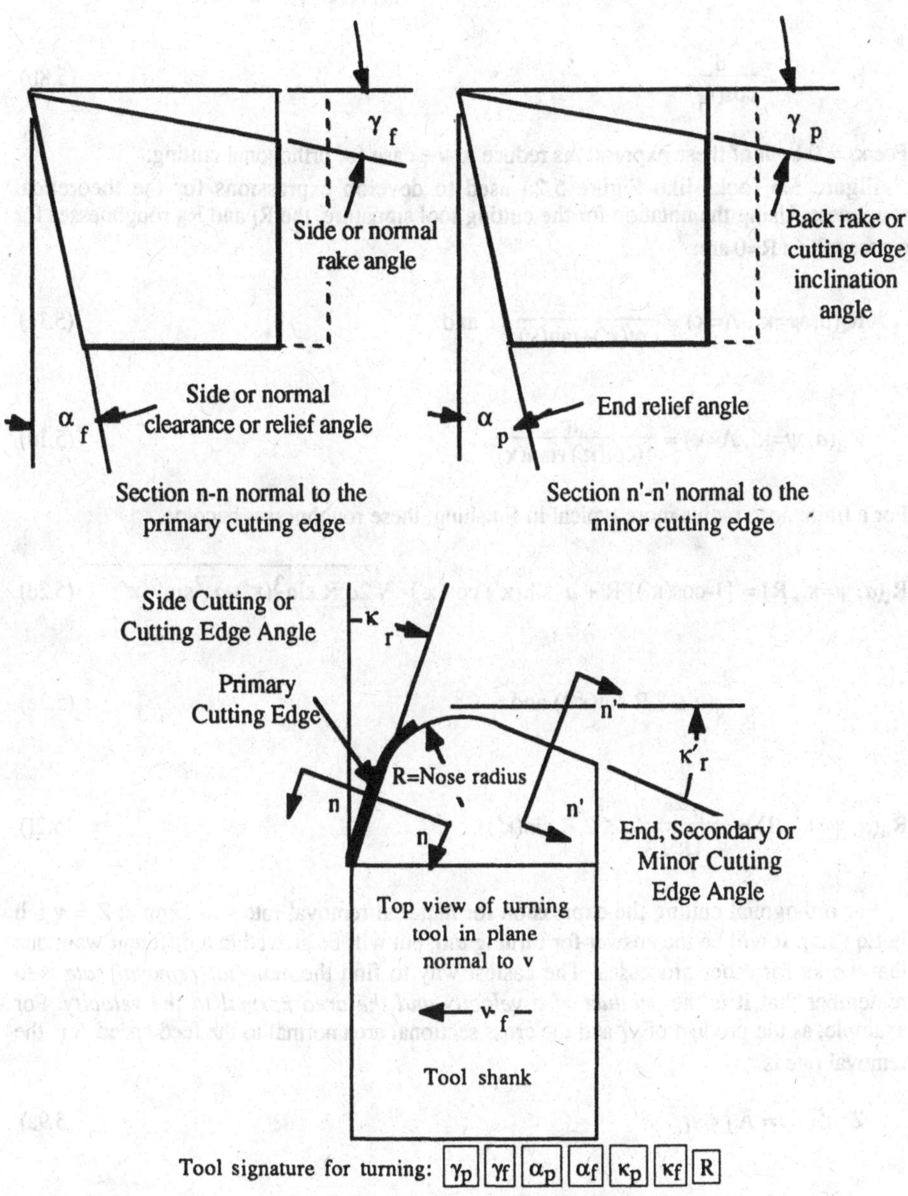

Tool signature for turning: $\boxed{\gamma_p}$ $\boxed{\gamma_f}$ $\boxed{\alpha_p}$ $\boxed{\alpha_f}$ $\boxed{\kappa_p}$ $\boxed{\kappa_f}$ $\boxed{R}$

Figure 5.5 Geometry for turning tools.

Quantities derived from the cutting conditions and edge geometry shown in Figure 5.6 include the uncut chip thickness, which is also the average uncut chip thickness

$$h \quad = h_{avg} = a \cos(\kappa_r) \tag{5.8a}$$

and the chip width

$$b = \frac{d}{\cos(\kappa_r)}. \tag{5.8b}$$

For $\kappa_r = 0$ both of these expressions reduce to the case for orthogonal cutting.

Figure 5.6 looks like Figure 5.2a used to develop expressions for the theoretical roughness. Using the notation for the cutting tool signature, the R_t and R_a roughnesses for a nose radius R=0 are:

$$R_t(a, \psi=\kappa', \Lambda=\kappa) = \frac{a}{\cot(\kappa')+\tan(\kappa)} \quad \text{and} \tag{5.1c}$$

$$R_a(a, \psi=\kappa', \Lambda=\kappa) = \frac{a}{4(\cot(\kappa')+\tan(\kappa))}. \tag{5.1d}$$

For a finite nose radius more typical in finishing, these roughnesses become

$$R_t(a, \psi=\kappa', R) = [1-\cos(\kappa')]\, R + a\ \sin(\kappa')\cos(\kappa') - \sqrt{2a\, R \sin^3(\kappa') - a^2 \sin^4(\kappa')} \tag{5.2d}$$

$$\approx \frac{a^2}{R\ 8} (a \leq 2\, R \sin(\kappa')) \text{ and} \tag{5.2e}$$

$$R_a(a, \psi=\kappa', R) \approx \frac{a^2}{R\ 18\sqrt{3}} (a \leq 2\, R \sin(\kappa')). \tag{5.2f}$$

For orthogonal cutting the expression for material removal rate was simple: $Z = v\ b\ h$ in Eq (3.5). It will be the answer for turning too, but will be arrived in a different way; one that works for other processes. The easiest way to find the *material removal rate* is to remember that it is the *product of a velocity and the area normal to the velocity*. For example, as the product of v_f and the cross sectional area normal to the feed speed $A_\perp$, the removal rate is

$$Z = A_\perp \bullet v_f \tag{5.9a}$$

$$= \pi \frac{D^2-(D-2d)^2}{4} v_f = \pi \frac{D^2-(D-2d)^2}{4} a\ N = \pi \frac{D^2-(D-2d)^2}{4} a \frac{v}{\pi \bullet D} \quad \text{or}$$

$$Z = v \bullet a \bullet d \bullet \left(1-\frac{d}{D}\right) \tag{5.9b}$$

$$\approx v \bullet a \bullet d \quad (d \ll D) \tag{5.9c}$$

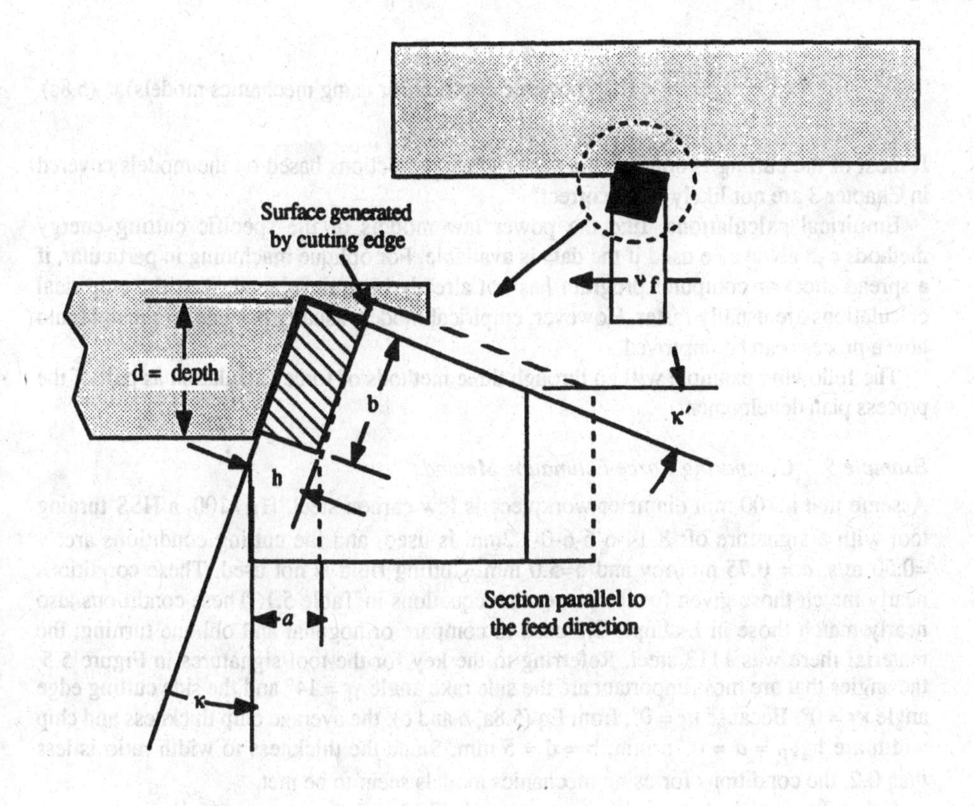

Figure 5.6 Chip cross section and surface finish in turning.

The same approximate result would be arrived at by finding the product of the cutting speed and the area of the chip normal to v using the results in Eq (5.8a and b).

One final calculation is the time to make a cut of length L, which is simply

$$t_c = \frac{L}{v_f} = \frac{L}{a \bullet N} = \frac{L \bullet \pi \bullet D}{a \bullet v}. \qquad (5.10)$$

5.2.3 Estimating Force and Power

Turning, facing and boring operations on a lathe come closest to the conditions for orthogonal or oblique machining covered in Chapter 3. Most research results to verify theoretical models are based on turning experiments.

For practical purposes of design, this means that several alternatives are available for estimating forces and power: the empirical methods of Section 5.1.2, and the orthogonal and oblique cutting models of Section 3.2. The problem with using the mechanics models is deciding which is appropriate, the orthogonal or oblique model, and then finding the data and models to predict shear plane and chip flow angles. With the right data, these models are appropriate if most of the cutting is done along the primary cutting edge, chip thickness is small relative to the chip width, so taking the ratio of Eq (5.8a) to Eq (5.8b) as a guide

$$\frac{h}{b} = \frac{a\ \cos^2(\kappa_r)}{d} < 0.2 \quad \text{(possible criterion for using mechanics models)} \qquad (5.8c)$$

If most of the cutting is done on the nose radius, predictions based on the models covered in Chapter 3 are not likely to be correct.

Empirical calculations, like the power law models or the specific cutting energy methods can always be used if the data is available. For oblique machining in particular, if a spread sheet or computer program has not already been developed, then the empirical calculations are usually faster. However, empirical models usually provide less insight into how a process can be improved.

The following example will go through three methods of force calculation as part of the process plan development.

Example 5.1 Comparing Force Estimation Methods

Assume that a 100 mm diameter workpiece is low carbon steel, H_B=100, a HSS turning tool with a signature of: 8-14-6-6-6-0-1.2mm is used, and the cutting conditions are: v =0.50 m/s, a = 0.75 mm/rev and d=5.0 mm. Cutting fluid is not used. These conditions nearly match those given for the power law equations in Table 5.1. These conditions also nearly match those in Example 3.1 used to compare orthogonal and oblique turning; the material there was 1113 steel. Referring to the key for the tool signatures in Figure 5.5, the angles that are most important are the side rake angle $\gamma_f = 14°$ and the side cutting edge angle $\kappa_f = 0°$. Because $\kappa_f = 0°$, from Eq (5.8a, b and c), the average chip thickness and chip width are $h_{avg} = a = 0.75$ mm, b = d = 5 mm. Since the thickness to width ratio is less than 0.2, the conditions for using mechanics models seem to be met.

Using the empirical power law expressions in Table 5.1 the force vector is

$$F_p = \begin{bmatrix} F_r \\ F_p \\ F_q \end{bmatrix} = \begin{bmatrix} 4.10(\text{kN})\left(\frac{0.75\ \text{mm/rev}}{25.4\text{mm/rev}}\right)^{0.56} \\ 592(\text{kN})\left(\frac{0.75\ \text{mm/rev}}{25.4\text{mm/rev}}\right)^{0.83}\left(\frac{5.0\ \text{mm}}{25.4\text{mm}}\right) \\ 150(\text{kN})\left(\frac{0.75\ \text{mm/rev}}{25.4\text{mm/rev}}\right)^{0.48}\left(\frac{5.0\ \text{mm}}{25.4\text{mm}}\right)^{1.45} \end{bmatrix} = \begin{bmatrix} 0.570\ \text{kN} \\ 6.26\ \text{kN} \\ 2.62\ \text{kN} \end{bmatrix}$$

The feed speed can be calculated as $v_f = a \bullet N = a \bullet v/(\pi \bullet D) = 1.19\text{x}10^{-3}$ m/s so that the power for a turning cut, based on Eq (3.13) is

$$P = v^T F_p = [0,\ 0.5\ \text{m/s},\ 1.19\text{x}10^{-3}\ \text{m/s}] \begin{bmatrix} 0.57\ \text{kN} \\ 6.26\ \text{kN} \\ 2.62\ \text{kN} \end{bmatrix} = 3.131\ \text{kW, or}$$

$$= 3.129\ \text{kW, neglecting } v_f.$$

This demonstrates why the specific cutting energy method assumes the power is due only to the force component parallel to the cutting speed; it is not because the forces in the other directions are small, but because the velocity components are small.

In Table 5.2 there is no entry for low carbon steel with H_B=100, but AISI 1020 is a fairly low carbon steel (remember the "20" stands for 0.2% carbon) and has E_c = 1.99 J/mm^3 for $H_B \approx 110$. The material removal rate from Eq (5.9b) is

$$Z = (0.50\ \text{m/s})\bullet(0.75\ \text{mm})\bullet(5\ \text{mm})\bullet(10^3\ \text{mm/m})\left(1 - \frac{5\ \text{mm}}{100\ \text{mm}}\right) = 1781\ \text{mm}^3/\text{s}$$

And from the graphs in Figure 5.3, the corrections are K_c(0.75 mm) = 0.80, C_c(0.75 mm) = 0.95, K_r(14°) = 0.75, C_r(14°) = 0.65. Since nothing is said about wear, assume that VB=0 so $K_w(0)=C_w(0)=1$. The estimated power and force vector are given by

$$P = (1069\ \text{mm}^3/\text{s})\bullet(1.99\ \text{J/mm}^3)\bullet(10^{-3}\ \text{kJ/J})[(0.80)\bullet(0.75)\bullet(1)] = 1.28\ \text{kW}$$

$$F_p = \begin{bmatrix} F_r \\ F_p \\ F_q \end{bmatrix} = \begin{bmatrix} 0 \\ \dfrac{1.28\ \text{kW}}{0.30\ \text{m/s}} \\ 4.25\ \text{kN}\bullet[(0.95)\bullet(0.65)\bullet(1)] \end{bmatrix} = \begin{bmatrix} 0\ \text{kN} \\ 4.25\ \text{kN} \\ 2.63\ \text{kN} \end{bmatrix}.$$

The force vectors and power components from this example, and the result from Example 3.1 are compared in the table that follows. The power law was the only one that predicted a radial force component. Actually, the predictions of F_p and P are within twenty percent by these three methods and the estimates of F_q differ by less than ten percent; this is reasonable considering the material property data is from 3 independent sources.

Comparing Results for Example 5.1				
	Power Law	Specific Cutting Energy	Mechanics from Example 3.1	Max-Min / 2 Avg
Force Vector F_r	0.570 kN	0 kN	0 kN	150.00%
F_p	6.26 kN	4.25 kN	4.84 kN	19.62%
F_q	2.62 kN	2.63 kN	2.22 kN	8.17%
Power P	3.13 kW	2.13 kW	2.42 kW	19.61%

Example 5.2 Design of a Process Plan for Turning

Figure 5.7 shows a part print for this example. Some other constraints on the process plan are:

- From Figure 5.7 the workpiece material is AISI 4340 steel that has been hardened to H_B = 300 and the turned surface should have a roughness of $R_a < 1\ \mu$m and a tolerance of ± 0.1 mm

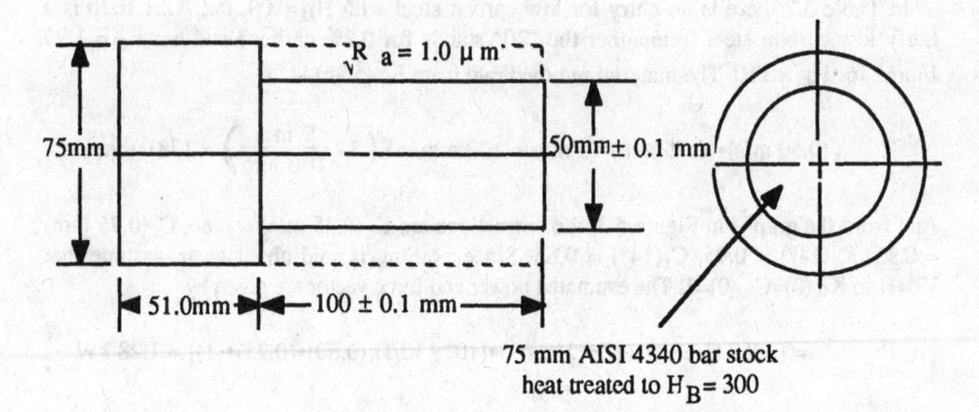

Figure 5.7 Part print for planning Example 5.1.

- Capacity and production constraints are assumed to be a function of the machine tool that is available; an engine lathe that has 10 kW available at the spindle. The machine tool itself has no stiffness limitations, so any compliance that would make it difficult to meet the tolerance specifications can be attributed to the workpiece.
- The cutting conditions should be selected so that at least two parts can be produced before a tool change and to assure that the power of the machine tool is not exceeded.

The Design or Planning Variables:

- Workholding, how should it be done? How a workpiece is fixtured can affect tolerances and the ability to resist chatter vibration. For simplicity, and not necessarily because it is the best technical solution, assume that the blank will be chucked.
- How many passes, and the cutting conditions for each pass?
- What tool material and tool signature for each pass?

The strategy for coming up with a plan illustrates one approach. It begins by assuming that it is possible to make a single pass. This determines the depth of cut. Then a cutting edge and feed are selected to meet the surface finish specification. Finally, the cutting speed is selected so the power is not exceeded when the cutting edge is worn out. Throughout this process, checks of the assumptions have to be made, for example, to see if at least two parts can be made.

- *Pick a depth of cut*, d, assuming the cut can be made in one pass. The depth is half the difference between the initial and final geometry

$$d = \frac{D - D_{final}}{2} = \frac{(75 - 50)\text{mm}}{2} = 12.5 \text{ mm}$$

- *Pick a cutting edge:* Many times this choice is made based on what a tool room stocks. Assume a carbide insert and tool holder combination that has a signature of: 0,5,5,5,15,15,3 mm, and a tool life equation, based on $VB^* = 0.5$ mm, of:

$$\left(\frac{T}{60\ s/edge}\right) = \left(\frac{1.5\ m/s}{v}\right)^{(1/0.25)}$$

- ***Pick a feed a,*** **to meet the finish requirements calculated by Eq (5.2f). From the tool signature, $\kappa' = 15°$ and R=3 mm:**

$$R_a(a, 15°, 3\ mm) \approx \frac{a^2}{3\ mm\ 18\sqrt{3}} = 1\ \mu m = 0.001\ mm$$

and solving for a

$$a = \sqrt{(0.001\ mm)\ (3\ mm)\ 18\sqrt{3}} = 0.177\ mm/rev.$$

Checking if Eq (5.2f) is valid, (0.177 mm/rev) < 2 (3 mm)sin(15°) = 1.55 mm so the approximation is valid.

- ***Use specific cutting energy to find v*****: From Table 5.2 for AISI 4340 in the hardness range 300 < H_B < 302, E_c = 4.12 J/mm^3. To estimate the average chip thickness the side cutting edge angle κ = 15° is need along with** a **,**

$$h_{avg} = (0.177\ mm)\cos(15°) = 0.171\ mm$$

so from Figure 5.3 K_c(0.171 mm) = 1.1. The side rake of γ_f = 5° determines K_r(5°) = 0.85. Because the tool life equation that will be used subsequently is based on VB*=0.5 mm, the correction for a worn tool based on Figure 5.3c is in the range 1.2 < K_w(VB=0.5 mm) < 1.4. From the hardness and the value of the specific cutting energy, this 4340 probably could be considered difficult to machine, so K_w(VB=0.5 mm) = 1.4 will be used. The cutting speed is found assuming all 10 kW is used at the cutting edge

$$v = \frac{(0.177mm)(12.5mm)\left(1-\frac{12.5mm}{75mm}\right)(4.12J/mm^3)(10^{-3}kJ/J)[(0.80)(0.85)(1.4)]}{10kW}$$

$$= 1.01\ m/s$$

- ***Check for at least 2 parts per cutting edge.*** **At the cutting speed of 1.01 m/s the predicted tool life is**

$$T = 60\ s/edge\left(\frac{1.5\ m/s}{1.01\ m/s}\right)^{(1/0.25)} = 294\ s/edge.$$

Comparing this to the cutting time from Eq (5.10) of

$$t_c = \frac{(100\ mm)\bullet\pi\bullet(75\ mm)}{(0.177\ mm/rev)\bullet(1.01\ m/s)(10^3\ mm/m)} = 132\ s/part$$

gives the number of parts per cutting edge of

$$\frac{T}{t_c} = \frac{294 \text{ s/edge}}{132 \text{ s/part}} = 2.23 \text{ parts/edge} > 2 \text{ parts/edge}.$$

This means that at the cutting conditions that have been calculated, it is possible to meet the finish requirements, power constraints and tool changing limits in a single pass.

What hasn't been done is check the ± 0.1 mm tolerance specifications on the diameter. This could be done by estimating the stiffness of the chucked workpiece, estimating the resultant force in the most compliant direction, and estimating the deflection. This would be an upper limit on the error or tolerance, because the deflection would cause h_{avg} to be smaller, reducing the forces. If the workpiece deflects too much, turning between centers might be the solution.

5.3 Face and End Milling

Milling machines are some of the most versatile and productive machine tools, in terms of their material removal rates. Part of this productivity is because most milling cutters use multiple, rather than the single, cutting edges. (There are exceptions; fly cutting uses a single cutter edge, but this is done for research investigations or for limited production with custom made tools.) The cutting speed is generated by rotating a cutter with the multiple edges and moving the workpiece in a plane normal to the axis of spindle rotation. This makes milling more complicated than turning because the geometry that the cutting edges "see" changes as the cut progresses. This means that both the thickness of the chip and the forces on an edge are constantly changing. While cutters can be designed and cutting conditions planned to even out the resultant force or torque on a cutter, the individual cutting edge is still subject to cyclic mechanical and thermal loading. This tends to shorten the tool life of an edge relative to comparable conditions in turning. Because the milling machine rotates the cutter, many times the machine tool's spindle or the cutter are the most compliant elements in the machine tool-workpiece system; there are notable exceptions, viz., when milling thin webs and ribs the workpiece is the most compliant element.

Milling processes are used most often to produce contour and planar surfaces as indicated by some of the more common types of milling in Figure 5.8. If the spindle axis of rotation is normal to the work surface the process is called *face* milling, but if the spindle rotation is parallel to the planar surface the process is called *slab* milling. Contour or planar surfaces are produced by *end* or *peripheral* milling with the spindle parallel to the generated surface, while a minor amount of cutting is done by the end of the cutter. In all of these cases, the feedrate is in a plane normal to the spindle rotation. Internal cylinders of moderate accuracy can be produced by *drilling* and high accuracy internal cylinders by *boring*; in either case, the feed motion is axial and the drill or boring tool rotates.

5.3.1 Specifying Cutting Conditions and Types of Milling

Terms for specifying cutting conditions in milling are more complicated than for turning, but there are still only three of them. The expression for the cutting speed is *exactly the same as for turning*

$$v = \pi \bullet D \bullet N, \qquad \text{(turning, boring, drilling and now milling)} \quad (5.6a)$$

but now D is the cutter diameter and N is the rotational speed of the spindle, the cutting speed is the peripheral speed of the cutting edge tip. The feedrate is

$$v_f = s_z \bullet N_t \bullet N. \quad (5.11)$$

The relative feed s_z, sometimes called the *chip load or feed per tooth*, is the amount one edge advances (mm/edge) as the spindle rotates relative to the next edge. N_t is the number of edges on a cutter, or to keep the units straight, the number of edges that advance in a revolution of the cutter (edges/rev). The uncut chip thickness which affects loads on individual cutting edges and the surface roughness of the workpiece are both functions of s_z. Also, Eq (5.11) points out that if the forces and roughness are related to s_z, then for the same cutting speed a way to increase the productivity in milling is by increasing N_t. This gives an additional degree of planning latitude. Two depths of cut are referred to in milling: The *axial depth of cut* or *back engagement* d is, as the name implies, the distance the cutter projects into the workpiece along the axis of spindle rotation; it generates the chip width and is never bigger than the cutter length or insert height. The projection of the cutter radially into the workpiece *a*, is termed the *radial depth of cut* or *working engagement*; it can never be greater than the cutter diameter D.

Conventional or up milling, down milling and slot milling depend on the way the chip thickness changes as the cutter rotates. As the cutter rotates from θ_0 by an angle θ, the path that an individual edge traces out is tricoidal, described by the expression

$$x(\theta) = \left(\frac{s_z N_t \theta}{2\pi}\right) + \left(\frac{D\cos(\theta + \theta_0)}{2}\right) \text{ and } y(\theta) = \left(\frac{D\sin(\theta + \theta_0)}{2}\right).$$

However, Martellotti in [Martel41] and [Martel45] showed that for values of s_z, N_t and D typical in milling, rather than using the tricoidal motion equations, this seemingly more complicated expression can be used

$$x(\theta) = s_z \,\text{Integer}\left(\frac{N_t \theta}{2\pi}\right) + \left(\frac{D\cos(\theta + \theta_0)}{2}\right) \text{ and } y(\theta) = \left(\frac{D\sin(\theta + \theta_0)}{2}\right).$$

This seems like a more complicated expression, but actually it is a much simpler geometric construction. It means that the edge path can be represented by a circular arc that has a constant center until a tooth passes. Then the center translates by the discrete amount given by the chip load s_z. This basic approximation will be used in all the subsequent calculations. It also aids in understanding Figure 5.9 which represents the three types of milling cuts: *down or climb milling*, *slot milling* and *up or conventional milling*. The dashed lines in Figure 5.9 represent the cross section of the chip that will be generated as the cutter rotates. In down milling the chip thickness is large when the edge enters the

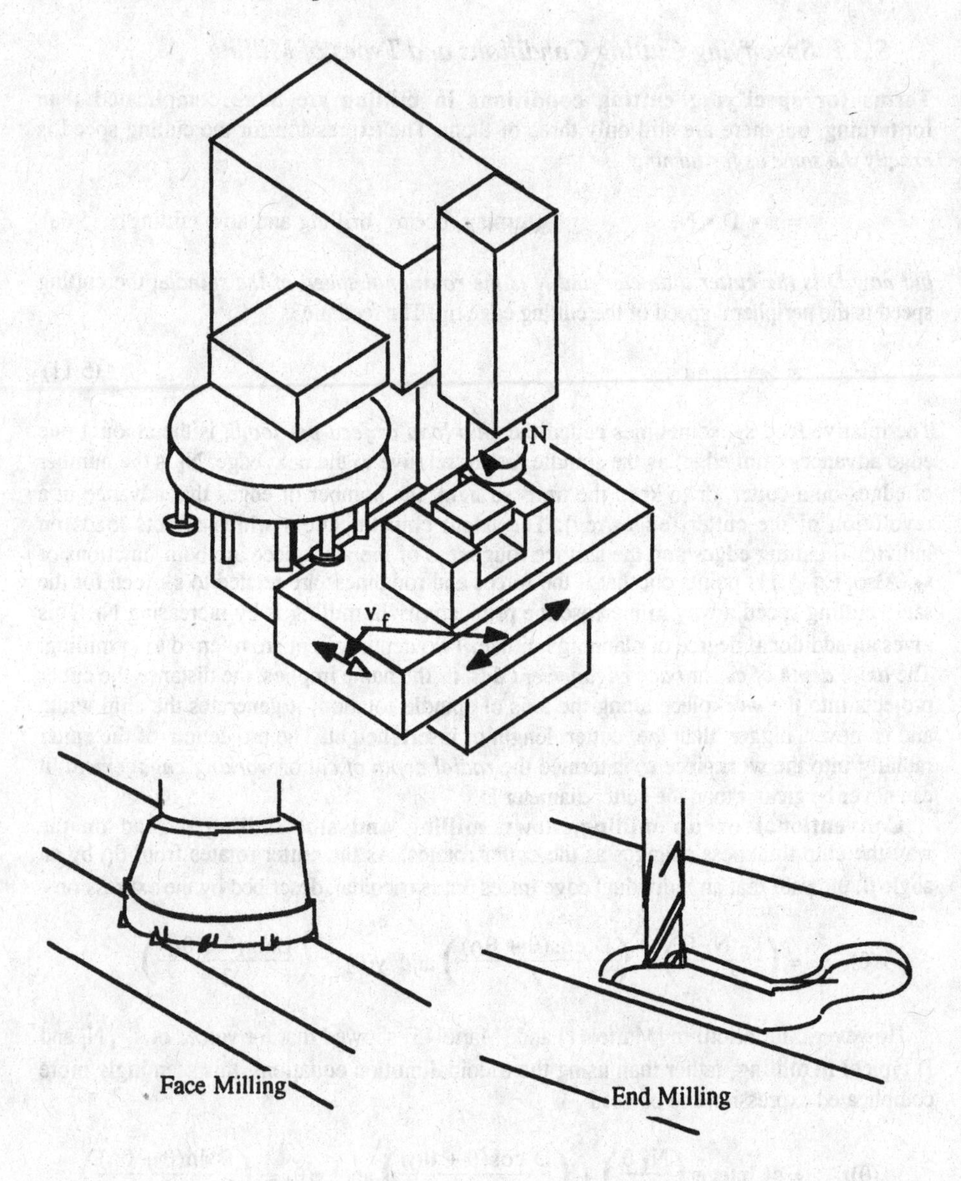

Figure 5.8 Common types of milling operations.

workpiece and leaves with a very thin cross section. The opposite occurs in up milling, with the chip thickness starting as a thin cross section and exiting with a large cross section. In slot milling, where the radial depth of cut is equal to the cutter diameter, half the cut is made up milling and half is made down milling.

Because the chip thickness plays an important role in forces, down milling tends to generate impact forces on the cutting edge which can be detrimental to both the cutting edge and the machine tool. Up milling, on the other hand, tends to produce thinner chips with higher specific cutting energy and a larger contact length between an edge an the workpiece, leading to higher temperatures. Before tough carbides became common tool materials, practical advice was to always use up milling; this was so common that the term "conventional" became synonymous with up milling. This was understandable for the less rigid machine tools and fixtures designed for HSS tools. Today's machine tools and carbide milling cutters favor down milling. The reasons are less heat causing less diffusion wear and the ability to have the initial contact with the workpiece be away from the cutting edges, reducing the chances of failure by chipping. Understanding these types of cuts also points out why, particularly for face milling carbide inserts, if the cutter diameter is larger than the width of the workpiece, the practice is not to center the cutter on the workpiece, but to have the greatest portion of the cut be down milling.

5.3.2 Cutting Edge Geometry and Kinematics for Face Milling

The geometry for face milling and angles for a face milling cutter are shown in Figure 5.10. Quantities derived from the cutting conditions and edge geometry are shown in Figure 5.11. Note that the coordinate system in Figure 5.11 uses a right hand coordinate system with the x-axis in the direction of feed motion, positive angles are measures from this axis in a counter clockwise direction (so that the normal clockwise spindle rotation is actually considered a negative

rotational direction). For example the detailed section in the upper right shows the cross section of the chip at its thickest point, where $\theta = 0$ and the cutting edge is parallel to v_f. The cross hatched chip width is

$$b = \frac{d}{\cos(\kappa)}. \tag{5.12a}$$

Looking at the top view of the cutter, it is apparent that the chip thickness, measured radially, changes throughout the cutter's rotation and the chip length depends on the cut geometry (This is different than the Type I,II and III chips in Section 3.1; there the Type I discontinuous chip was due to inherent properties of the material being cut.) The contact length is

$$l_c = \frac{D\,\Delta\theta}{2} = \frac{D}{2}\left(\sin^{-1}\left(\frac{2a_2}{D}\right) + \sin^{-1}\left(\frac{2a_1}{D}\right)\right) \quad \text{if } a_1 \neq a_2 \tag{5.12b}$$

$$= \frac{D}{2}\sin^{-1}\left(\frac{a}{D}\right) \quad \text{if } a_1 = a_2. \tag{5.12c}$$

using the coordinate system in Figure 5.11, where θ_1 is the exit angle, θ_2 is the entrance angle, and $\theta_1 < \theta_2$ with the sign convention in Figure 5.11. This figure is the way a face mill would usually be located over a workpiece of width w, i.e., not located symmetrically but so the chip thickness is large on entry and smaller on exit. This variation in chip thickness will be defined by two values. The *maximum chip thickness* in the upper right of Figure 5.11 is

$$h_{max} = s_z \cos(\kappa), \tag{5.12d}$$

so if the inclination or lead angle is zero, then $h_{max} = s_z$.

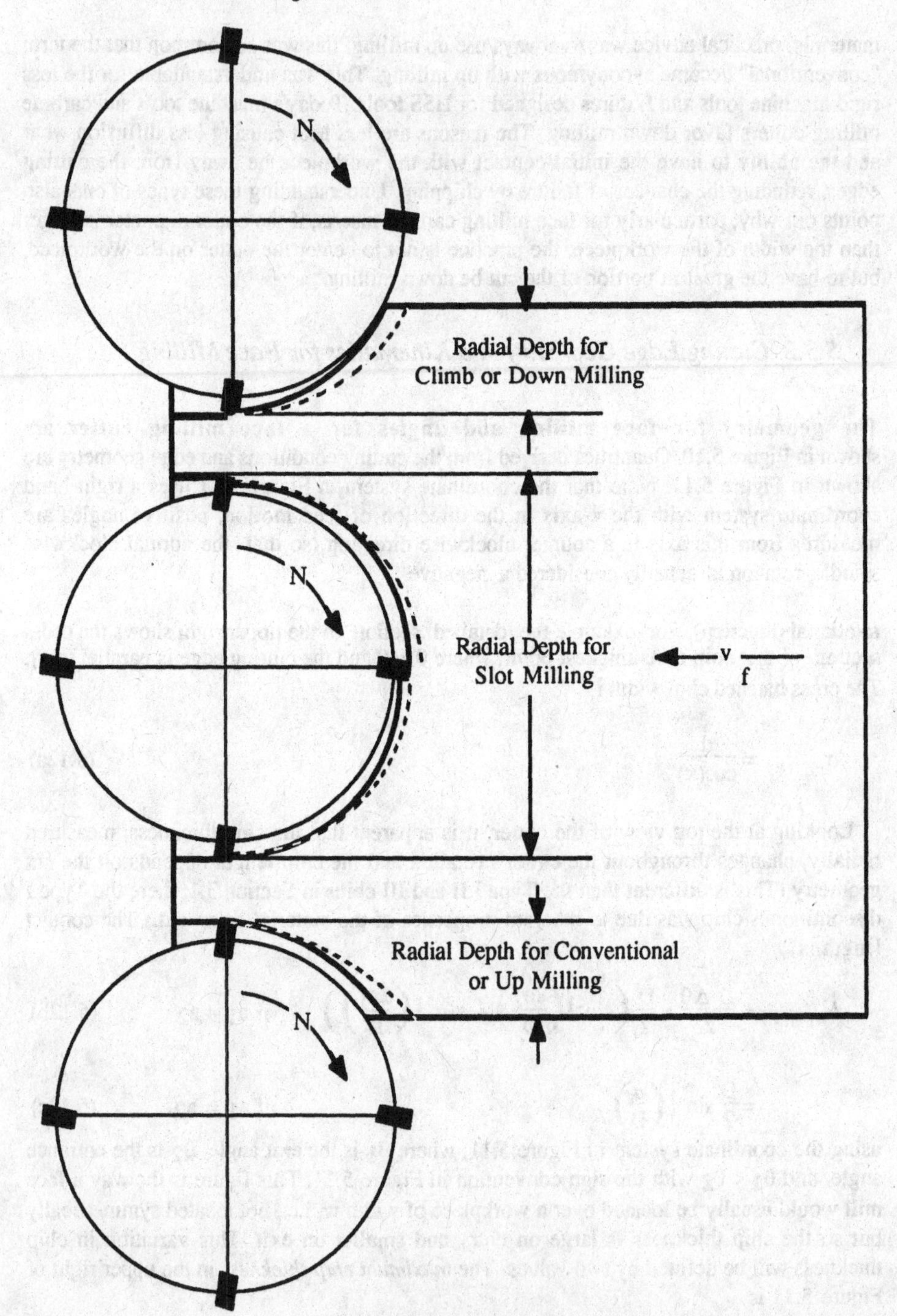

Figure 5.9 Types of milling cuts.

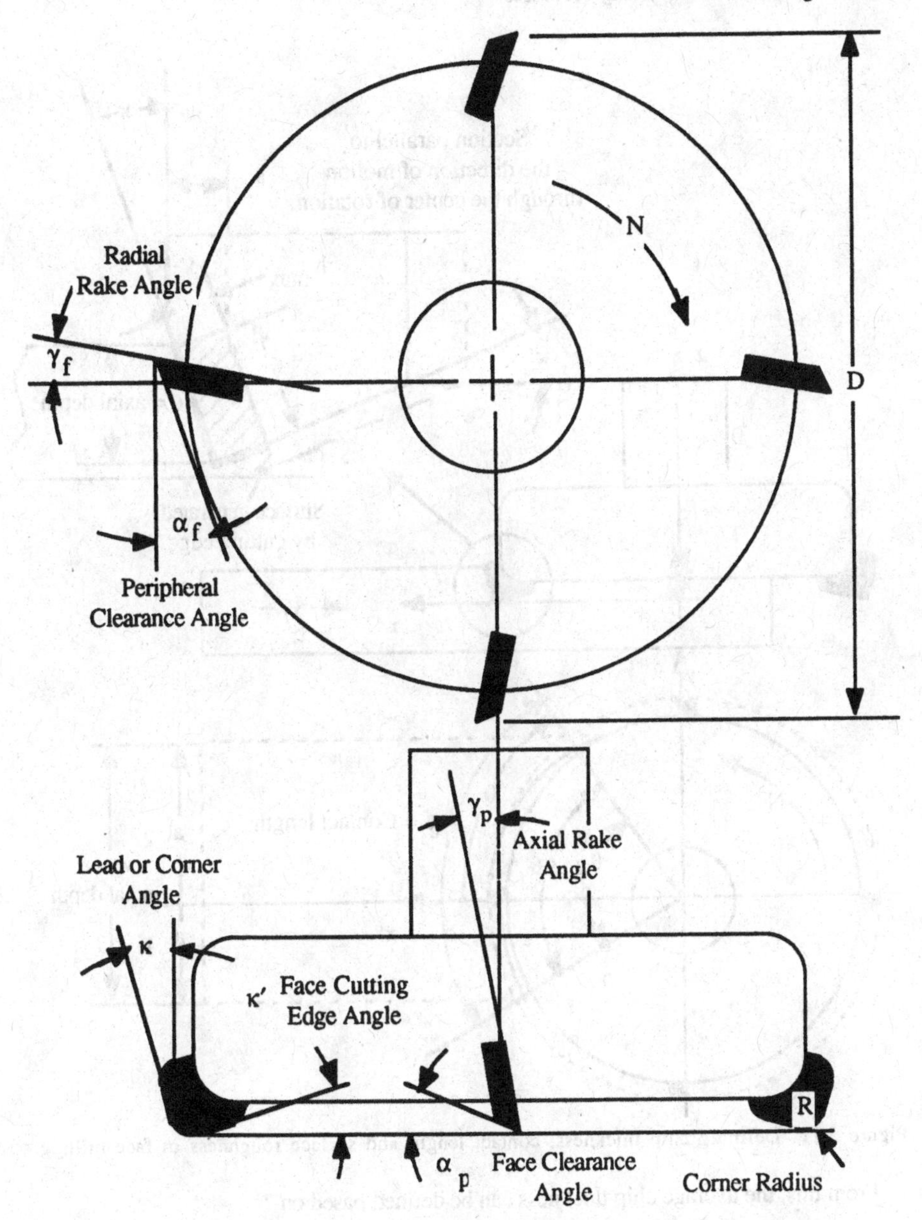

Figure 5.10 Geometry of a face mill.

It can be shown (and it will be later on when end milling is considered) that for both face milling and end milling the chip thickness is

$$h(\theta) = h_{max} \cos(\theta), \quad \frac{-\pi}{2} < \theta < \frac{\pi}{2}. \tag{5.13}$$

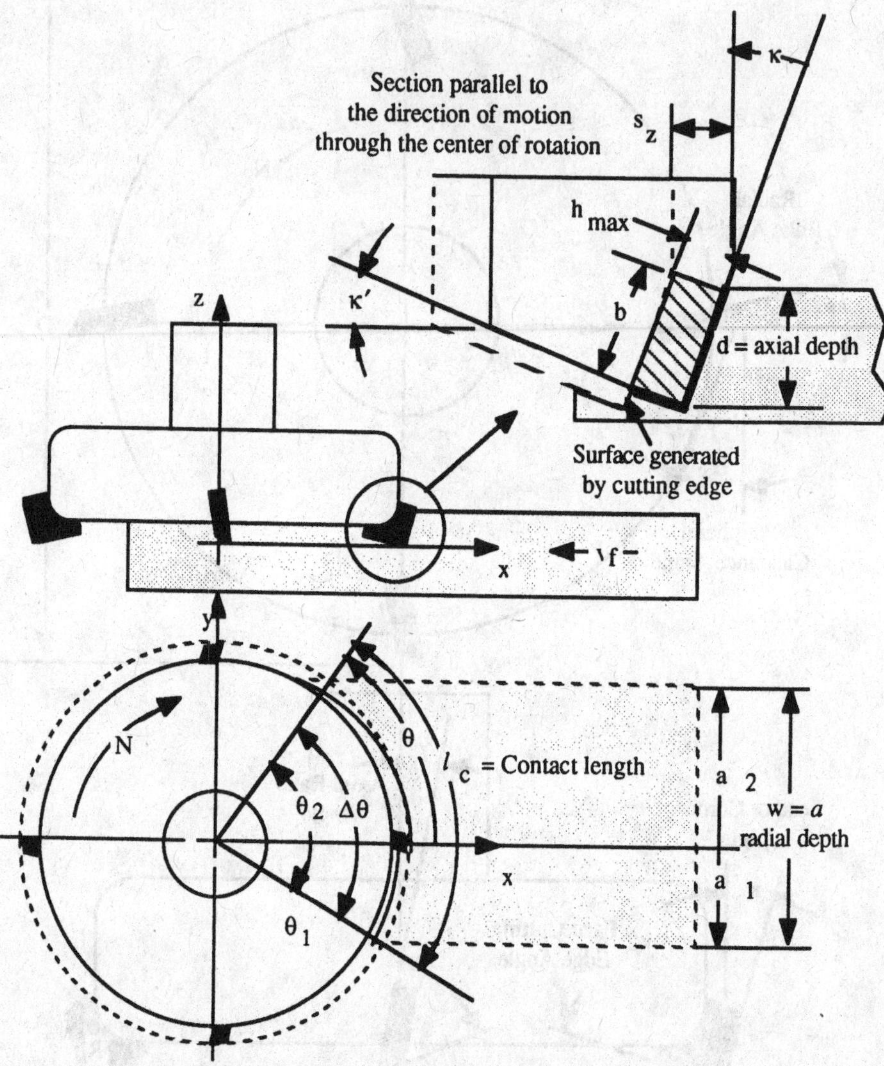

Figure 5.11 Defining chip thickness, contact length and surface roughness in face milling.

From this, the average chip thickness can be defined based on

$$h(\theta) = h_{max} \cos(\theta), \quad \frac{-\pi}{2} < \theta < \frac{\pi}{2}. \tag{5.13}$$

From this, the instantaneous chip thickness can be defined based on

$$h_{avg} \bullet l_c \equiv \int_{\theta_1}^{\theta_2} h(\theta) \frac{D}{2} d\theta = = \frac{D\, h_{max}}{2} \int_{\theta_1}^{\theta_2} \cos(\theta)\, d\theta = \frac{D\, h_{max}}{2} [\sin(\theta_2)+\sin(\theta_1)]$$

or eliminating θ_1 and θ_2

$$h_{avg} = h_{max} \frac{2(a_2+a_1)}{D\left(\sin^{-1}\left(\frac{2a_2}{D}\right) + \sin^{-1}\left(\frac{2a_1}{D}\right)\right)} \quad \text{if } a_1 \neq a_2 \tag{5.12e}$$

$$= h_{max} \frac{2\,a_1}{D \sin^{-1}\left(\frac{2a_1}{D}\right)} \quad \text{if } a_1 = a_2. \tag{5.12f}$$

Being able to estimate the average chip thickness is of particular importance when using the specific cutting energy method to estimate power and forces. Note that Eq (5.12e and f) are always such that $0 < h_{avg} < h_{max}$.

The theoretical surface profile measured parallel to v_f and coincident with the cutter axis is like the profile parallel to the axis of rotation in turning. For example, compare the surface in the upper right hand of Figure 5.11 with Figure 5.2a. Therefore, when the cutter corner radius R=0, the R_t and R_a roughnesses, in terms of the nomenclature for milling cutters, are:

$$R_t(a = s_z, \psi=\kappa', \Lambda=\kappa) = \frac{s_z}{\cot(\kappa')+\tan(\kappa)} \text{ and} \tag{5.1e}$$

$$R_a(a = s_z, \psi=\kappa', \Lambda=\kappa) = \frac{s_z}{4(\cot(\kappa')+\tan(\kappa))}. \tag{5.1f}$$

In a similar way, if the nose radius on the cutting edges are not zero, the roughnesses are:

$$R_t(a = s_z, \psi=\kappa', R) = [1-\cos(\kappa')]R + s_z\sin(\kappa')\cos(\kappa') - \sqrt{2s_zR\sin^3(\kappa') - s_z^2\sin^4(\kappa')} \tag{5.2g}$$

$$\approx \frac{s_z^{\,2}}{R\,8} \quad (s_z \leq 2\,R\,\sin(\kappa')) \text{ and} \tag{5.2h}$$

$$R_a(a = s_z, \psi=\kappa', R) \approx \frac{s_z^2}{R\,18\sqrt{3}} (s_z \leq 2\,R\,\sin(\kappa')). \tag{5.2i}$$

To reiterate, these roughnesses are theoretical values based on kinematics. However, these theoretical values will not be correct if

- One edge projects down into the plane of the workpiece, this is called *axial runout*, and will produce a rougher surface with an upper limit on the roughness predicted by using $a = s_z \bullet N_t$.
- Cutting occurs as the cutter rotates across the back or previously cut surface, which is called *back cutting* and is often remedied by putting a slight *tilt*, toward the front side of the cutter. (This changes most of the other angles if the tilt angle is too large.)

Kinematics to estimate cutting time and tool life for face milling is a bit more complicated than for turning because there is a transient portion at the beginning and end of a cut where an edge is not continuously cutting like it is in turning. Referring to Figure 5.11, to cut a workpiece of length L, the cutting time spent producing a chip is

$$t_c = \frac{\frac{D}{2}\left(1 + \sqrt{1 - \text{Max}\left(\left(\frac{2a_1}{D}\right)^2 \left(\frac{2a_2}{D}\right)^2\right)}\right) + L}{v_f}. \tag{5.14a}$$

A simpler and more common expression estimates the time to move across and completely clear the workpiece

$$t_c = \frac{D + L}{v_f} = \frac{D + L}{s_z \bullet N_t \bullet N} = \frac{(D + L) \bullet \pi \bullet D}{s_z \bullet N_t \bullet v}. \tag{5.14b}$$

It is assumed that the wear on each cutting edge on the face mill can be described by the Taylor's tool life equation. However, the tool life equation is based on continuous cutting. For milling, an edge spends at least half of the time not cutting, so the tool life expression to handle the kinematics of milling is

$$T = \left(\frac{2\pi}{\Delta\theta}\right) T_R \left(\frac{v_R}{v}\right)^{(1/n)} = \left(\frac{2\pi}{\sin^{-1}\left(\frac{2a_2}{D}\right) + \sin^{-1}\left(\frac{2a_1}{D}\right)}\right) T_R \left(\frac{v_R}{v}\right)^{(1/n)} \tag{5.15}$$

Remember that Eq (5.15) increases the tool life of an edge because not all of the time is spent making a chip; it does not take into account tool life reduction caused by intermittent cutting action and thermal cycling discussed in Section 4.3.

Example 5.3 Simulation of Face Milling

The calculations so far have all been used to predict average values. However, milling produces periodic variation in forces just from the kinematics of the cut. This example, which is based on a simplified version of the simulation in [Fu*etal*84]. Fu, DeVor and Kapoor model the more difficult cases of axial and radial *runout or throw* and a specific cutting pressure that is a function of chip thickness.

Figure 5.12 defines the face milling cut and a set of world coordinate axes that are fixed to the 150 mm x 200 mm x 50 mm workpiece. The 200 mm, 7 insert face mill uses the same set of coordinates shown in Figure 5.11, but during the course of the simulation the center of the cutter translates from an initial position of {X(0), Y(0), Z(0)} = {-100 mm, 100 mm, 40 mm} to a final position {X(final), Y(final), Z(final)} = {300 mm, 100 mm, 40 mm}. The cutter geometry is $\kappa = \kappa' = 15°$, $\gamma_f = \gamma_p = \alpha_f = \alpha_p = 5°$. The cutting conditions are N = 6 rev/s, v_f = 25 mm/s and d = 10 mm.

The specific cutting pressure method, Eq (5.4), is used to estimate the force on a particular cutting edge, and for the simulation k'_p = 0.7 kN/mm^2 and k'_q = 0.4. Then at a particular angular position θ, the specific cutting pressure and force ratio and the cutting edge geometry can be used to define a cutting stiffness vector that is a function of the chip thickness h(θ) in Eq (5.13), see [Fu*etal*84]. In R-θ-z coordinates the force and stiffness vectors for the ith cutting edge are:

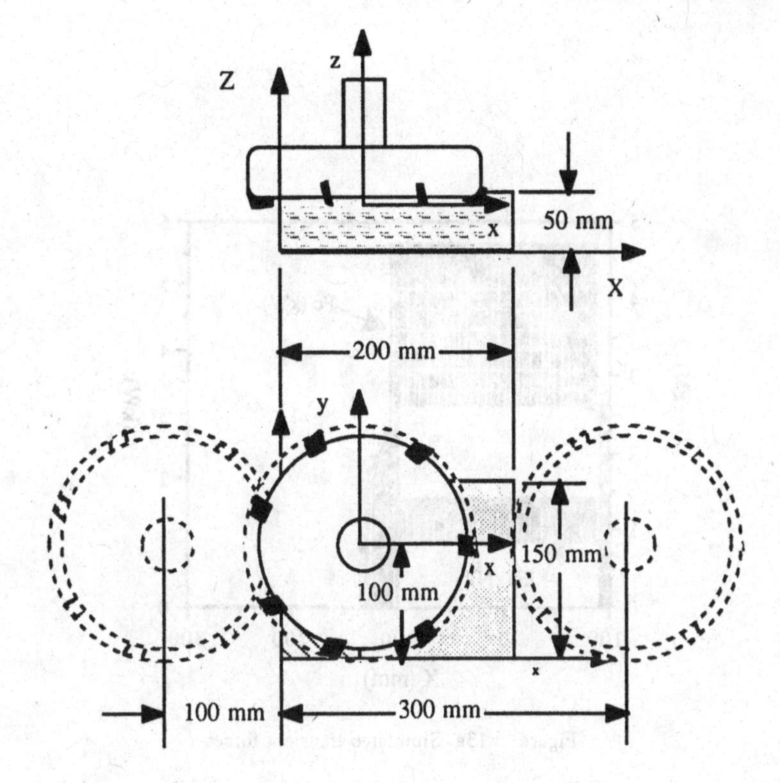

Figure 5.12 Face milling cut for Example 5.3.

$$\mathbf{F}(\theta^i(t)) = \begin{bmatrix} F_t(\theta^i(t)) \\ F_R(\theta^i(t)) \\ F_z(\theta^i(t)) \end{bmatrix}$$

$$= \begin{bmatrix} k_t \\ k_R \\ k_z \end{bmatrix} h(\theta^i(t)) = k'_p\, d \begin{bmatrix} \left(1 + \dfrac{k'_q \cos(\kappa) \tan(\gamma_f)}{\cos(\gamma_f)}\right) \\ \left(-\tan(\gamma_f) + \dfrac{k'_q \cos(\kappa)}{\cos(\gamma_f)}\right) \\ \left(\dfrac{-\tan(\gamma_p)}{\cos(\gamma_f)} + \dfrac{k'_q \sin(\kappa) \tan(\gamma_f)}{\cos(\gamma_p) \cos(\gamma_f)}\right) \end{bmatrix} h(\theta^i(t))$$

and $h(\theta^i(t)) = s_z \cos(\kappa) \cos(\theta^i(t))\, \delta(\theta^i(t))$, where $\delta(\theta^i(t)) = 1$ if the edge is engaged or $= 0$ otherwise. Then the total force vector is the sum of the forces on each cutting edge resolved in the x-y-z directions,

$$\mathbf{F}(t) = \begin{bmatrix} F_x(t) \\ F_y(t) \\ F_z(t) \end{bmatrix} = \sum_{i=0}^{N_t-1} \begin{bmatrix} \cos(\theta^i) & \sin(\theta^i) & 1 \\ -\sin(\theta^i) & \cos(\theta^i) & 0 \\ 0 & 0 & 1 \end{bmatrix} \mathbf{F}(\theta^i(t))\delta(\theta^i(t)))$$

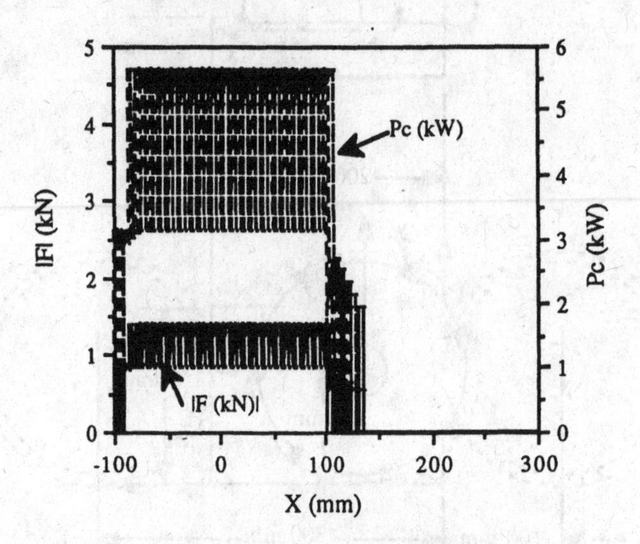

Figure 5.13a Simulated transient forces.

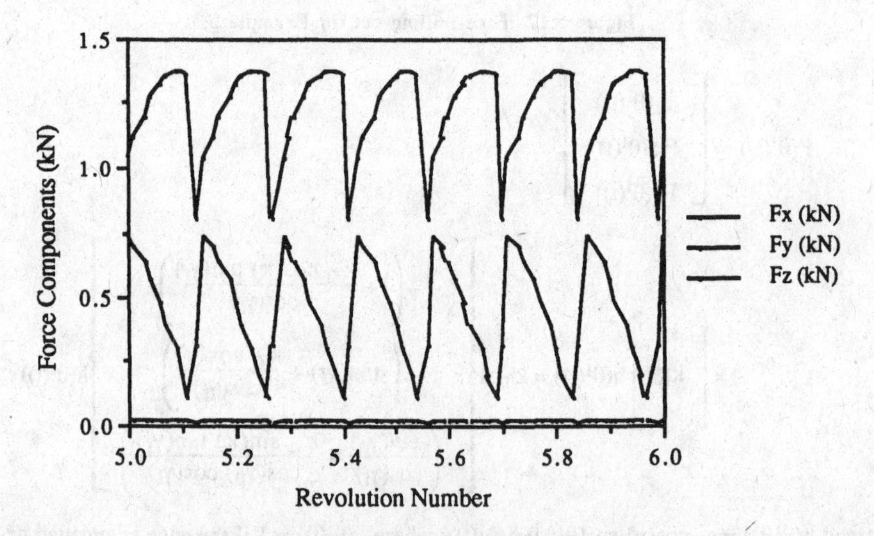

Figure 5.13b Simulated forces on the fifth revolution.

Figure 5.13 Simulation results for Example 5.3.

The kinematics that drive the translation and rotation of the cutter are simple: $X(t) = X(t-\Delta t) + v_f \bullet \Delta t$ and $\theta^i(t) = \theta^i(t-\Delta t) - 2\pi \bullet N \bullet \Delta t$.

Figure 5.13 shows the force history using the data in this example, with Figure 5.13a showing the entire force history, and Figure 5.13b on the fifth revolution, when the cutter is fully into the cut. The forces are always periodic, but as the cutter first contacts and finally leaves the workpiece the forces actually drop to zero or may even reverse direction.

In the Appendix , the function Face_Mill used to do this simulation is given. Similar approaches are used for end milling. The the cutter is broken down into axial disks, each treated like a face mill. Then the forces are summed both radially and axially. Some papers to consult on this include [FusSri89a], [Kli*etal*82a], [KolDeV91] and [SutDeV86].

5.3.3 Cutting Edge Geometry and Kinematics for Peripheral End Milling

Figure 5.14 shows the geometry of an end mill. The length to diameter ratio of most end mills classifies them as slender elements that are subject to deflection when cutting. The primary cutting edges are along the *flutes* of the cutter, with the secondary cutting edges along the bottom. As with the face mills, the radial rake γ_f is the major cutting edge angle with an effect on forces, while the axial rake angle γ_f affects cutting on the face of the cutter. As with the face mills, the *peripheral clearance* (α_f) and the axial clearance are to minimize rubbing of the cutting edge on the newly cut surface. The helix angle κ, serves many of the functions of the inclination angle in oblique cutting, i.e., it directs chip flow and distributes the forces along a longer edge. But the primary function of the helix angle is to even out the force variation that the cutter sees, by having a cutting edge enter the workpiece gradually. The *concavity angle* (κ') which defines the secondary cutting edge, can have a major effect on the roughness of the floor surface. The wall surface, on the other hand, develops a roughness profile that is a function of the cutter diameter and the feed per tooth s_z.

The chip geometry in peripheral end (or slab) milling is usually idealized with an approximately triangular cross section like that in Figure 5.15. Actually Figure 5.15 is the same geometry as the down or up milling cases in Figure 5.11, but the radial depth is much smaller than the cutter diameter. The cutting is done in the region where the chip thickness grows slowly from zero for the up milling case considered here. The total chip width is given by

$$b = \frac{d}{\cos(\kappa)}, \tag{5.13a}$$

essentially the same expression as Eq (5.12a), but κ represents the helix angle of the end mill, rather than a face mill's lead angle. Figure 5.15 represents a cross section along the cutter's length, and the chip contact length: from the geometry in Figure 5.15 is

$$l_c = \frac{D\,\Delta\theta}{2} = \frac{D}{2}\cos^{-1}\left(1 - \frac{2a}{D}\right) \text{ or} \tag{5.13b}$$

$$= \sqrt{a\,D} \qquad \text{if } 2a \ll D. \tag{5.13c}$$

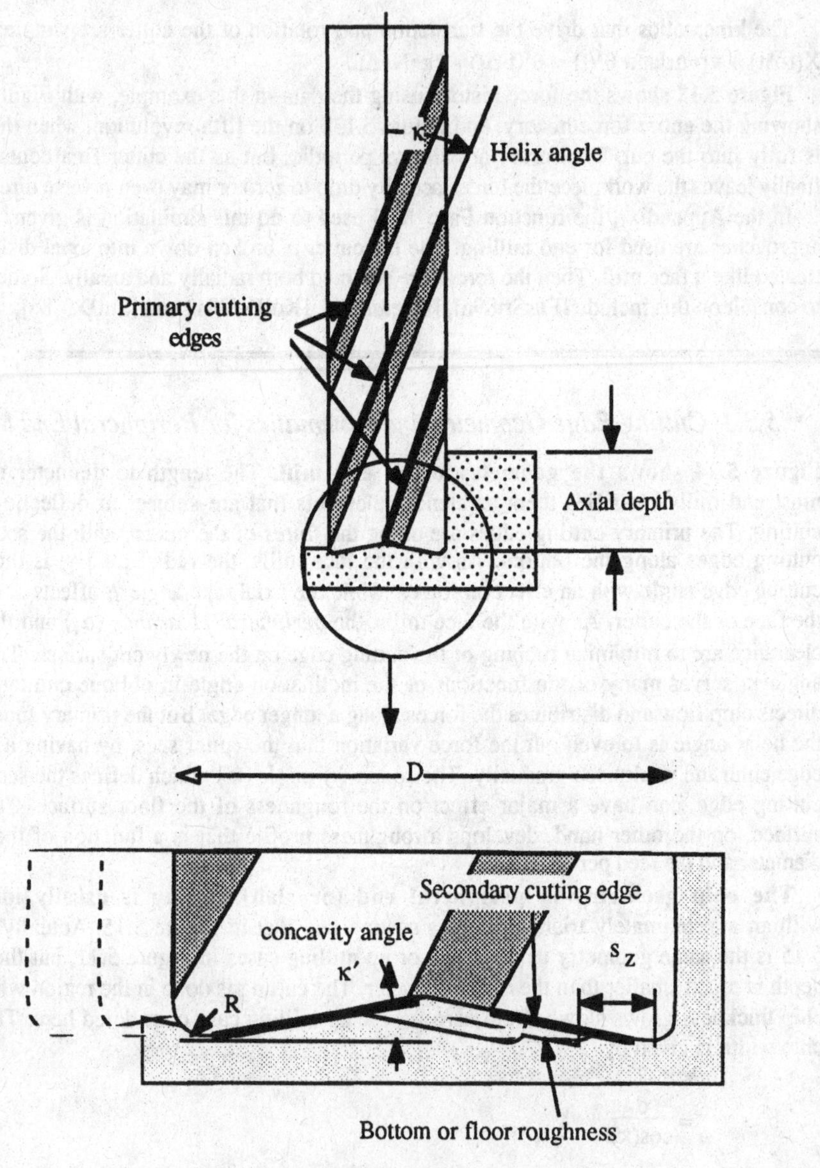

Figure 5.14 Geometry of an end mill.

Since the chip cross section is nearly a triangle, the maximum chip thickness is measured radially from the obtuse angle to the opposite side, this becomes

$$h_{max} = s_Z \sin(\Delta\theta') = s_Z \sin\left(\cos^{-1}\left(1 - \frac{2a}{D}\right)\right) \text{ or} \qquad (5.13d)$$

$$= 2\, s_Z \sqrt{\frac{a}{D}} \qquad \text{if } 2a \ll D. \qquad (5.13e)$$

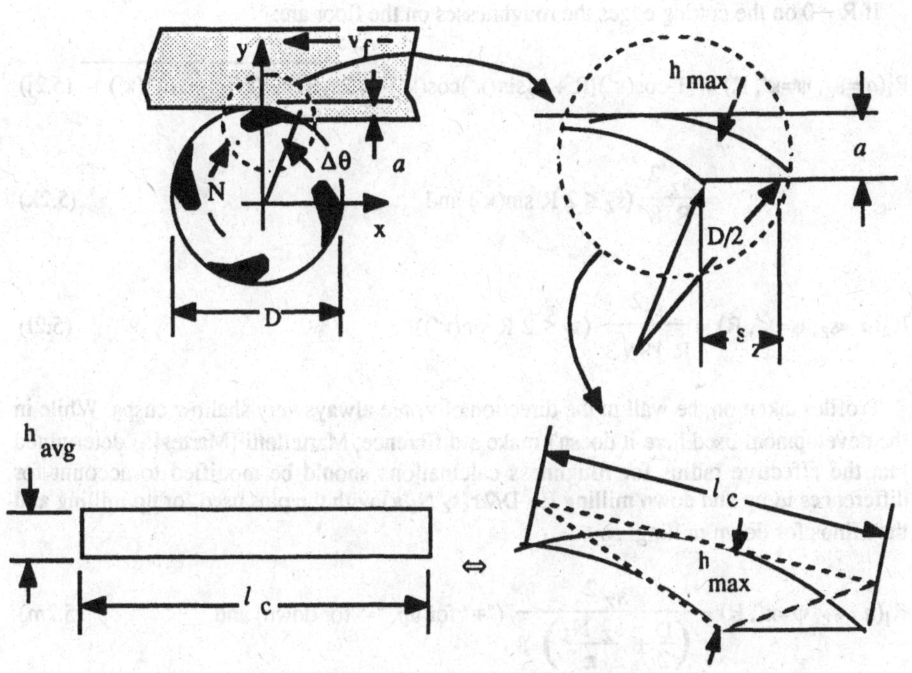

Figure 5.15 Chip geometry in end milling.

The average chip thickness is the thickness of a rectangular cross section of length l_c that has the same cross section as the triangular cross section, i.e., $h_{avg} \bullet l_c = (h_{max} \bullet l_c)/2$ so that

$$h_{avg} = \frac{h_{max}}{2} \tag{5.13f}$$

This expression for h_{avg} can be used in calculation using specific cutting energy methods to compensate for chip thicknesses with K_c or C_c in Figure 5.3.

The theoretical roughnesses for both the floor and wall in end milling can both be estimated using Eqs (5.1a-c). Beginning at the floor, if the corner radius R=0, the roughness of the floor depends only on the concavity angle κ' and s_z, i.e., $\kappa = 0$

$$R_t(a = s_z, \psi = \kappa', \Lambda = 0) = \frac{s_z}{\cot(\kappa')} \text{ and} \tag{5.1g}$$

$$R_a(a = s_z, \psi = \kappa', \Lambda = 0) = \frac{s_z}{4\cot(\kappa')}. \tag{5.1h}$$

If $R \neq 0$ on the cutting edges the roughnesses on the floor are:

$$R_t(a = s_z, \psi = \kappa', R) = [1-\cos(\kappa')]R + s_z\sin(\kappa')\cos(\kappa') - \sqrt{2 s_z R \sin^3(\kappa') - s_z^2 \sin^4(\kappa')} \quad (5.2j)$$

$$\approx \frac{s_z^{\ 2}}{R\ 8} \ (s_z \leq 2\,R\,\sin(\kappa') \text{ and} \quad (5.2k)$$

$$R_a(a = s_z, \psi = \kappa', R) \approx \frac{s_z^2}{R\ 18\sqrt{3}} \ (s_z \leq 2\,R\,\sin(\kappa')). \quad (5.2l)$$

Profiles taken on the wall in the direction of v_f are always very shallow cusps. While in the development used here it doesn't make a difference, Martellotti [Martel45] determined that the effective radius for roughness calculations should be modified to account for differences in up and down milling $R = D/2 \pm (s_z N_t/\pi)$ with the plus used for up milling and the minus for down milling, so

$$R_t(a = s_z, \psi = \kappa', R) \approx \frac{s_z^{\ 2}}{\left(\frac{D}{2} \pm \frac{s_z N_t}{\pi}\right) 8} \ (\text{"+" for up, "-" for down) and} \quad (5.2m)$$

$$R_a(a = s_z, \psi = \kappa', R) \approx \frac{s_z^{\ 2}}{\left(\frac{D}{2} \pm \frac{s_z N_t}{\pi}\right) 18\sqrt{3}} \ (\text{"+" for up, "-" for down}). \quad (5.2n)$$

Estimating the cutting time and tool life for end milling is similar in approach to Section 5.3.2 for face milling. The cutting time spent producing chips is

$$t_c = \frac{\frac{D}{2}\sqrt{\frac{4a}{D}\left(1 - \frac{a}{D}\right)} + L}{v_f} \quad (5.14a)$$

or the simpler expression that allows for the cutter to clear the workpiece is

$$t_c = \frac{\left(\frac{D}{2} + L\right)}{v_f} = \frac{\left(\frac{D}{2} + L\right)}{s_z \bullet N_t \bullet N} = \frac{\left(\frac{D}{2} + L\right) \bullet \pi \bullet D}{s_z \bullet N_t \bullet v} \quad (5.14b)$$

To account for the time not cutting, the Taylor's tool life equation for an end milling cut like that shown in Figure 5.15 is

$$T = \left(\frac{2\pi}{\Delta\theta}\right) T_R \left(\frac{v_R}{v}\right)^{(1/n)} = \left(\frac{2\pi}{\cos^{-1}\left(1 - \frac{2a}{D}\right)}\right) T_R \left(\frac{v_R}{v}\right)^{(1/n)} \quad (5.15b)$$

Again, remember that Eq (5.15b) does not take into account the reduction in the life of a cutting edge due to thermal or mechanical shock; it only accounts for the reduced time spent in the cut.

5.3.4 Estimating Average Forces and Power in Milling

Forces in both face and end milling are time varying because the chip geometry is always changing. As a result, either an excellent intuitive feel for how the forces change during a cutter rotation or simulation techniques like those illustrated in Example 5.3 are necessary to understand the nature of the milling forces.

Many times average forces, torques or power are sufficient for making process planning decisions. Specific cutting energy methods are usually sufficient for estimating approximate force levels that can be used to size a machine tool or to determine power requirements. As was the case for turning, this means finding the material removal rate, which for both face and end milling is

$$Z = v_f \bullet \text{radial depth} \bullet \text{axial depth} = v_f \bullet \text{Min}[a, D] \bullet d, \tag{5.16}$$

where the radial depth depends on if the cutter is taking a full width slotting cut or is up or down milling. The corrections for edge geometry, chip thickness and wear in Figure 5.3 are estimated in a way similar Example 5.1 for turning. The radial rake γ_f should be used to estimate the angle correction factor, the average chip thickness from either Eq (5.12e) for face milling or Eq (5.13f) for peripheral end milling can be used, even though the instantaneous chip thickness is always changing. The wear correction can be used to account for peripheral wear on the face or end mill. Then the *average* power is

$$P_{avg} = E_c \bullet Z \bullet K_r(\gamma_f) \bullet K_c(h_{avg}) \bullet K_w(VB) \text{ or} \tag{5.17a}$$

$$= F_p \bullet v = T_{avg} \bullet 2\pi \bullet N \tag{5.17b}$$

to calculate an *average* torque or *average* tangential resultant force.

Example 5.4 Elegant Ways to Increase Milling Productivity

A station on a transfer line needs to face mill 4.5 mm off of a part made of 1045 steel that is 175 mm wide and 500 mm long as shown in Figure 5.16. This will be done with a 250 mm cutter that has 16 carbide inserts that have a radial rake angle of -6°, axial rake angle of 0° and a lead angle of 30°. This station has a spindle that has a maximum rotational speed of 5 rev/sec and can provide no more that 40 kW to the cutter. The total time allocated to this operation is 45 seconds, which includes the time cutting and clearing the workpiece.

Make a calculation using the specific cutting energy method to determine if this cut is possible, and also be sure that the cutters don't have to be changed all the time, which means that making only 5 or 10 parts between cutter changes is unacceptable. Begin the calculations by assuming:

- The spindle speed is at its upper limit to estimate the feedrate v_f (mm/s) and the feed per tooth s_t (mm/tooth) needed to make this cut.
- Estimate the power required to make this cut when the flank wear on the insert reaches 0.5 mm.

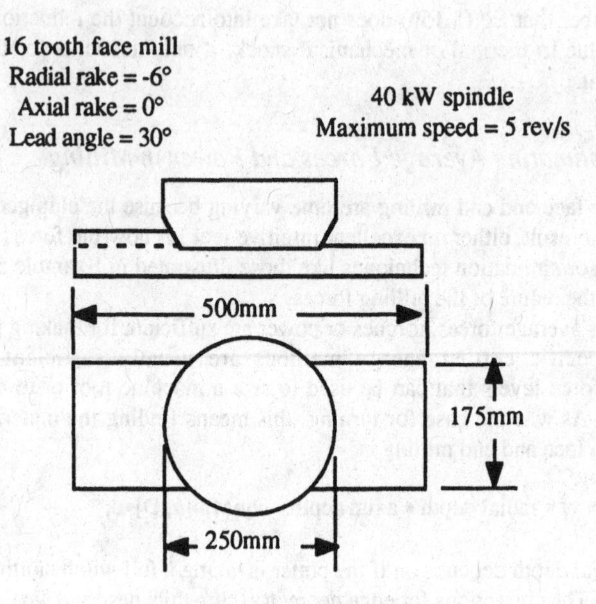

Figure 5.16 A face milling station.

- If each cutting edge has a Taylor tool life equation given by:

$$\left(\frac{T}{60 \text{ s/edge}}\right)^{0.25} = \left(\frac{5.0 \text{ m/s}}{v}\right)$$

and all the edges wear out at the same rate, estimate the number of parts that can be machined before the cutter has to be changed.

If N=5 rev/s, for the 250 mm cutter the cutting speed is $v = \pi(250 \text{ mm})(10^{-3} \text{ m/mm})(5 \text{ rev/s}) = 3.93$ m/s. The feed rate and s_z can be estimated from Eqs (5.12 and 12b), where the $t_c = 45$ s, so

$$v_f = \frac{(250 \text{ mm} + 500 \text{ mm})}{45 \text{ s}} = 16.7 \text{ mm/s} = s_z \text{ (16 edges/rev)(5 rev/s), or}$$

$$s_z = 0.208 \text{ mm/edge.}$$

The material removal rate can be estimated because all the cutting conditions are available. In this case it is the product of the axial depth d = 4.5 mm, the workpiece width of 175 mm, and the feed rate that was just calculated, Z = (4.5 mm)(175 mm)(16.7 mm/s) = 13125 mm^3/s. To check if the spindle power is not exceeded when the cutting edge wears out, the specific cutting energy method can be used. For 1045 with $190 < H_B < 200$, from Table 5.2 $E_c = 2.26$ J/mm^3. The corrections for rake angle and flank or peripheral wear can be read from Figure 5.3a and c; $K_r(-6°) = 1.0$ and $K_w(0.5 \text{ mm}) = 1.3$ - half way between the easy and difficult to machine materials. Determining the chip thickness correction requires estimating h_{avg} using Eqs (5.12d and e) for $a_1 = 125$ mm (θ_1=-90°) and $a_2 = 50$ mm (θ_2=23.6°). Then for the lead angle $\kappa = 30°$, h_{max} = (0.208 mm/edge) cos(30°) = 0.180 mm

$$h_{avg} = (0.180\text{ mm})\frac{2(50\text{ mm}+125\text{ mm})}{250\text{ mm}\left(\sin^{-1}\left(\frac{2(50\text{ mm})}{250\text{ mm}}\right)+\sin^{-1}\left(\frac{2(125\text{ mm})}{250\text{ mm}}\right)\right)}$$

$$= 0.127\text{ mm}.$$

From Figure 5.3b, $K_c(0.127\text{ mm}) = 1.2$ so the estimated average power when the cutting edges are worn out is

$$P_{avg} = (13125\text{ mm}^3/\text{s})\,(2.26\text{ J/mm}^3)(1.0)(1.3)(1.2) = 46.3\text{ kW}$$

and from Eq (5.15) used to predict the life of the edges

$$T = \frac{2\pi}{\left(\sin^{-1}\left(\frac{2(50\text{ mm})}{250\text{ mm}}\right)+\sin^{-1}\left(\frac{2(125\text{ mm})}{250\text{ mm}}\right)\right)}(60\text{ s/edge})\left(\frac{5.0\text{ m/s}}{3.93\text{ m/s}}\right)^{(1/0.25)}$$

$$= 500\text{ s/edge or with the 45 second cutting time 11.1 parts/edge.}$$

As proposed, this is not a particularly good process plan. The power will probably be exceeded when the cutters are worn and the tool life is not very good either.

Several remedies could be proposed, such as a reducing the metal removal rate, but then the allocated time at this station would be exceeded. Looking at the power calculations, the most elegant way would be to reduce the chip thickness correction K_c to get on the nearly flat portion of the curve in Figure 5.3b. This means increasing, not decreasing, h_{avg}. From Eq (5.12) there are three way to do this: increasing v_f, decreasing the number of inserts N_t or decreasing the rotational speed from the maximum used in these calculations. The last is probably the most elegant, i.e., decrease N to say 2.5 rev/s, which has the same effect on chip thickness as removing every other insert in the cutter. Then $h_{avg} = 0.254$ mm so $K_c(0.254\text{ mm})$ drops to 1.0 and $P_{avg} = 38.6$ kW. Even more significant is that the reduced speed of v = 1.92 m/s increases to tool life to 8000 s/edge so that approximately 178 parts can be machined before a cutter change.

5.4 Drilling

Drilling is a process for producing internal cylindrical surfaces - holes - of moderate accuracy in terms of position, *roundness* and *straightness*. As has been alluded to earlier, drilling can be done on engine lathes or milling machines, in addition to specialized drill presses, hand held drills or special stations on a transfer line. Drilled holes are often used for mechanical fasteners like bolts or rivets, drilling a hole is a preliminary step for processes like tapping, boring or reaming, drilling is used in many other places. Drilled holes are probably the most common manufactured feature specified on process plans.

In terms of the basic mechanics of chip formation, drills have multiple cutting edges (usually two, but a single edge is possible as in a spade drill or more than two edges for special new drill designs) at the drill tip. Because the cutting speed is due to the rotation about the drill's axis and the nature of the tip geometry, the cutting speed and the basic geometry of the cutting edge changes along the drill radius, making chip formation difficult. Since the chip is formed at the bottom of the hole, it makes removal of the chip difficult and more important than was the case for either turning or milling processes. These things make predicting forces and power for drilling difficult unless empirical methods are used.

5.4.1 Drill Tip Geometry

Of the drills used, the twist drill is the most common. The tip geometry of such a drill is shown in Figure 5.17. While two cutting edges are the most common, the number of *flutes* N_f, determines the number of cutting edges. The flutes on a twist drill are helical (formed by twisting, hence the name), with the helix angle given by κ'. The helix on a drill is not designed for cutting, as was the case for the helix on an end mill. The point angle κ, defines the profile of the tip of the drill and serves many of the functions of the inclination angle in oblique cutting, i.e., defining the primary cutting edge and increasing the chip width to distribute cutting forces. These features of the drill are easy to see and produce. On a drill the helix provides a way to remove chips from the hole. Cutting occurs along the *lips* of the drill, but the geometry varies radially, i.e., negative rake and large clearance adjacent to the chisel edge and positive rake and reduced clearance at the circumference.

For example normal rake and clearance angles, as well as the inclination angles have been derived, [Oxford55], [CowPeg58],

$$\gamma_n(x=\sin^{-1}(w/r)) = \tan^{-1}\left(\frac{\tan(\kappa')}{\sin(\kappa/2)}\left(\cos(x)+(w/r)\tan(x)\cos^2(\kappa/2)\right) - \tan(x)\cos(\kappa/2)\right) \quad (5.18a)$$

$$\alpha_n(x=\sin^{-1}(w/r)) = \frac{\cos(\kappa/2)\left(\cos(\sin^{-1}(2w/D))\tan(x)-\left(\tan(\sin^{-1}(2w/D))-\tan(\alpha_p)\tan(\kappa/2)\right)\right)}{\cos(\sin^{-1}(2w/D))+\tan(x)\cos^2(\kappa/2)\left(\tan(\sin^{-1}(2w/D))-\tan(\alpha_p)\tan(\kappa/2)\right)}$$

(5.18b)

$$\lambda(x=\sin^{-1}(2r/D)) = \sin^{-1}(\sin(x)\,\sin(\kappa/2)) \,. \quad (5.18c)$$

Figure 5.18 shows how these angles vary radially and for two different web thicknesses. It shows how the region near the *chisel edge* has large negative rakes and positive clearance and inclination angles, which become nearly constant at the drill periphery. This is why specifications on the drill geometry are given at the periphery of the drill, for example the rake (γ_f) and lip relief (γ_f) angles. Drills with a more complex geometry than the twist drill analyzed here have been developed, for example Racon™ drills have a compound parabolic, rather than conical, point angle. The difficulty with these drills (as well as more conventional geometry) is being able to consistently produce and re-grind the proper geometry.

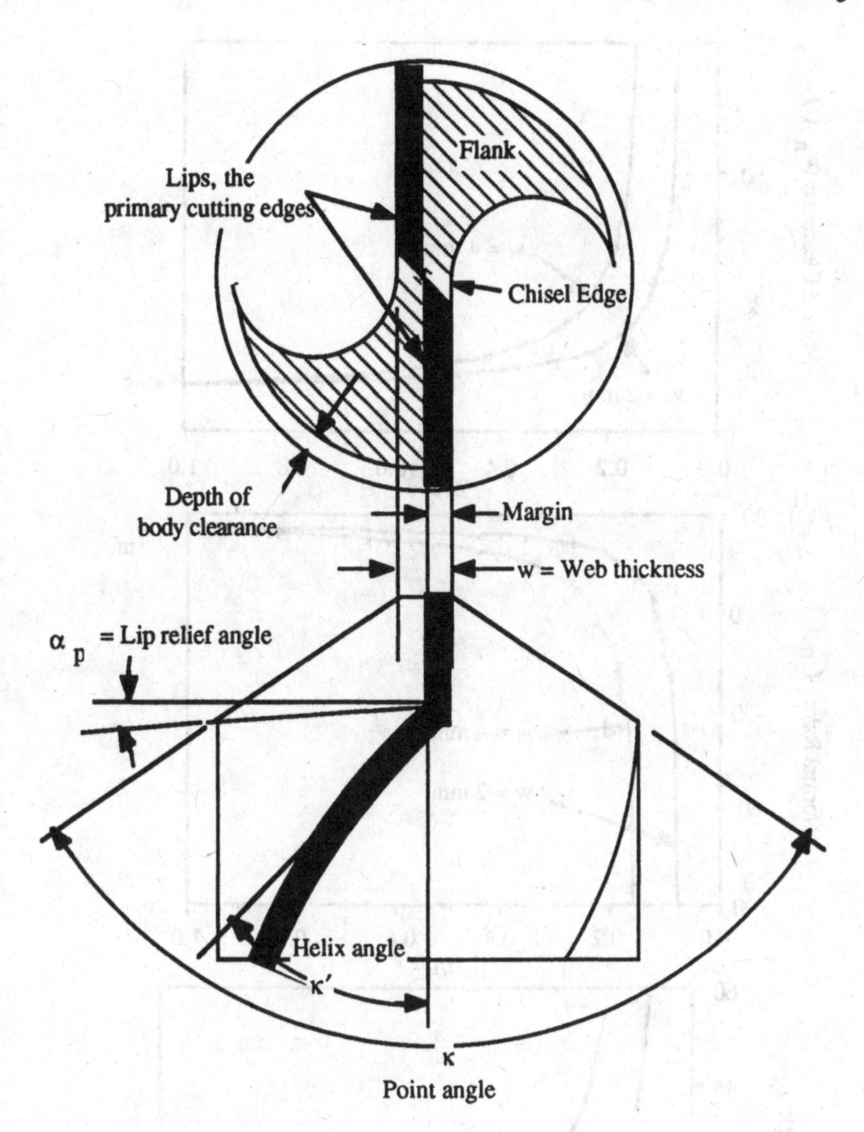

Figure 5.17 . Twist drill point geometry.

Figure 5.18 shows why the common shop practice of "thinning" a drill works. Thinning reduces the web thickness. The thinner web causes the large negative rakes to become positive quicker leading to lower forces and heat build up. The negative side of drill point thinning is it weakens the chisel edge where the relative velocity is zero so just rubbing or pushing material occurs. The compliance of the long and slender shape and the chisel edge can make it difficult to start and produce a quality hole. As a result, to produce a straight round drilled hole, many times *drill bushings* are needed to start a straight hole.

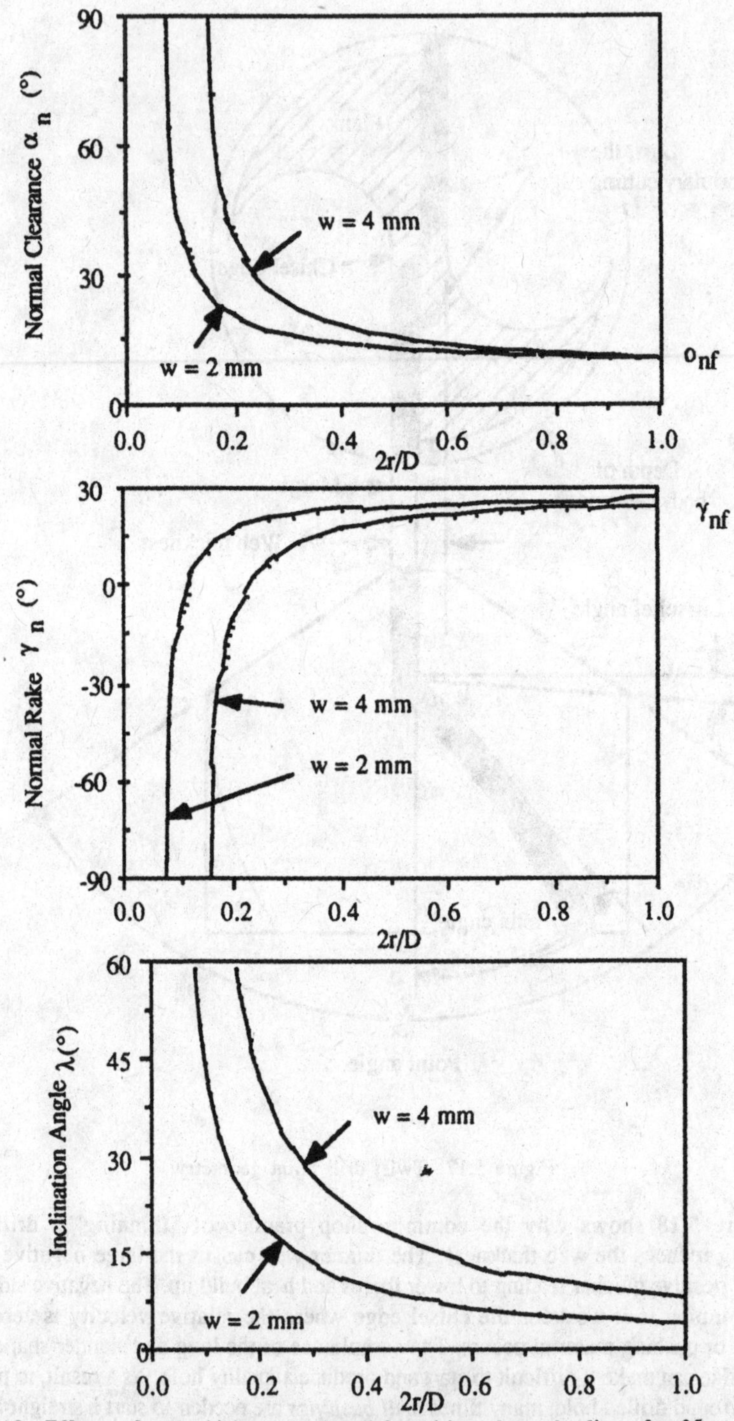

Figure 5.18 Effects of web thickness and radial position along the lips of a 25 mm twist drill with a helix angle = 25° and κ= 118° on clearance, rake and inclination angle.

5.4.2 Empirically Estimating Forces and Power

Specifying the cutting conditions for drilling is easier than defining the drill point geometry. The cutting speed is defined at the periphery of the drill, however, as was

$$v = \pi \cdot D \cdot N, \qquad (5.19)$$

pointed out earlier, this decreases in proportion to the drill radius with zero velocity at the chisel edge. Feed for drilling is the axial advance in one revolution of the spindle (mm/rev), and the feed rate or feed speed is

$$v_f = a \cdot N \qquad (5.20)$$

where N is the rotational speed of the spindle. The average chip thickness can be estimated as a function of the feed, the point angle and the number of drill flutes,

$$h = h_{avg} = \left(\frac{a \cos(\kappa/2)}{N_f}\right). \qquad (5.21)$$

The chip width is actually the determined by the drill diameter, from which the chip width can be estimated as,

$$b = \left(\frac{D - w}{2 \cos(\kappa/2)}\right) \qquad (5.22)$$

where w is the web thickness. From the geometry, the machining time calculation for a hole L units deep with a drill of diameter D and a point angle, κ is:

$$t_c = \frac{L + \frac{D \cdot \cot(\kappa/2)}{2}}{a \cdot N} = \frac{L + \frac{D \cdot \cot(\kappa/2)}{2}}{v_f} = \frac{\pi \cdot D \cdot \left(L + \frac{D \cdot \cot(\kappa/2)}{2}\right)}{a \cdot v} \qquad (5.23)$$

Estimating forces and power in drilling is usually done empirically, however some mechanics and semi-mechanistic models have been reported in the literature [StepWu88b] [Oxley89]. The methods reviewed here are empirical, using the specific cutting energy method and power law models.

For the specific cutting energy method, the material removal rate is

$$Z = v_f \cdot \frac{\pi D^2}{4}. \qquad (5.24)$$

Because the geometry along the lips of the drill is constantly changing, it is difficult to determine the angle to use for the rake corrections, but the most readily available is γ_f. The average chip thickness can be estimated by Eq (5.21). Then the power is

$$P = Z \cdot E_c \cdot [K_c(h_{avg}) \cdot K_r(\gamma_f) \cdot K_w(VB)] \qquad (5.25a)$$

from which the drilling torque can be estimated

$$T_p = \left(\frac{Z \bullet E_c \bullet [K_c(h_{avg}) \bullet K_r(\gamma_f) \bullet K_w(VB)]}{2\pi \bullet N}\right). \tag{5.25b}$$

The force to feed the drill is more difficult to estimate. If it is assumed that the resultant forces act at the drill periphery, then the force on each flute in the plane normal to the axis of rotation is

$$F_p = \left(\frac{Z \bullet E_c \bullet [K_c(h_{avg}) \bullet K_r(\gamma_f) \bullet K_w(VB)]}{v \bullet N_f}\right). \tag{5.25c}$$

The resultant force along the axis of rotation is

$$F_z = N_f \bullet F_p \bullet [C_c(h_{avg}) \bullet C_r(\gamma_f) \bullet C_w(VB)]. \tag{5.25d}$$

Another empirical method is to use a power law expression like one in Table 5.1. If the conditions for a particular drilling application match those in the table this is probably the most accurate.

Example 5.5 Comparing Methods For Estimating Drilling Torque and Thrust

Consider the problem of drilling a 25 mm hole in a cast iron workpiece (H_B=163). A HSS twist drill with a 31° helix, 121° point angle, 136° chisel edge and 5° relief angle will be used. The cutting conditions are: v=0.30 m/s, a = 0.25 mm/rev with an emulsion of soluble oil 1:16 water. These conditions match those in Table 5.1b, so the power law models predict

$$T = 515(\text{N-m})(0.25 \text{ mm/rev} / 25.4\text{mm/rev})^{0.60}(25\text{mm} / 25.4\text{mm})^2 = 31.2 \text{ N-m}$$

$$F_z = 65.5(\text{kN})(0.25 \text{ mm/rev} / 25.4\text{mm/rev})^{0.60}(25/25.4\text{mm}) = 4.03 \text{ kN}$$

and the power would be

$$P = (31.2 \text{ N-m}) \frac{0.30 \text{ m/s}}{(25 \text{ mm})(10^{-3} \text{ m/mm})} = 0.748 \text{ kW}.$$

Using the specific cutting energy method, start by assuming this is a new drill without any wear so K_w(VB=0) = C_w(VB=0) = 1. In Table 5.1 there is no specific cutting energy listed for cast iron with a hardness of H_B = 163; the closest is E_c = 1.39 J/mm^3 for H_B = 150. From Eq (5.21) the average chip thickness for a two flute drill is

$$h_{avg} = \left(\frac{0.30 \text{ mm/rev} \cos(121°/2)}{2}\right) = 0.062 \text{ mm}.$$

The values from Figure 5.3a are K_c(0.062 mm) = 1.4 and C_c(0.062 mm) = 2.0. The rake is not given, but can be estimated from Eq (5.18a). For the given values of κ = 121°, κ' = 31° and α_f = 5° the normal rake at the periphery $\gamma_n(x=\sin^{-1}(4\text{mm}/12.5\text{mm}))$= 31.2°.

Because the rake keeps decreasing toward the center, assume that the rake to use is half the value at the periphery, so from Figure 5.3b $K_r(15.6°) = 0.75$ and $C_r(15.6°) = 0.65$. For the cutting conditions specified, $Z = 469\ mm^3/s$, so the power is

$$P = (469\ mm^3/s)\bullet(1.39\ J/mm^3)\bullet[(1.4)\bullet(0.75)\bullet(1)](10^3\ kW/W) = 0.684\ kW$$

From Eq (5.25b) the torque is

$$T_p = \left(\frac{\frac{(0.684\ kW)(10^3\ W/kW)}{2(0.30\ m/s)}}{(25\ mm)(10^{-3} m/mm)}\right) = 28.5\ N\text{-}m.$$

The resultant force is estimated from Eq (5.25c) is $F_p = 1.140$ kN, or solving for the thrust force

$$F_z = 2\bullet(1.140\ kN)\bullet[(2.0)\bullet(0.65)\bullet(1)] = 2.96\ (kN).$$

For calculations based on empirical data from different sources, the simplifications in the modeling and approximations made because complete data was not available, these are quite comparable results. The power and torque are within ten percent, while the thrust force which depends on the power estimate had an error of about twenty five percent.

Example 5.6 Is It Possible to Drill All These Holes?

The bolt hole pattern shown in Figure 5.19 has to be drilled in aluminum. Standard 12 mm, 2 flute high speed steel twist drills will be used to make these 50 mm deep blind holes. The drill tip geometry is such that the average rake angle is 5°. In trying to design the process to accomplish this, there are several constraints: the maximum power available to drive the drills is 2 kW, the drilling time to produce *all* the holes is 48 s (don't worry about the time to retract and move to the next hole), the criterion for a worn out drill is 0.5 mm of wear on the flank, but the transfer station must make 1200 holes or 100 blocks before it can be stopped to change drills, and empirically it has been found that the thrust force cannot exceed 1.5 kN or the drill may break.

In this problem, assume that the tool life for a drill is given by:

$$\left(\frac{T}{60\ s/edge}\right) = \left(\frac{0.4\ m/s}{v}\right)^{(1/0.1)}$$

The specific cutting energy for drilling aluminum takes on two values that are related to the uncut chip thickness,

$$E = 1.0\ J/mm^3,\ h_{avg} \le 0.25\ mm$$

$$= 0.5\ J/mm^3,\ h_{avg} > 0.25\ mm.$$

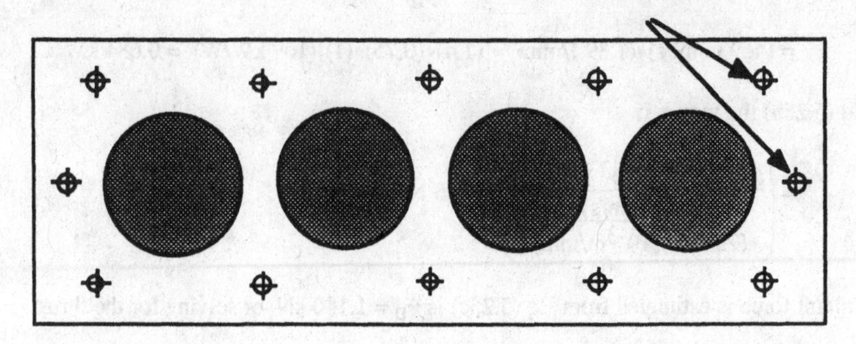

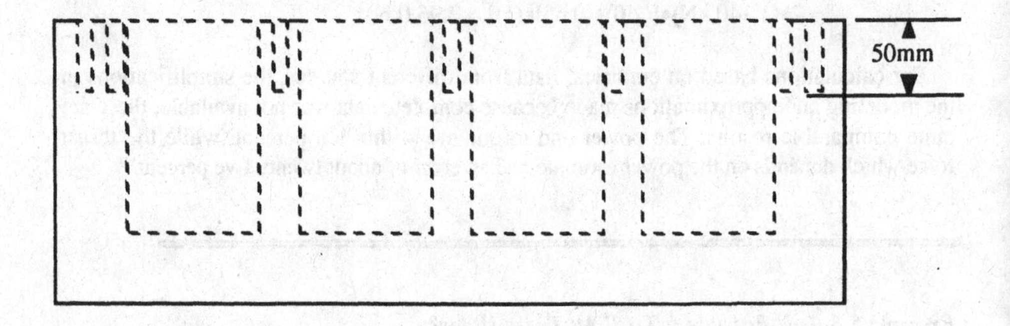

Figure 5.19 Bolt hole pattern for Example 5.6

Analyze this problem to determine cutting conditions for feed rate and rotational speed to see it it is possible to meet the constraints on force, power, time and tool life.

To meet the time limits, the feed rate per hole needs to be calculated first. The time to drill one hole is (48 s/block)/(12 holes/block) or t_c = 4 s/hole. Assuming that the standard drill also has a standard" point angle of 118°, so from Eq (5.23) the feed rate to drill a 50 mm hole in 4 seconds is:

$$v_f = \frac{50 \text{ mm} + \frac{12 \text{ mm} \cot(118°/2)}{2}}{4 \text{ s/hole}} = 13.40 \text{ mm/s}$$

To meet the tool changing requirements, the cutting speed and then the rotational speed can be estimated based on the required tool life of T = (4 s/hole) (1200 holes/edge) = 4800 s/edge so

$$v = \frac{0.4 \text{ m/s}}{\left(\frac{4800 \text{ s/edge}}{60 \text{ s/edge}}\right)^{0.1}} = 0.258 \text{ m/s} = \pi \text{ (12 mm)}(10^{-3} \text{ m/mm}) \text{ N, or}$$

$$N = 6.85 \text{ rev/s.}$$

With these cutting conditions specified, the average chip thickness and the material removal rate can be estimated from Eqs (5.20, 5.21 and 5.24). Substituting the feed expression for the feed in Eq (5.20) into the Eq (5.21), the average chip thickness is

$$h_{avg} = \left(\frac{(13.40\ \text{mm/s})\cos(118°/2)}{(6.85\ \text{rev/s})(2\ \text{flutes/rev})}\right) = 0.97\ \text{mm},$$

which should be recognized as a fairly large chip thickness; possible but not very common. It also means that the specific cutting energy is 0.5 J/mm^3. The material removal rate is

$$Z = (13.40\ \text{mm/s})\bullet\frac{\pi\,(12\ \text{mm})^2}{4} = 1516\ \text{mm}^3/\text{s}.$$

The correction factors for chip thickness, the average rake angle of 5° and flank wear of 0.5 mm for easy to machine materials are: $K_c(0.97\ \text{mm}) = 0.80$, $K_r(5°) = 0.85$, $K_w(0.5\ \text{mm}) = 1.2$, $C_c(0.97\ \text{mm}) = 0.95$, $C_r(5°) = 0.90$ and $C_w(0.5\ \text{mm}) = 1.5$. The estimated power at the spindle when the drill is worn out is

$$P = (1516\ \text{mm}^3/\text{s})(0.5\ \text{J/mm}^3)(0.80)(0.85)(1.2) = 0.618\ \text{kW}.$$

The tangential force on each flute F_p = 1.20 kN/flute is used to estimate the thrust force,

$$F_z = (2\ \text{flutes/rev})(1.20\ \text{kN/flute})(0.95)(0.90)(1.5) = 3.07\ \text{kN}.$$

Based on these results, the thrust force of F_z = 3.07 kN when the drill is worn exceeds the limit of the 2 kN constraint, while the power, time and tool life constraints are not exceeded.

Given the constraints on the time allocated to make the holes, the best solution to this problem is "gang" drilling the holes using a drill head with multiple spindles so that several holes can be drilled simultaneously at a reduced feed rate. Changing tool material or increasing the number of flutes may have a beneficial effect, but the most direct way to reduce the thrust force is by reducing the material removal rate. Gang drilling is how holes are often drilled on a transfer line and this example points out why drilling is often referred to as a bottleneck operation.

5.5 References

[AmeMac79] Edited by American Machinist, (1979), *Metalcutting: Today's Techniques for Engineers and Shop Personnel*, American Machinist/McGraw-Hill, New York.

[ArmBro69] Armarego, E.J.A, and R.H. Brown, (1969), *The Machining of Metals*, Prentice Hall, Inc., Englewood Cliffs, NJ.

[ASME52] American Society of Mechanical Engineers (ASME), (1952), *Manual on Cutting of Metals, Second Edition*, The American Society of Mechanical Engineers.

[ASTE50] American Society of Tool Engineers (ASTE), (1950), *Practical Design of Manufacturing Tools, Dies, and Fixtures, First Edition*, McGraw-Hill.

[ASTME60] American Society of Tool and Manufacturing Engineers, (1960) *Metal Cutting Research Reports*, (1954-1960), ASTME Research Fund, Detroit, MI.

[BabSut86] Babin, T. S., J. W. Sutherland, and S. G. Kapoor, (1986), "On the Geometry of End Milled Surfaces," *Proc. 14th NAMRC*, Minneapolis, MN, May 1986, pp. 168-176.

[Boston41] Boston, Orlan W., (1941), *Metal Processing*, John Wiley & Sons, New York.

[CowPeg58] Cowie, R.J., and J.O.M. Pegler, (1958) "Some Factors Affecting the Performance of Twist Drills and Taps," *Proceedings of the Institution of Mechanical Engineers*, Conference on Technology of Engineering Manufacture, pp. 463.

[Fu*etal*84] Fu, H.J., R.E. DeVor and S.G. Kapoor, (1984), "A Mechanistic Model for the Prediction of the Force System in Face Milling Operations," *ASME Journal of Engineering for Industry*, Vol. 106, Feb., pp. 81-88.

[Fujetal70a] Fujii, S., M.F. DeVries, and S.M. Wu, "An Analysis of Drill Geometry for Optimum Drill Design by Computer, Part I - Drill Geometry Analysis," *ASME Journal of Engineering for Industry*, Series B, Vol. 92, No. 3,. pp. 647-656.

[FusSri89a] Fussell, B. K., and Srinivasan, K., (1989a), "An Investigation of the End Milling Process Under Varying Machining Conditions," *ASME Journal of Engineering for Industry*, Vol. 111, Feb., pp. 27-36.

[Hatsch79] Hatschel, R.L., (1979), "Fundamentals of Drilling," *American Machinist*, Vol. 122, No. 2, pp.107-130.

[Kalpak84] Kalpakjian, S., (1984), *Manufacturing Processes for Engineering Materials*, Addison-Wesley, Reading, MA.

[KenMet84] *Kennametal Metalcutting Tools*, (1984), Kennametal Inc., Latrobe, PA.

[KenMil86] *Kennametal Milling Systems*, (1986), Kennametal Inc., Latrobe, PA.

[Kli*etal*82a] Kline, W. A., DeVor, R. E., and Lindberg, J. R., (1982a), "The Prediction of Cutting Forces in End Milling with Application to Cornering Cuts," *Int. J. of Machine Tool Design and Research*, Vol. 22, No. 1, pp. 7-22.

[KliDev83] Kline, W. A., and DeVor, R. E., (1983), "The Effect of Runout on Cutting Geometry and Forces in End Milling," *Int. J. of Machine Tool Design and Research*, Vol. 23, No. 2/3, pp. 123-140.

[KolDeV91] Kolarits, Francis M., and Warren R. DeVries, (1991), "A Mechanistic Dynamic Model of End Milling for Process Controller Simulation," *ASME Transactions, Journal of Engineering for Industry*, Vol. 113, No. 2, May, pp.176-183.

[Ludetal87] Ludema, Kenneth C., Robert M. Caddell and Anthony G. Atkins, *Manufacturing Engineering: Economics and Processes*, Prentice Hall, Inc. Englewood Cliffs, NJ, 1987.

[MachDa80] *Machining Data Handbook, Third Edition*, (1980), Vol.s 1 and 2, Machinability Data Center, Metcut Research Associates, Inc., Cincinnati, OH .

[Martel41] Martellotti, M.E., (1941), "An Analysis of the Milling Process," *ASME Transactions*, Vol. 63, pp. 667.

[Martel45] Martellotti, M.E., (1945), "An Analysis of the Milling Process - Part II: Down Milling," *ASME Transactions*, Vol. 67, pp. 233.

[Nakaya78] Nakayama, K., (1978), *The Metal Cutting in its Principles*, (in Japanese) Corona Publishing Co., Ltd., Tokyo, Japan.

[Oxford55] Oxford, C.J., Jr., (1955) "On the Drilling of Metals, I: Basic Mechanics of the Process," *ASME Transactions*, Vol. 77, pp. 103.

[Oxley89] Oxley, P.B.L., (1989), *Mechanics of Machining: An Analytical Approach to Assessing Machinability*, Ellis Horwood Limited, Chichester, England.

[Shaw84] Shaw, Milton C., (1984), *Metal Cutting Principles*, Clarendon Press, Oxford, England.

[SmiTlu91] Smith S., and Tlusty, J., (1991), "An Overview of Modelling and Simulation of the Milling Process," *ASME Transactions, Journal of Engineering for Industry*, Vol. 113, No. 2, May, pp. 169-175.

[StepWu88b] Stephenson, D.A., and S.M. Wu, (1988b), Computer Models for the Mechanics of Three-Dimensional Cutting Processes-Part II: Results for Oblique End Turning and Drilling," *ASME Transaction, Journal of Engineering for Industry*, Vol. 110, February, pp. 38-43.

[SutDeV86] Sutherland, J. W., and R. E. DeVor, (1986), "An Improved Method for Cutting Force and Surface Error Prediction in Flexible End Milling Systems," *ASME J. of Engineering for Industry*, Vol. 108, November, pp. 269-279.

[Zorev66] Zorev, N.N., (1966), *Metal Cutting Mechanics*, translated from Russian by H.S.H. Massey and edited by Milton C. Shaw, Pergamon Press, Ltd., London.

5.6 Problems

5.1. Assume that you have to cut a thread on a bar 6 mm in diameter. Cutting a thread means a fairly large relative feed, for example, $a = 1.27$ mm/rev (this is turning your own 1/4-20 stock). Other things to assume for this problem are:

$$v = 0.25 \text{ m/s}, \; b = 0.5 \text{ mm}$$

and the applicable empirical relationships for the cutting, feeding and radial forces in turning are:

$$F_p = 5 \text{ kN} \left(\frac{a}{1 \text{ mm/rev}}\right)^{0.8} \left(\frac{b}{1 \text{ mm/rev}}\right)^{0.5}$$

$$F_q = 4 \text{ kN} \left(\frac{a}{1 \text{ mm/rev}}\right)^{0.5} \left(\frac{b}{2.5 \text{ mm}}\right)^{0.8}$$

$$F_r = 4 \text{ kN} \left(\frac{a}{1 \text{ mm/rev}}\right)^{0.5} \left(\frac{b}{2.5 \text{ mm}}\right)^{0.5}$$

a. Determine the velocity vector **v** and the force vector $\mathbf{F_p}$ for this operation.
b. Determine the total power and the fraction of the the power contributed by each component, i.e., (P_p/P), (P_q/P) and (P_r/P).

5.2 The purpose of this problem is to plan for a facing operation to produce a flat and smooth ($R_a = 0.5\ \mu m$) surface on the 2024-T4 Aluminum cylinder shown in Figure 5.20. Before being turned, the length of the cylinder is 55 mm and the diameter is 64 mm. This will be done on a lathe that has 1.5 kW available at the spindle and with a left hand HSS cutting edge with an in hand tool signature of 10°, 12°, 6°, 6°, 6°, 20°, 1.5 mm. This cut will be made with a constant rotational speed N.

a. Choose a cutting speed and relative feed that will meet the surface finish specification and make the cut as fast as possible without exceeding the power available at the spindle.

b. Estimate the time it will take to make the cut that you have planned.

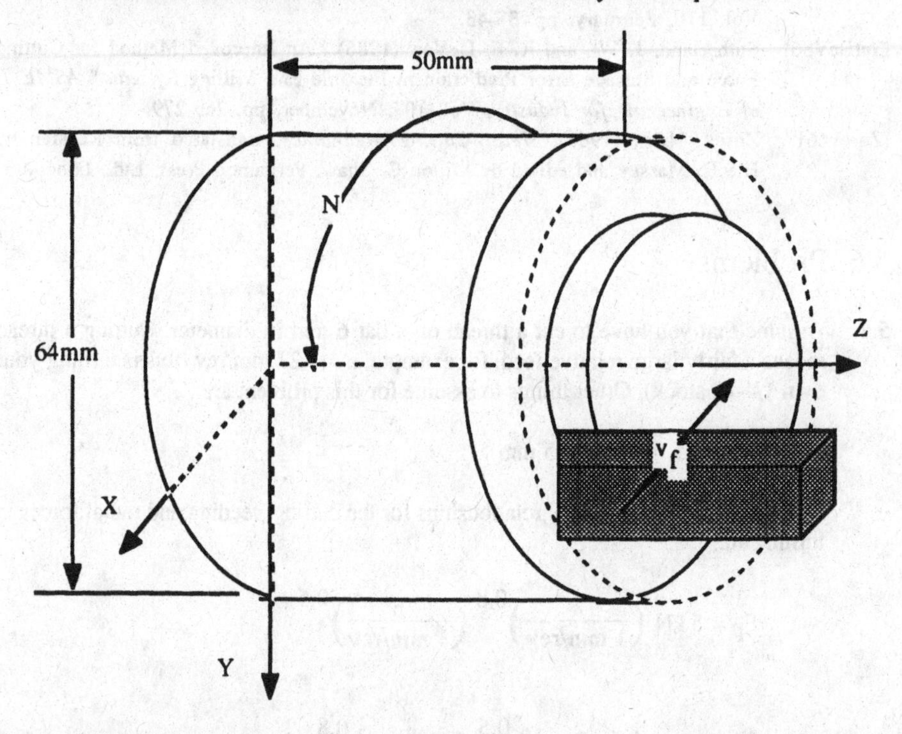

Figure 5.20 Determining cutting conditions for a facing cut in Problem 5.2.

5.3 The slab milling operation illustrated in Figure 5.21 has the following specifications:

Cutter: 100 mm diameter, 10 flute HSS cutter with a 10 degree normal rake and 15 degree helix angle.

Workpiece: 1045 steel with H_B = 190, 150 mm wide, 500 mm long and 90 mm thick.

Cutting Conditions: Climb milling with N = 100 rev/min, a = 15 mm, v_f = 4 mm/s

a. Determine the average chip thickness, h_{avg}.

b. Estimate the power, in kW, to make this cut with a new cutter using the specific cutting energy method.

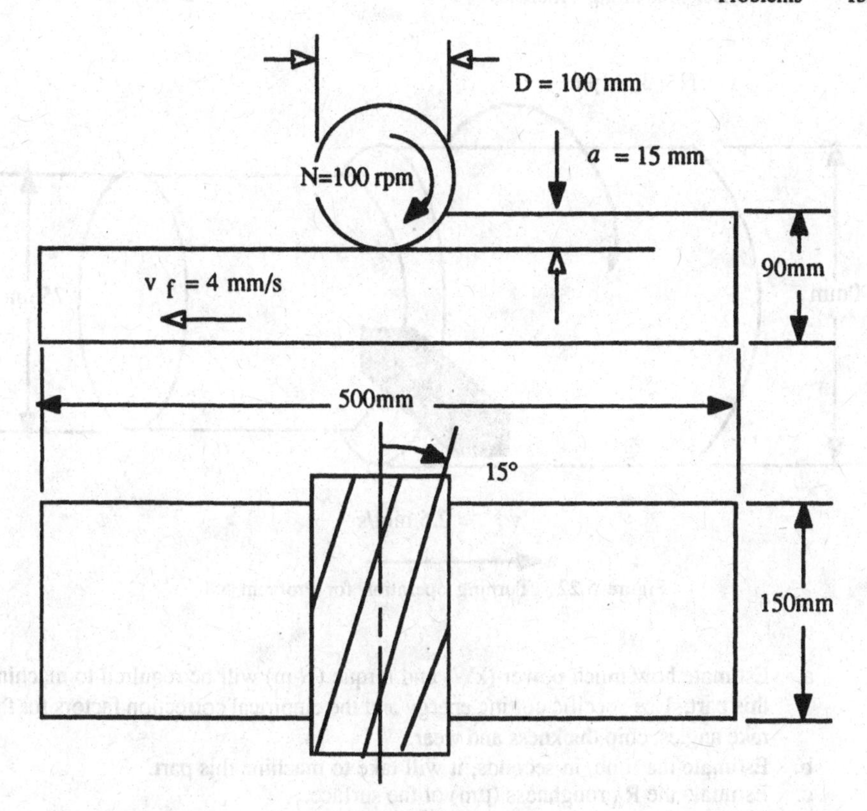

Figure 5.21 Estimating the power for the slab milling operation in Problem 5.3.

5.4 Figure 5.22 shows a turning operation with the following conditions:
Cutting Tool: A carbide cutting tool with a signature: 0, -10°, 5°, 5°, 15°, 15°, 3 mm that will be used until the cutting edge has 0.25 mm of flank wear.
Workpiece: 1045 steel bar 90 mm in diameter with H_B = 200, that is 200 mm long and will be turned down to a 75.0 mm diameter.
Cutting Conditions: N = 300 rev/min and v_f = 1.5 mm/s

a. Estimate the theoretical R_a roughness, in μm, for a new cutting edge.
b. Use the oblique cutting model with τ_s = 200 MPa to estimate the forces needed for power calculations, but use the correction factors for the specific cutting energy method to account for flank wear,
c. Estimate the power requirements in kW, for this operation.

5.5 Face milling is to be done on the heads of an engine using a 12 tooth cutter. In Figure 5.23 the idealizations of the cutter and the engine head are shown. (Note that the cylinders are "idealized away" so no preference for 4, 6 or 8 cylinders is expressed by the author.) The specifics of this proposed operation are:
Cutting Tool: Twelve 25 mm square carbide inserts with a 1 mm corner radius will be used. The insert holder and insert geometry give a radial and axial rake angle of 0° and a corner cutter angle of 60°. These inserts are used until VB=0.75 mm.
Workpiece: Cast iron with H_B = 100, 500 mm long and 200 mm wide that will have 14 mm of the material face milled off the top.
Cutting Conditions: N = 200 rev/min and the chip load, s_z= 0.5 mm/tooth.

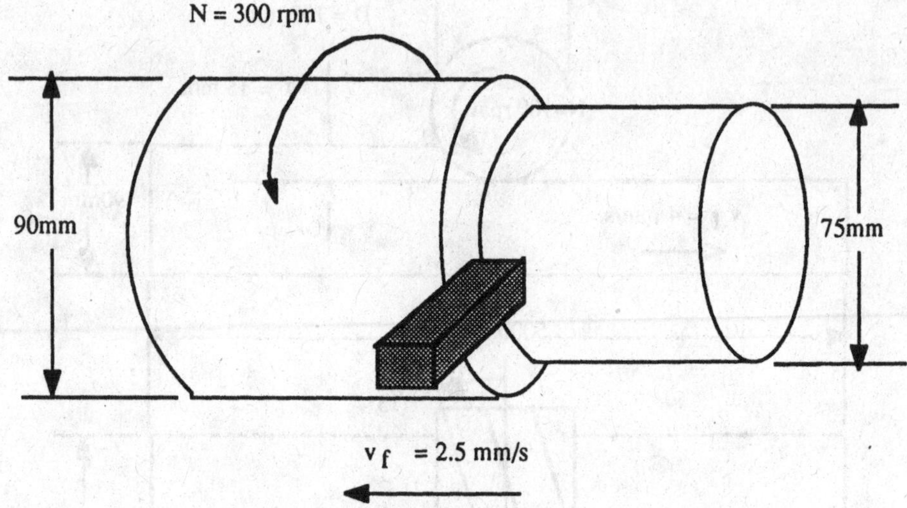

Figure 5.22 Turning operation for Problem 5.4.

a. Estimate how much power (kW) and torque (N m) will be required to machine this part. Use specific cutting energy and the empirical correction factors for the rake angles, chip thickness and wear.
b. Estimate the time, in seconds, it will take to machine this part.
c. Estimate the R_a roughness (µm) of the surface.

5.6 This problem is aimed at comparing face milling with carbide inserts and conventional slab milling with high speed steel on the same 2024-T4 aluminum workpiece; the surface finish $R_a \leq 0.5$ µm, s_z and depth are the same in both cases. Estimate power with the specific cutting energy method for new cutting edges.

Table of Cutting Conditions and Cutter Geometry to Compare Face and Slab Milling in Figure 5.24	
Common conditions: $s_z = 0.25$ mm/tooth and $a = 10$ mm	
Face Mill	Slab Mill
v = 3.0 m/s	v = 1.0 m/s
D = 62.5 mm	D = 62.5 mm
4 square inserts	10 helical flutes
3 mm corner radius	50 mm wide
5° lead angle	15° helix angle
5° radial rake	5° radial rake

a. For the *face mill* estimate, the average uncut chip thickness h_{avg}, the power at the spindle, and the average torque, i.e.,
b. Repeat these calculations, but for the *slab mill*, i.e., estimate the average uncut chip thickness h_{avg}, power at the spindle, and the average torque.
c. Which of these cuts: meets theoretical surface finish requirements? takes the shortest time, taking into account the cutter clearing the workpiece?

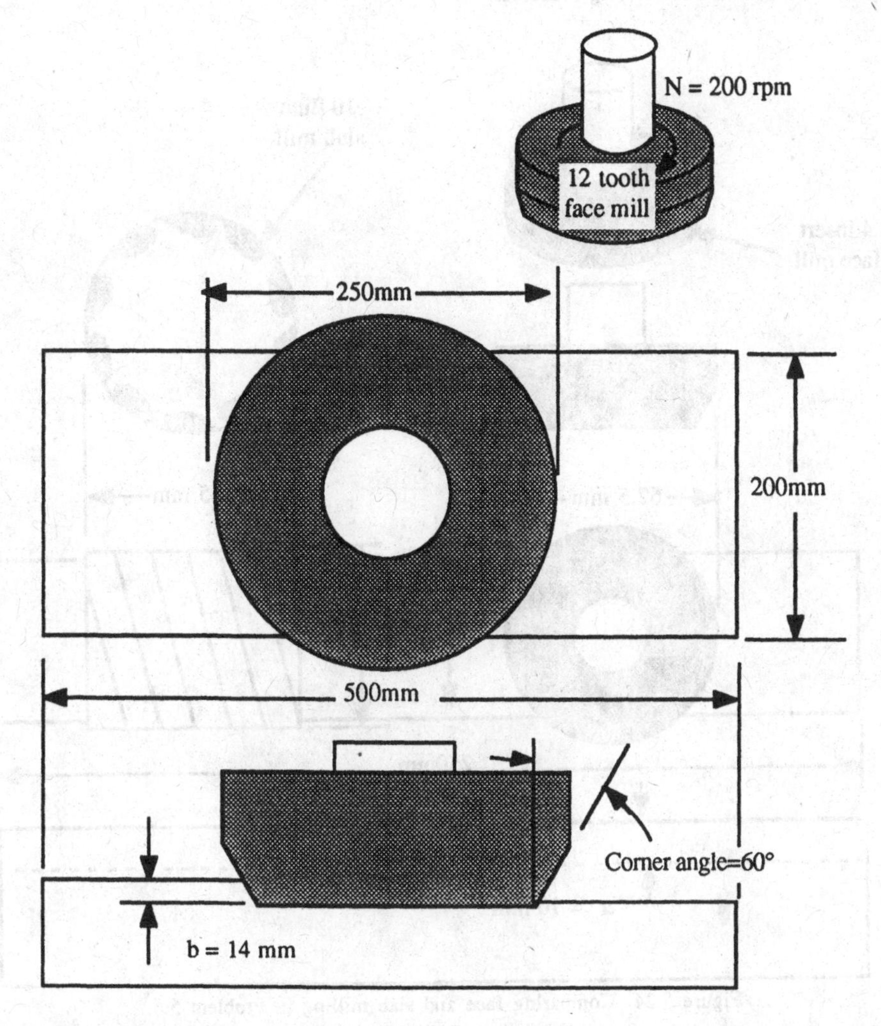

Figure 5.23 Idealization of face milling an engine block head in Problem 5.5.

5.7 Assume that a 10 mm diameter two flute twist drill with the conventional point angle of 118° will be used to make a number of holes in cast iron with a Brinell hardness of H_B = 100. Assume that the cutting speed is v = 0.2 m/s and the relative feed is 0.2 mm/rev. Empirical equations have been developed for this grade of cast iron that relate the relative feed *a*, and drill diameter D (mm), to the drilling force and thrust as given below:

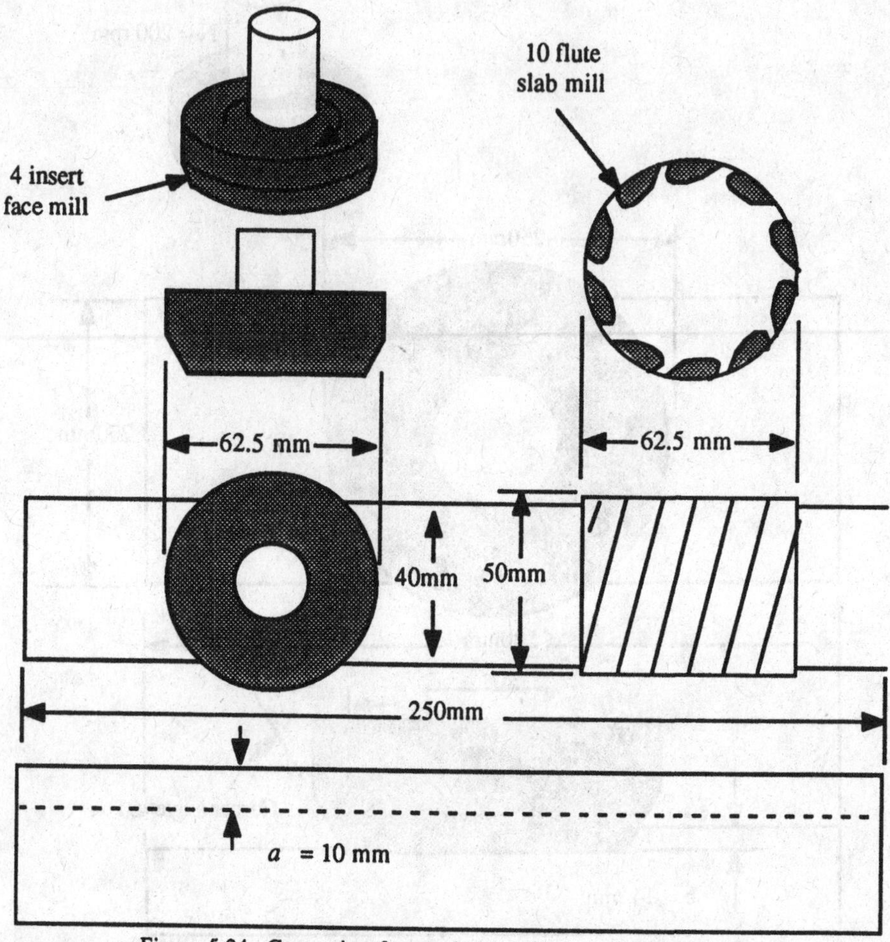

Figure 5.24 Comparing face and slab milling in Problem 5.6.

$$T_{cutting} = 8.0\text{ N-m}\left(\frac{a}{25.4\text{ mm/rev}}\right)^{0.6}\left(\frac{D}{1\text{ mm}}\right)^{2}$$

$$F_{thrust} = 2.6\text{ kN}\left(\frac{a}{25.4\text{ mm/rev}}\right)^{0.6}\left(\frac{D}{1\text{ mm}}\right)$$

a. Determine the rotational speed N, and the time t_c to make a hole through 75 mm of material.
b. Estimate the torque, thrust force and power needed to make a hole using the power law equations given with this problem.
c. It is also possible to estimate the torque and power using the specific cutting energy method, so repeat the calculations using this approach.

5.8 This problem is to select one of the cutting conditions for turning, viz. the depth of cut b, based on a maximum value for the radial force $F_{r,max}$. This means the orthogonal cutting or specific cutting energy methods will not work. The other things that are specified are: the rotational speed N= 10 rev/s, the feedrate $v_f = 2$ mm/s, and a new carbide cutting tool with a signature of 0°, 10°, 5°, 5°, 15°,45°, 0.5 mm. The workpiece that has to be turned down is 2024-T4 aluminum that is 200 mm long and 110 mm in diameter. Based on this information and Figure 5.25, find the depth of cut b, that causes the radial force to just equal $F_{r,max} = 0.75$ kN.

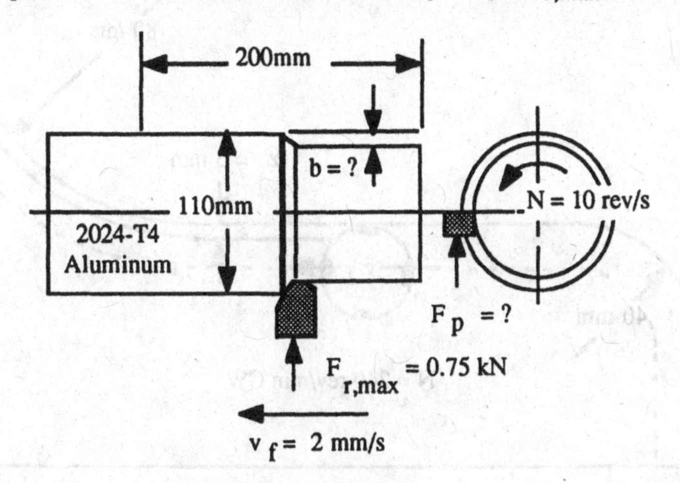

Carbide tool with signature
0°, 10°, 5°, 5°, 15°, 45°, 0.5mm

Figure 5.25 For planning based on radial force limits in Problem 5.8.

5.9 Peripheral end milling with an inside corner, a straight section and an outside corner are shown in Figure 5.26. This will done on a CNC machine tool where the programmed feedrate is $v_f = 4$ mm/s , the velocity of the center line of the cutter, and the rotational speed is programmed at N = 240 rev/min. Experiments found that the specific cutting energy is related to the average chip thickness by $E_c(h_{avg}) = 2.0$ J/mm^3 $(h_{avg}/0.25 \text{ mm})^{-0.25}$.

a. For the straight sections of this cut, estimate the average chip thickness, material removal rate and power.
b. Repeat the calculations in 5.9a, but for the inside corner section.
c. Repeat the calculations in 5.9a, but for the outside corner section.
d. Based on your calculations, what is an appropriate way to program the CNC machine tool to keep the chip load constant when making corners?

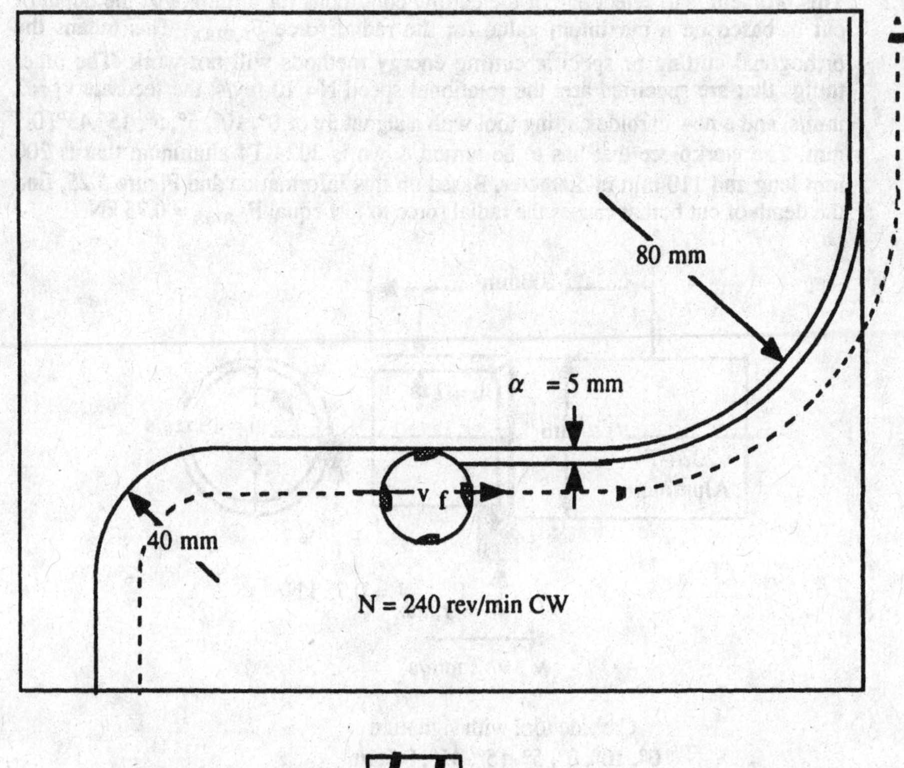

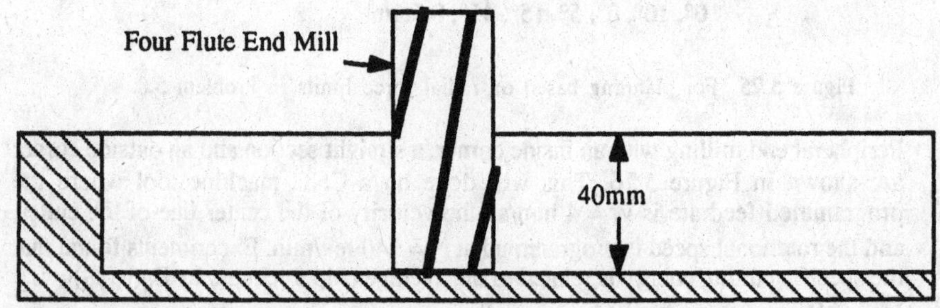

Figure 5.26 Contour end milling cut for Problem 5.9.

6
Grinding Processes

Grinding processes differ enough from cutting processes that they should be treated in a separate chapter. Whole books have been written on grinding, [KinHah86] and [Malkin89] are recent examples, and they should be consulted for more technical and practical details. This chapter will concentrate on the basics of grinding for the analysis that goes into process planning, and will compare and contrast grinding with cutting.

Many of the inputs to the grinding process are similar to the ones for cutting processes, as a comparison of Figure 5.1 and Figure 6.1 illustrates. For example workpiece material, geometry and tolerances as well as many of the work holding techniques are common to both grinding and cutting processes. One of the major differences between grinding and cutting processes is the number and geometry of the cutting edges. Grinding uses an abrasive wheel with many randomly oriented edges, while cutting uses a known number of cutting edges with controlled geometry. In most cases the constraints on the stiffness and power of the grinding machines is important because grinding is considered a precision material removal process where deflections take on added significance. More of the energy provided at the spindle is converted into heat, making the possibility of thermal damage to the workpiece an important technical limitation. Analyzing the grinding process still relies on mechanics, kinematics and heat transfer. But the use of abrasive wheels leads to a new terminology to describe the grinding wheel and how it wears. The technological results of this analysis pertinent to planning a grinding process are indicated on the right side of Figure 6.1. Force, temperature and wear predictions are used to select the grinding process and conditions needed to meet surface finish and tolerance specifications.

6.1 Comparing Grinding With Cutting Processes

A way of explaining the differences between cutting and grinding processes is the cutting edge geometry and relative scale of the chips produced [Shaw90]. Figure 6.2 highlights these differences. Cutting processes have controlled geometry, usually with positive or "moderately" negative rake angles like that in Figure 6.2.a. Grinding processes, on the other hand, use small *abrasive grits* with random orientation and geometry as Figure 6.2.b shows, that gives rise to very negative "rake" angles.

Large negative rake angles and thin chips are produced by grinding. The previous background on the mechanics of machining explains some of the differences between grinding and cutting that need to be considered when planning a grinding operation. In Section 3.2.1 an expression for the specific cutting energy that included an empirical power law expression for strain hardening and size effect was proposed:

$$E_c = \frac{\cos(\beta-\gamma_o)\,\tau_0 \left(\frac{h}{h_R}\right)^a}{\sin(\phi_o)\cos(\phi_o+\beta-\gamma_o)}$$

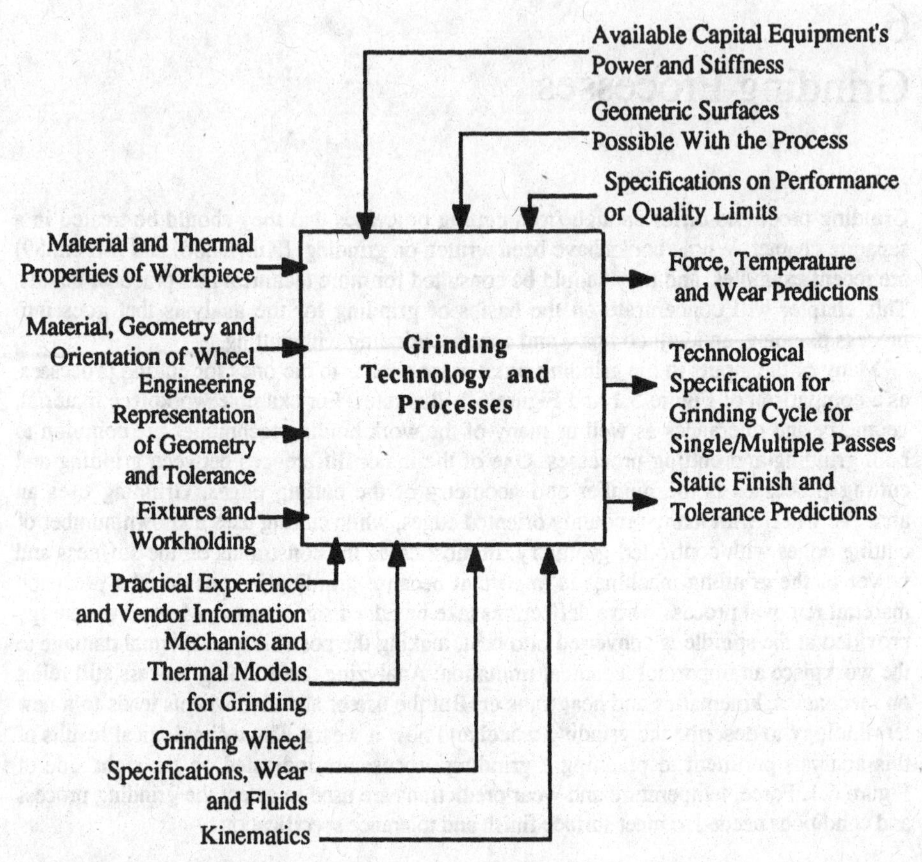

Figure 6.1 The Inputs, Constraints, Outputs and Mechanisms considered when analyzing and modeling a grinding process.

In cutting, the average chip thickness is $h_{avg} \approx 0.1\text{-}1.0$ mm and rake angles are about $-5° < \gamma_o < 30°$ producing specific cutting energies on the order of 1-2 J/mm^3. Grinding produces very thin chips or *swarf*, described later on a macro scale by the equivalent chip thickness h_{eq} or on a smaller scale by the maximum chip thickness. Due to the grit geometry, large negative rakes of $\gamma_o << 5°$ are common. These facts, in light of the expression for E_c, can explain the increases in the specific energy in grinding E_g, of 2 to 5 times that in cutting. However *cutting* is not the only mechanism that produces forces on the grits. In addition to cutting, there is *rubbing* of what is the flank of the cutting edge formed by the grit, as well as *plowing* or plastic flow to the side of the grit rather than ahead and out of the contact zone. All three increase the specific energy in grinding and are modeled as:

$$E_g = E_{cutting} + E_{plowing} + E_{rubbing} \,. \tag{6.1}$$

These three mechanisms generate forces that lead to higher specific energies in grinding than in cutting as Table 5.2 shows for some selected materials where both specific grinding and cutting energies are available. This means that the heat source intensity in grinding is greater than in cutting, affecting temperatures. In cutting, because the heat can be conducted away from the workpiece, usually elevated workpiece temperatures are not a

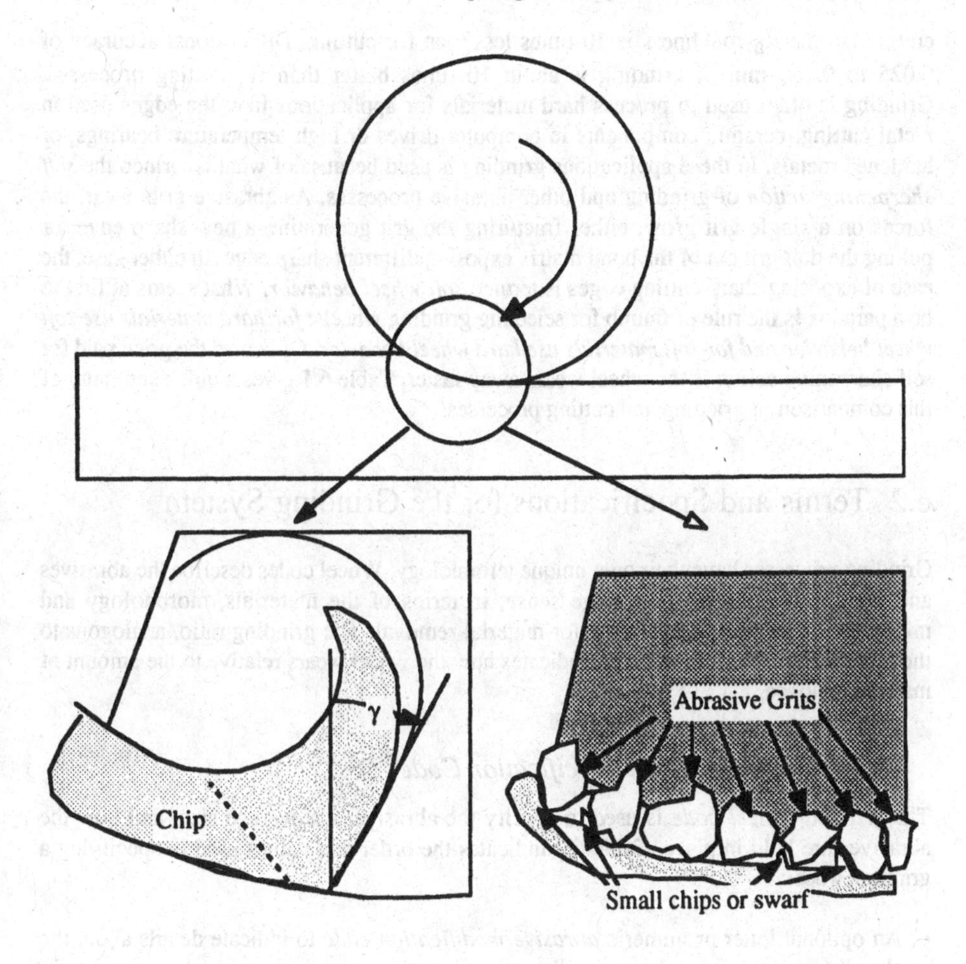

Figure 6.2 Comparing chip formation in cutting and grinding.

great problem; instead it is the cutting edge that experiences the elevated temperatures. Because most abrasives are insulators, more of the heat generated in grinding goes into the workpiece. This leads to elevated workpiece temperatures that can cause metallurgical changes and sub-surface damage often termed *burning*.

While these are some of the negative aspects of grinding, fortunately they can be controlled by selecting the wheel and grinding fluid, as well as the grinding conditions. For slab or peripheral end milling, the axial depth of cut is on the order of 10 mm; in grinding this is reduced by a factor of 10 or more. The wheel surface speed in grinding is often in the range of $10 < v_s < 25$ m/s, while the corresponding cutting speed is ten to twenty percent of these values.

When grinding conditions are selected correctly, several measures of workpiece quality improve an order of magnitude over what can be achieved in cutting processes. First is surface finish - one of the reasons to grind - depends on the feed or chip load. In milling processes this is limited by the number of cutting edges, which are on the order of tens, producing finishes of $R_a \approx 1$ µm. In grinding the grits provide the cutting edges, so the number of cutting edges increases by several orders of magnitude when compared to cutting, so the R_a roughness is 10 times less than for cutting. Dimensional accuracy of 0.025 to 0.100 mm in grinding is about 10 times better than for cutting processes. Grinding is often used to process hard materials for applications like: the edges used in metal cutting, ceramic components in computer drives or high temperature bearings, or hardened metals. In these applications grinding is used because of what is termed the *self sharpening action* of grinding and other abrasive processes. As abrasive grits wear, the forces on a single grit grow, either fracturing the grit generating a new sharp edge, or pulling the dull grit out of the bond matrix exposing different sharp edge. In either case, the ease of exposing sharp cutting edges is termed *soft wheel behavior*. What seems at first to be a paradox is the rule of thumb for selecting grinding wheels; *for hard materials use soft wheel behavior and for soft materials use hard wheel behavior*. Of course the price paid for self sharpening action is the wheel wears away faster. Table 6.1 gives a quick summary of this comparison of grinding and cutting processes.

6.2 Terms and Specifications for the Grinding System

Grinding processes have their own unique terminology. Wheel codes describe the abrasives and bonding matrix in an average sense, in terms of the materials, morphology and mixtures in a grinding wheel used for material removal. The grinding ratio, analogous to the tool wear relations for cutting, indicates how the wheel wears relative to the amount of material removed.

6.2.1 Grinding Wheel Specification Code

The *grinding wheel code* is used to specify the abrasive, it's average size, and how the abrasives are held in place. Table 6.2 indicates the order and values used in specifying a grinding wheel:

- An optional letter or numeric *abrasive modification code* to indicate details about the abrasive.
- The *types of abrasives;* consists of a menu of: (A)luminum oxide, the most common abrasive for ferrous materials is relatively soft and tough. Silicon (C)arbide, the most common abrasive for non-ferrous materials like aluminum is harder and more expensive than aluminum oxide. (Z)erconia alumina is one of the toughest abrasives and is used for high impact rough grinding where self sharpening is important, i.e., snagging and cutoff. Note that the next two abrasives are classified as *super abrasives* and have a slightly different code than the conventional abrasives listed up to this point. This is because these abrasives are usually bonded in a thin layer to a metallic hub, due to their high cost. Cubic (B)oron nitride, usually referred to as CBN, is a super abrasive suitable for grinding ferrous materials. (D)iamond is another super abrasive that can be used for non-ferrous materials and ceramics.
- *Grit size*, is indicated by number that is inversely proportional to the size, i.e., 46 grit is larger than 100 grit. (This sizing is based on statistical sieve classification. For example, 100 grit means that the abrasives pass through a screen with 100 wires per *inch* and do not pass through the next finest size, e.g., 120 grit. As a result, an agreed upon average grit diameter is

Table 6.1 A Simplified Comparison of Cutting and Grinding Processes

Quantity	Cutting	Grinding
Specific Energy (J/mm^3)	$1 < E_c < 2$	$2 < E_g < 10$
Tolerances (mm)	0.1 - 1.0 mm	0.01 - 0.10 mm
Surface Roughness (μm)	$1.0 < R_a < 2.0$ μm	$0.10 < R_a < 0.20$ μm
Cutting Speed (m/s)	$0.5 < v < 5.0$ m/s	$5.0 < v_s < 50$ m/s
Chip Thickness (mm)	$0.1 < h_{avg} < 1.0$ mm	$0.01 < h_{eq} < 0.1$ mm
Normal Rake Angles (°)	$-5° < \gamma_n < 30°$	$-30° < \gamma_n < 5°$

$$d_g = \frac{0.6\ (25.5\ \text{mm/in})}{\text{grit size}}. \tag{6.2}$$

- The *grade or hardness* is a letter code indicating how difficult it is to remove a grit from the wheel, i.e., the self sharpening action. The "softest" wheel is designated by A, the "hardest" by Z, and the L and N in Table 6.2 for the conventional and super abrasives are moderately hard wheels. Hardness is a function of several things: the *friability* or how easily the abrasive fractures, the bond strength, and the void space.
- A numeric code for *structure or openness* indicates the relative porosity or approximate spacing between grits. For conventional abrasives, a relative ranking scale is used (0-25), with low numbers indicating tightly packed abrasives. The super abrasives use a vendor designated code.

- A letter code indicates the *type of bond* used to hold the grits in the wheel matrix(V)itrified is the most common inorganic bond. (B) designates a resinoid bond, where grits are removed from the bond matrix primarily by heat, not mechanical action.

(R)ubber is the most resilient and toughest bond. The (E) code is used for shellac bonds which are more heat sensitive than vitrified bonds. As a result, if temperatures rise a little, the shellac bonded grits pull out and tend to prevent workpiece burning. (M)etal bonds are used almost exclusively for the super abrasives - CBN or diamond - because of their cost. These bonds are strong and designed so that these expensive abrasives do not pull out.

- (For super abrasives, two further specifications are made. The first is the vendor's *bond modification code*. The second is the the *depth of abrasive* into the bond material. This last specification is usually a few millimeters because of the cost of these high performance abrasives.)
- A final *optional bond modification* code can be assigned by the vendor.

As another example, a 32A46 - H8VBE wheel can be decoded by looking at Table 6.2. This is a 46 grit aluminum oxide wheel of soft to medium hardness and density, with a vitrified-resin-shellac combination bond, probably designed to prevent workpiece burning. The vendors proprietary abrasive code is 32. Note the pattern on these codes: number-letter-number-letter.

Table 6.2 Identification Codes for Standard Commercial and Super Abrasives
(See ANSI B74.13-1977 and ISO 525-1975E)
Identification Code for Grinding Wheels and Other Bonded Abrasives

Abrasive Type	Grade			Bond Type
	Soft	Medium	Hard	
A - Aluminum Oxide	A E	I M	Q V	B - Resinoid
C - Silicon Carbide	B F	J N	R W	BF - Resinoid Reinforced
Z - Zerconia Alumina	C G	K O	S X	E - Shellac
	D H	L P	T Y	O - Oxy-chloride
			U Z	R - Rubber
				RF - Rubber Reinforced
				S - Silicate
				V - Vitrified
				Mg - Magnesia

↓ ↓ ↓

51 A 36 L 5 B ___

↑ ↑ ↑ ↑

Prefix	Abrasive Grit Size				Structure		Manufacturer's Record
	Coarse			Very Fine	Dense	Open	
Manufacturer's Symbol indicating exact kind of abrasive (use optional)					1	9	Manufacturer's identification Symbol (use optional)
	8	30	90	280	2	10	
	10	36	100	320	3	11	
	12	46	120	400	4	12	
	14	54	150	500	5	13	
	16	60	180	600	6	14	
	20	70	220		7	15	
	24	80	240		8	16	

6.2.2 Grinding Ratio and Dressing Cycle

The grinding ratio G, can be considered an overall measure of the grinding system performance. It is defined as the ratio of the workpiece material removal rate Z_W (mm^3/s) to the wheel wear rate, Z_S (mm^3/s),

$$G \equiv \frac{\text{Volume of material removed}}{\text{Volume of wheel used}} = \frac{Z_W}{Z_S}. \tag{6.2a}$$

The material removal rate will largely depend on the operating conditions for the process. But the wheel wear rate depends on a number of things such as: characteristics specified by the wheel code in Section 6.2.1, the grinding fluid used, temperatures in the grinding zone or the reactivity of the workpiece-abrasive combination. Many of these are hard to predict and can be affected by operating conditions and the process design. Figure 6.3 shows a curve similar in shape to the tool wear figures in Section 4.3.2, but it represents how a grinding wheel wears, represented by Z_S, as a function of the volume of material removed Z_W. The instantaneous grinding ratio is the inverse of the slope of this curve. Like tool

Table 6.2 (continued)
Identification Codes for Standard Commercial and Super Abrasives
(See ANSI B74.13-1977 and ISO 525-1975E)
Identification Code for Diamond, CBN and Other Bonded Super Abrasives

Abrasive Type	Grade			Bond Type	Depth of Abrasive
	Soft	Medium	Hard		
B - CBN D - Diamond	A E B F C G D H	I M J N K O L P	Q V R W S X T Y U Z	B - Resin M - Metal V - Vitrified	Working depth of abrasive in millimeters or inches (millimeters illustrated)

M D 120 N 100 B 77 3 —

Prefix	Abrasive Grit Size				Concentration	Bond Modification	Manufacturer's Record
Manufacturer's symbol indicating exact kind of abrasive (use optional)	Coarse 8 10 12 14 16 20 24	30 36 46 54 60 70 80	90 100 120 150 180 220 240	Very Fine 280 320 400 500 600	Manufacturer's designation. May be number or symbol	Manufacturer's notation of special bond type or modification	Manufacturer's identification symbol (use optional)

wear rates, when a single G-ratio is reported, it refers to the region with a nearly constant slope. G-ratios on the order of 10 to 100 are reported for conventional abrasives, while for super abrasives the values range from 100 to 1000. As might be expected based on the definition of soft versus hard wheel behavior, soft wheels tend to have the lower G-ratios than wheels exhibiting hard wheel behavior.

Dressing and truing are both treatments that consume the grinding wheel so they add to the cost of a grinding process. Both may be done with a dressing tool, either a *diamond stick* that is fed across the wheel at a very small radial depth, or a *crush dresser* which is a free turning wheel that is fed radially into the wheel so as to crush the abrasives. *Dressing* is applying these treatments to expose sharp new cutting edges and should just fracture existing grits. This is needed for hard wheels because the abrasives tend to wear on the surface tangent to the velocity, just like flank wear in cutting. Figure 6.3 indicates that immediately after dressing the G-ratio may drop for a short time, then reach an approximately steady state value where the most productive grinding is done. Then the G-ratio drops, either because of wear on the abrasives or because workpiece material bonds to the abrasive or the bond matrix, making another dressing cycle necessary. ***Truing*** is a deeper form of dressing applied to soft wheels aimed at controlling the geometry of the wheel - remember they sharpen themselves - by removing lobing or circumferential grooves. Of course, when using a dressing tool, both dressing and truing are occurring.

6.3 Kinematics of Grinding

As in cutting, the conditions set on a grinding machine play a major role in determining fundamental quantities like chip thicknesses or contact lengths. These are primarily kinematic calculations that depend on the particular type of grinding process. This section will consider only two grinding arrangements: surface and cylindrical grinding. Then ways to compare results for these two processes will be considered in terms of the equivalent chip thickness and equivalent wheel diameter.

6.3.1 Surface Grinding

Surface grinding, while the least common production grinding process, is the easiest to understand because it's a close analog to slab or peripheral milling as can be seen in Figure 6.4. This type of grinding is designed to produce high tolerance, low surface roughness, flat planar surfaces. A slight variation on the notation used in milling is introduced for grinding: The subscript "s" is added to quantities that pertain to the wheel, particularly its surface or circumference. Similarly, the workpiece has a subscript "w" to define the *workpiece velocity* v_W. For surface grinding v_W changes direction at the end of a stroke, and at the same time the workpiece is indexed along the axis of spindle rotation by an

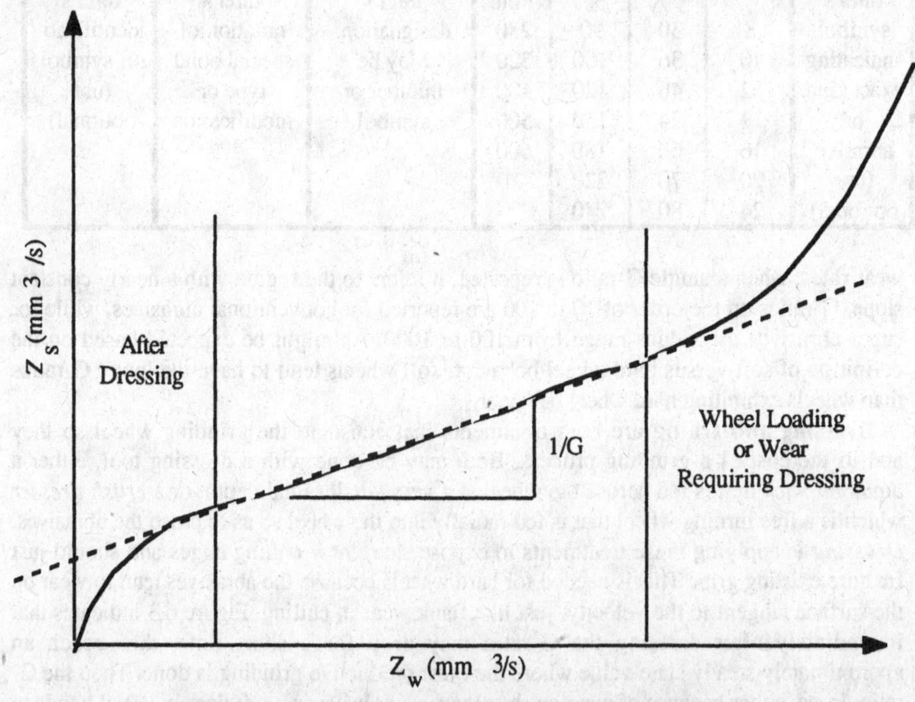

Figure 6.3 The grinding ratio G, and the dressing cycles.

amount b, the *axial feed*. This value is usually several times smaller than the wheel width, w. Because the direction of v_w changes, surface grinding is constantly switching between conditions of up and down grinding. The *depth, radial depth or radial infeed a*, defines the distance that the wheel projects below the unground surface of the workpiece. The radial infeed is less than or equal to the *grinding allowance*, the amount of excess material to be removed by grinding, on the order of 0.1 to 1 mm.

With these terms defined, by analogy with the results in Section 5.3.3 the following geometric and kinematic quantities can be defined. The kinematic contact length (the actual length is a combination of this value and the elastic "flattening" of the wheel) is

$$l_c = \frac{D_s}{2}\cos^{-1}\left(1 - \frac{2a}{D_s}\right) \text{ or} \tag{6.3a}$$

$$\approx \sqrt{a\,D_s} \qquad \text{if } 2a \ll D_s. \tag{6.3b}$$

If the circumferential spacing of active grits K (grits/mm) can be determined, then the feed per grit is defined as

$$s_g = \frac{v_w}{K \bullet v_s} \tag{6.4}$$

and from this, the maximum swarf or chip thickness due to a single grit can be estimated by

$$h_{max} = 2\,s_g\sqrt{\frac{a}{D_s}} = 2\left(\frac{v_w}{K \bullet v_s}\right)\sqrt{\frac{a}{D_s}} \tag{6.5a}$$

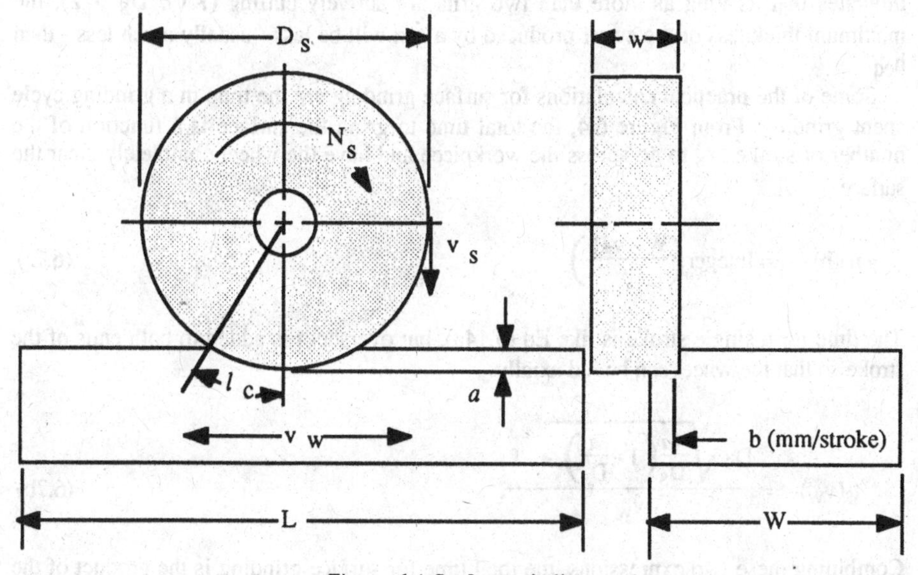

Figure 6.4 Surface grinding.

If a chip cross section, for example, a rectangle or triangle, is assumed, the average chip thickness can be estimated just as in Section 5.3. It will not be done here. The cross section of a chip produced by an individual grit is assumed to have a length given by Eq (6.3a or b), and a maximum thickness given by Eq (6.5a).

Forces on an individual grit, the same ones mentioned earlier in the discussion of the self sharpening action of a grinding wheel, are a strong function of the chip thickness. This means that if a material is more difficult to grind than expected, but the wheels have already been purchased, Eq (6.5a) suggests how the self sharpening action characteristic of a soft wheel can be achieved: increase the radial depth a, or increase the ratio (v_W/v_S). Both increase the chip thickness so the forces to remove an individual grit increase.

Writing the material removal rate for surface grinding in terms of the stock entering the grinding zone and the material leaving as swarf

$$b \bullet a \bullet v_W = b \bullet h_{eq} \bullet v_S$$

defines the *equivalent chip thickness*

$$h_{eq} = a\left(\frac{v_W}{v_S}\right). \tag{6.6}$$

This is not a measurable physical quantity. Instead it is the thickness of an idealized continuous ribbon of material removed by grinding. It is an aggregate quantity that has been used to correlate a number of experimental measurements such as surface roughness, specific grinding energy, etc. The relationship between h_{eq} and h_{max}

$$h_{max} = h_{eq}\left(\frac{2}{K\sqrt{a\,D_S}}\right) \tag{6.5b}$$

indicates that as long as more than two grits are actively cutting ($K\sqrt{a\,D_S} > 2$), the maximum thickness of the swarf produced by a grit will be less - usually much less - than h_{eq}.

Some of the practical calculations for surface grinding are the time in a grinding cycle spent grinding. From Figure 6.4, the total time to grind the surface is a function of the number of strokes n_S to go across the workpiece and have the wheel completely clear the surface

$$n_S(b) = \text{Integer}\left(\frac{W+w+b}{b}\right). \tag{6.7a}$$

The time for a single stroke is like Eq (5.14a), but overshoot is added to both ends of the stroke so that the wheel can be fed axially

$$t_S(v_W) = \frac{D_S\sqrt{\frac{4a}{D_S}\left(1-\frac{a}{D_S}\right)} + L}{v_W}. \tag{6.7b}$$

Combining these two expressions, the total time for surface grinding is the product of the number of strokes and the time per stroke,

$$t_g(b,v_w) = \text{Integer}\left(\frac{W+w+b}{b}\right)\left(\frac{D_s\sqrt{\frac{4a}{D_s}\left(1-\frac{a}{D_s}\right)}+L}{v_w}\right) \qquad (6.7c)$$

When the grinding wheel wears radially, i.e., the radius decreases at a rate v_r, then the radial wheel wear rate is

$$Z_s(t) \quad = b \bullet v_r \bullet \pi \bullet D_s\left(1 - \frac{2v_r t}{D_s}\right) \qquad (6.8a)$$

$$\approx b \bullet v_r \bullet \pi \bullet D_s \qquad (2v_r t \ll D_s). \qquad (6.8b)$$

The approximation works when the diameter reduction ($2v_rt$) is small relative to the wheel diameter D_s. Eqs (6.8) are *valid for both surface and cylindrical grinding if the wheel wear is radial*. The removal rate for *surface grinding* can be written in a similar manner, but the radial wear reduces the infeed so that

$$Z_w(t) \quad = b \bullet v_w \bullet a\left(1 - \frac{v_r t}{a}\right) \qquad (6.9a)$$

$$\approx b \bullet v_w \bullet a = b \bullet v_s \bullet h_{eq} \qquad (v_r t \ll a). \qquad (6.9b)$$

Because Eq (6.9) indicates how the infeed is reduced, this also indicates how the accuracy of the part will be reduced by v_r. Using the approximate expressions for Z_w and Z_s, the grinding ratio for *surface grinding* is

$$G \quad \approx \frac{a \bullet v_w}{\pi \bullet D_s \bullet v_r}. \qquad (6.2b)$$

Eq (6.2b) can be thought of as either an experimental way of finding the G-ratio by measuring v_r, or for a given G-ratio it provides a way to estimate the radial wear on the grinding wheel and the reduction in a.

6.3.2 Cylindrical Grinding and D_{eq} to Relate to Surface Grinding

The workpiece, as well as the grinding wheel, rotate in cylindrical grinding. This type of grinding, along with centerless grinding which is not covered in this book, are the most common production grinding processes used to make bearings and bearing surfaces. Figure 6.5 represents an *external cylindrical grinding* process and Figure 6.6 an *internal cylindrical grinding* process. Both of these processes generate surfaces of revolution like an engine lathe. Variations on the cylindrical grinding are when $v_a = 0$ and the radial depth is generated by the infeed v_f,

$$a \quad = \frac{v_f}{N_w}. \qquad (6.10a)$$

This is termed external or internal cylindrical *plunge* grinding, where the grinding width b is equal to the wheel width w. For internal or external cylindrical *traverse* grinding, $v_a \neq 0$ and the radial feed motion is intermittent and generates the radial depth. Grinding width is determined by

$$b = \frac{v_a}{N_w}, \tag{6.10b}$$

the feed per revolution of the workpiece. While the surfaces generated by these processes are different than surface grinding, it is the contact conditions that are of fundamental importance for analyzing grinding performance. For example, when visualizing the contact lengths for Figure 6.5 and 6.6, it is easy to imagine that l_c is greater for internal than for external cylindrical grinding. This, along with the reduced stiffness of a spindle for a small wheel makes internal grinding particularly prone to chatter.

Removal rate calculations for cylindrical grinding are similar to those for surface grinding. In cylindrical *plunge* grinding the radial feed rate and wheel diameter are both affected by the radial wheel wear rate v_r: the radial feed rate v_f is reduced by (v_r/v_f) and the wheel diameter is reduced by $(2v_rt/D_s)$ so

$$Z_w(t) = b \bullet \pi \bullet D_w \left[1 \pm \frac{2v_rt}{D_w}\left(1 - \frac{v_r}{v_f}\right)\right] \bullet v_f \left(1 - \frac{v_r}{v_f}\right) \text{ (- for external, + for internal)} \tag{6.11a}$$

$$\approx b \bullet \pi \bullet D_w \bullet v_f (2v_rt \ll D_w \text{ and } v_r \ll v_f) \tag{6.11b}$$

Then using Eq (6.11b) and (6.8b), the approximate G-ratio for cylindrical plunge grinding is

$$G \approx \frac{D_w \bullet v_f}{D_s \bullet v_r}. \tag{6.2c}$$

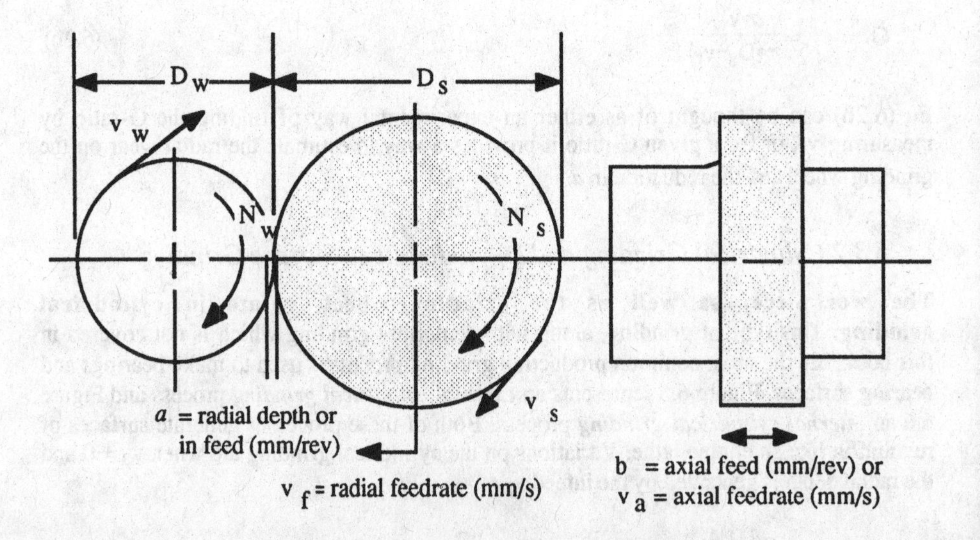

Figure 6.5 External cylindrical grinding.

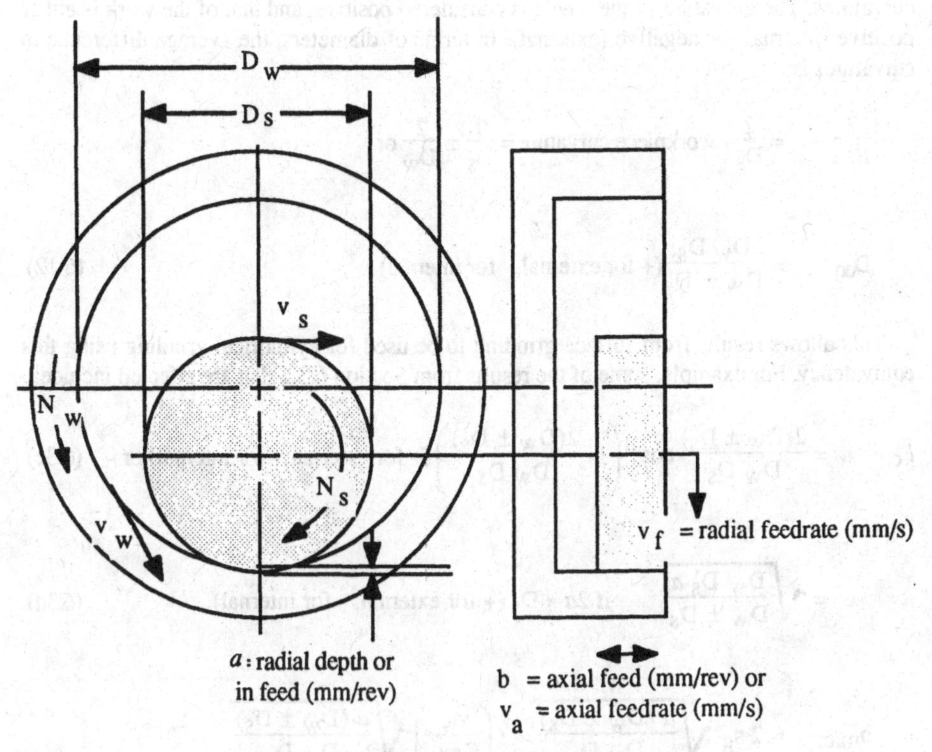

Figure 6.6 Internal cylindrical grinding.

Traverse grinding differs because the axial feed rate v_a and the radial or depth feed determine the material removal rate. Again, the radial wear rate reduces the radial depth a by the ratio
($v_r t/a$) so

$$Z_w(t) = v_a \bullet \pi \bullet D_w \left[1 \pm \frac{a}{D_w}\left(1 - \frac{v_r t}{a}\right)\right] \bullet a \left(1 - \frac{v_r t}{a}\right)$$

(- for external, + for internal) (6.11c)

$$\approx a \bullet \pi \bullet D_w \bullet v_a = a \bullet \pi \bullet D_w \bullet b \bullet N_w \quad (v_r t \ll a \text{ and } a \ll D_w) \tag{6.11d}$$

The second approximate expression uses b, the axial feed per revolution of the workpiece, from which the approximate G-ratio for cylindrical traverse grinding is

$$G \approx \frac{D_w \bullet v_a \bullet a}{D_s \bullet v_r \bullet b} = \frac{D_w \bullet a \bullet N_w}{D_s \bullet v_r} \tag{6.2d}$$

The equivalent wheel diameter D_{eq} is based on matching the contact conditions in cylindrical grinding to those in surface grinding. This is done by matching the contact curvatures. The curvature of the wheel is considered positive, and that of the work is either positive (internal) or negative (external). In terms of diameters, the average difference in curvatures is:

$$\frac{2}{D_{eq}} = \frac{2}{D_s} - \text{workpiece curvature} = \frac{2}{D_s} \pm \frac{2}{D_w} \text{ or}$$

$$D_{eq} = \frac{D_w D_s}{D_w \pm D_s} \text{(+ for external, - for internal)} \tag{6.12}$$

This allows results from surface grinding to be used for cylindrical grinding using this equivalency. For example, some of the results from Section 6.3.1 that are affected include:

$$l_c = \frac{2(D_w \pm D_s)}{D_w D_s} \cos^{-1}\left(1 - \frac{2(D_w \pm D_s)}{D_w D_s}\right) \text{(+ for external, - for internal) or} \tag{6.3c}$$

$$= \sqrt{\frac{D_w D_s a}{D_w \pm D_s}} \quad \text{if } 2a \ll D_s \text{ (+ for external, - for internal).} \tag{6.3d}$$

$$h_{max} = 2\, s_g \sqrt{\frac{a\,(D_w \pm D_s)}{D_w D_s}} = 2\left(\frac{v_w}{K \bullet v_s}\right)\sqrt{\frac{a\,(D_w \pm D_s)}{D_w D_s}}$$

$$\text{(+ for external, - for internal)} \tag{6.5c}$$

6.4 Calculations for Planning a Grinding Cycle

In planning a grinding process, common calculations are to check the power and force limits of a particular grinding setup and determine the grinding cycle which is comprised of the feed and sparkout time. Empirical methods are the most common for these purposes, and several approaches will be outlined. The forces and compliances are needed in the development of the sparkout time calculations, and this is the first time compliance is introduced for models of machining or grinding.

6.4.1 Estimating Power, Forces

Figure 6.7 serves as a reference for the force calculations, with the notation and location for the forces in Figure 6.7b. The forces are assumed to act along the line through the centers of the wheel and workpiece; this is not exactly correct but quite close.

Table 6.3 Approximate Empirical Values for $\frac{F_t}{F_n}$

Wheel Condition	Sharp or Freshly Dressed	Constant G-Ratio	Dull or Loaded Wheel
Approximate Value for μ_g	0.7	0.5	0.3

Table 6.4 Typical Grinding Conditions Comparing Conventional and Creep Feed

Type of Grinding	v_s(m/s)	v_w(m/s)	*a* (mm/rev or stroke)	Grinding Allowance (mm)
Conventional	20-60	0.01 - 1.0	0.012-0.050	0.10-0.75
Creep Feed	20-60	0.0002-0.025	1.0-25	2-20

Power and forces can be estimated using the specific grinding energy which is like the specific cutting energy used for cutting. First the power is estimated as a function of the specific grinding energy and the material removal rate, then the tangential force F_t is estimated by assuming all the power is used to rotate the spindle, and finally the normal force F_n is calculated. The power at *grinding wheel* spindle is

$$P = E_g Z_w \tag{6.13a}$$

$$= F_t v_s = T_s (2\pi N_s). \tag{6.13b}$$

Values for the specific grinding energy E_g are listed in Table 5.2. E_g is assumed to be a property of the material and the size of the swarf. Examples of how to estimate the material removal rate for some of the different grinding processes were given in Section 6.3. Next, the normal grinding force F_n, is estimated using the empirical force ratio coefficient μ_g

$$F_n = \frac{F_t}{\mu_g}. \tag{6.13c}$$

Average values for μ_g are related to the wheel dressing condition described in Figure 6.3 and are given in Table 6.3. This method of calculation is very much an approximation, but many times it is sufficient for initial planning. Referring back to Chapter 5, the constants to determine corrections for edge geometry, chip thickness and wear are not used in a quantitative way. Table 6.3 is the only place where the grinding ratio is taken into account, and in a very approximate way.

The material and wheel removal parameter method assumes that the normal force F_n drives the grinding process and causes material to be removed. As a result, this method starts with F_n and then calculates F_t. The defining relationship with this method is really another way of writing the grinding ratio

$$G = \frac{Z_w}{Z_s} = \frac{\Lambda_w(F_n - F_{n0})}{\Lambda_s(F_n - F_{n0})}. \tag{6.14a}$$

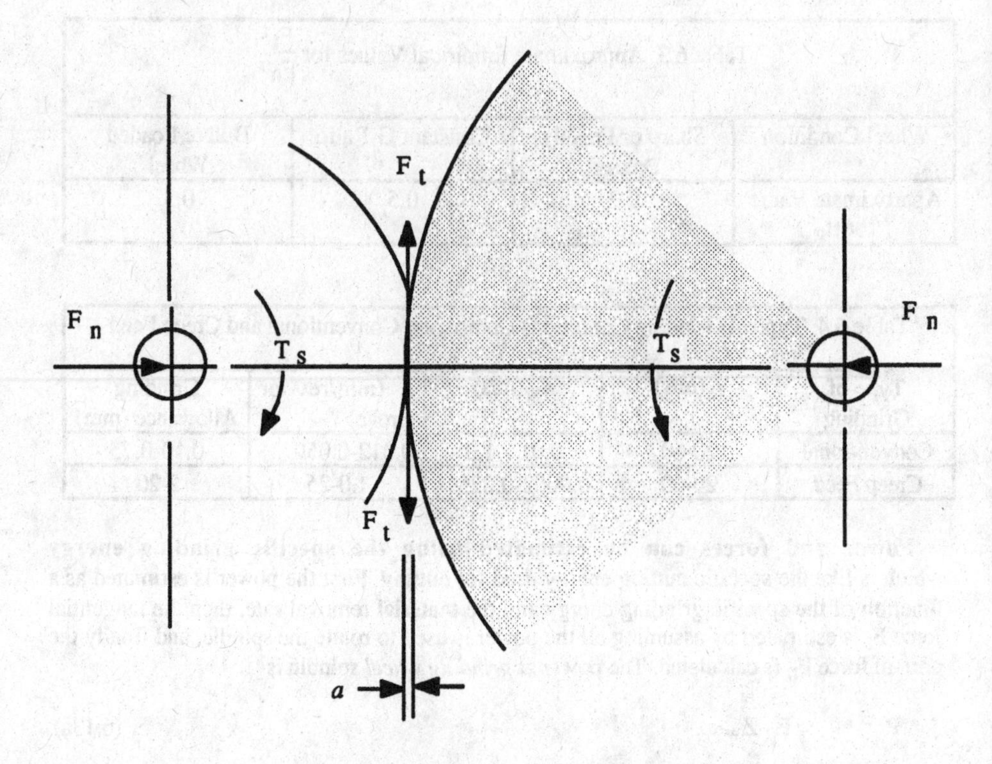

Figure 6.7a Grinding forces and torques.

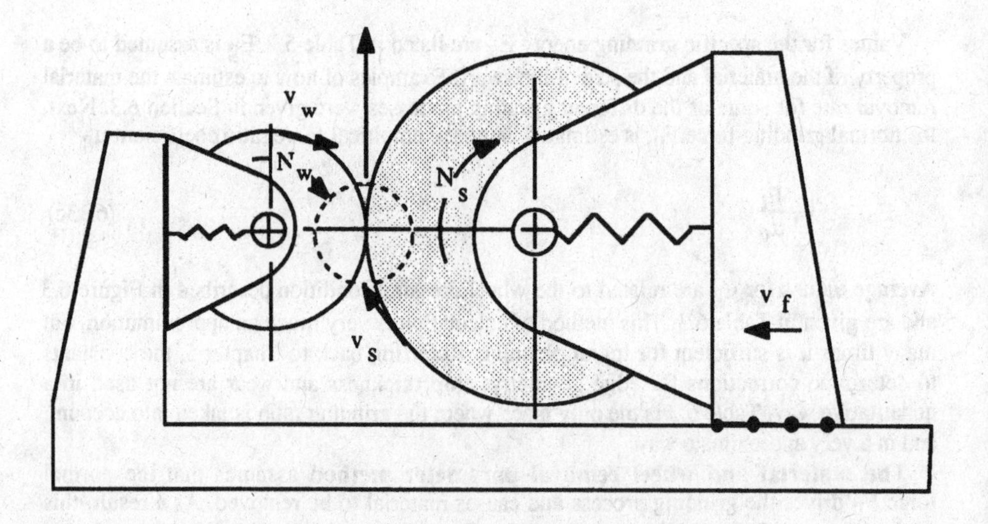

Figure 6.7b Cylindrical Plunge grinding use to analyze forces.

Figure 6.7 Idealization used to analyze grinding forces.

Λ_W is the workpiece removal parameter that indicates how the material removal rate increases for a unit increase in normal force, Λ_S is the wheel removal parameter of proportionality between normal force and the volumetric wheel wear rate, and F_{no} is the normal force threshold that has to be exceeded before material is removed from the workpiece. Going back to Eq (6.1), this force threshold is the same as saying that $E_{plowing}$ and $E_{rubbing}$ contribute to the energy consumed but not to the material removed. This approach requires at least three empirical parameters: Λ_W, Λ_S, and F_{n0} to find F_n from

$$F_n = \frac{Z_W}{\Lambda_W} + F_{n0}, \tag{6.14b}$$

and then Eq (6.13c) to find F_t

$$F_t = \mu_g F_n. \tag{6.14c}$$

Power and torque at the wheel can be estimated this way using Eq (6.13b). The specific grinding energy E_g and the material removal parameter Λ_W are related by

$$\Lambda_W \approx \frac{v_S}{\mu_g E_g} \tag{6.14d}$$

so Λ_W is not a material property. Because the spindle speed and wheel size for most production grinders is fixed so that v_S is nearly a constant, this approach is useful nevertheless.

The grinding stiffness approach assumes that the normal force is proportional to the radial depth,

$$F_n = k_g a \tag{6.15a}$$

Once F_n is known, the tangential force can be found from Eq (6.14c). This approach will be used in the next section to estimate sparkout time. A similar approach for cutting processes was introduced in Chapter 5 and will be used again in Chapter 8 on dynamics and vibration. The term stiffness is used because k_g has dimensions of and is idealized as a spring when used in modeling the grinding interface. As an empirical quantity, the problem is always finding values for k_g. Ways to estimate it from either E_g or Λ_W are

$$k_g \approx \frac{E_c\, b\, v_W}{\mu\, v_S} \approx \frac{b\, v_W}{\Lambda_W} \tag{6.15b}$$

6.4.2 Sparkout Time

Because the normal force F_n acts to separate the wheel from the work, if there is compliance in the system the normal force and material removal rate do not go to zero immediately when the $v_f = 0$. Instead the force decays until any elastic deformation in the system is relieved by removing workpiece material. While the wheel is removing material it is generating sparks so this part of the grinding cycle is termed *sparkout*. The time for sparkout to occur has to be planned in the grinding cycle, because if it isn't, the dimensions, shape and surface finish for a part will not meet technical specifications.

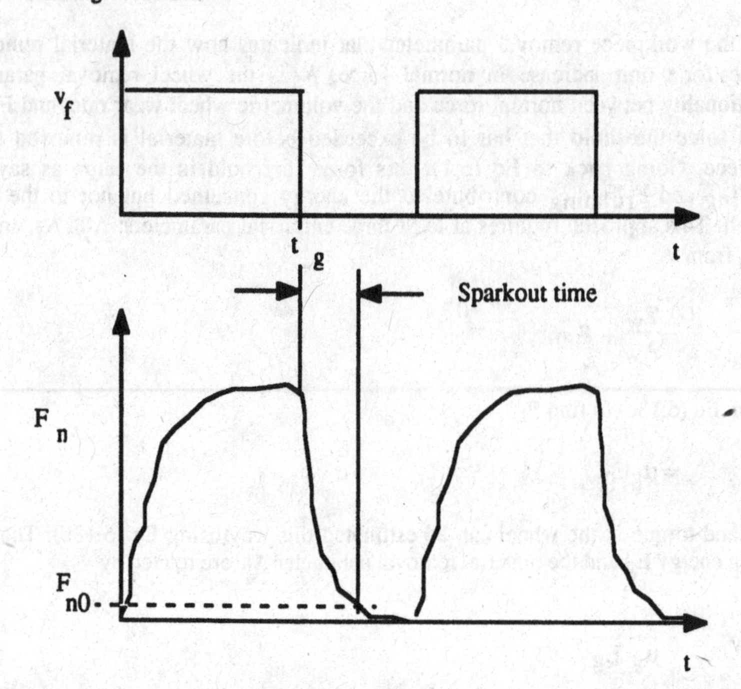

Figure 6.8a Grinding cycle.

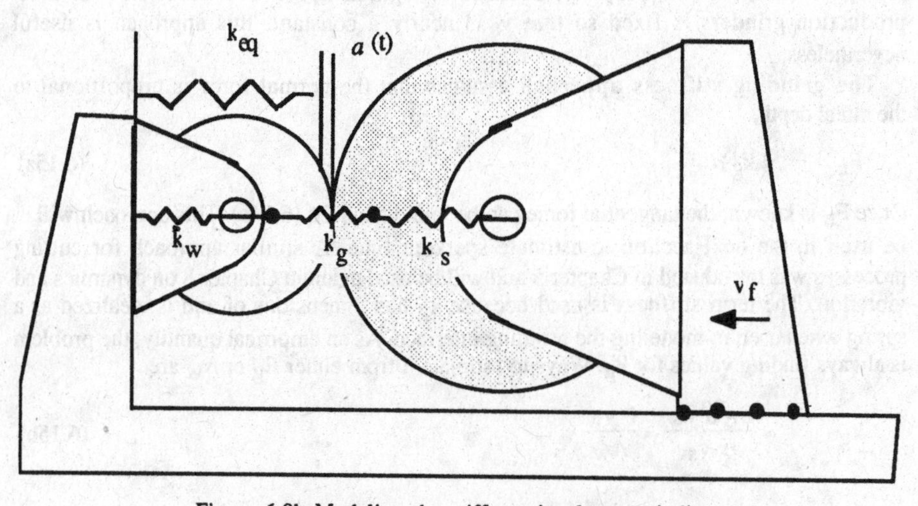

Figure 6.8b Modeling the stiffness in plunge grinding.

Figure 6.8 Model to estimate sparkout.

Figure 6.8 shows an idealization used to model the sparkout for cylindrical plunge grinding and how v_f and F_n would appear as a function of time.

The first step is to determine the equivalent stiffness made up of the workpiece stiffness k_w, the wheel and machine stiffness k_s, and what is termed the contact stiffness at the interface k_a. The equivalent stiffness is

$$\frac{1}{k_{eq}} = \frac{1}{k_s} + \frac{1}{k_w} + \frac{1}{k_a} \tag{6.16a}$$

and is lumped together as in Figure 6.8b. The normal grinding force, modeled by Eq (6.15a) causes the the workpiece to deflect by an amount $\Delta r(t)$, so the *instantaneous normal force* is

$$F_n(t) = k_g\, a\,(t) = k_{eq}\,\Delta r(t). \tag{6.16b}$$

The instantaneous radial depth $a(t)$, is the relative velocity between the wheel advance v_f and the rate that the radius of the workpiece reduces $(d\Delta r(t)/dt)$, divided by N_w or

$$a\,(t) = \frac{v_f(t) - \frac{d\Delta r(t)}{dt}}{N_w} \tag{6.16c}$$

When Eq (6.16b) is used to eliminate $a(t)$, the first order differential equation for the workpiece radius reduction is

$$\frac{d\Delta r(t)}{dt} = -N_w \frac{k_{eq}}{k_g}\Delta r(t) + v_f(t) = \frac{-1}{\tau}\Delta r(t) + v_f(t),\ \text{where}\ \tau = \frac{k_q}{k_{eq}\,N_w} \tag{6.16d}$$

The particular solution for Eq (6.16d) for a grinding cycle, as Figure 6.8a shows, has a segment where the infeed velocity is constant, i.e., $v_f(t) = v_{fo}$, $0 < t < t_g$ and the sparkout time where $v_f(t) = 0$, $t \geq t_g$. For estimating sparkout time, the solution to the free response that describes the decreasing part of the curve is

$$\Delta r(t) = \frac{k_q\, v_{fo}}{k_{eq}\, N_w} \exp\left(\frac{-k_{eq}\, N_w\,(t-t_g)}{k_q}\right) \text{or}\ F_n(t) = \frac{k_q\, v_{fo}}{N_w} \exp\left(\frac{-k_{eq}\, N_w\,(t-t_g)}{k_q}\right) \tag{6.16e}$$

Finding the sparkout time, $(t-t_g)$ requires a criterion for either the radial deflection Δr^* - for example the tolerance on the diameter - or a specification on the normal force $F_n{}^*$ - F_{n0} in Eq (6.14a) is often used because forces below that level don't remove any material. As an example using the force threshold criteria, the sparkout times from Eq (6.16e) is

$$t_{sparkout} = (t-t_g) = -\tau \ln\left(\frac{k_q\, v_{fo}}{N_w\, F_{n0}}\right)$$

This reveals that reduced sparkout times require low grinding stiffness k_q - a workpiece material characteristic, high equivalent stiffness k_{eq} - a grinding machine-workpiece-workholding characteristic dominated by the weakest link in the design, and the rotational speed of the workpiece N_s, which is an operating variable that can be adjusted.

A similar difference equation solution can be developed for surface grinding,

$$\Delta r(n) = \frac{k_q a}{k_{eq}}\left(\frac{1}{1+\frac{k_{eq}}{k_q}}\right)^n \quad \text{or } F_n(n) = k_q a\left(\frac{1}{1+\frac{k_{eq}}{k_q}}\right)^n. \tag{6.16f}$$

where a is the radial infeed that is set and n is the number of strokes of the wheel. These expressions can be solved to find the number of strokes needed for the deflection or normal force to reach a specified threshold for plunge surface grinding. As should be the case, the way to reduce the number of strokes for sparking out is a high k_{eq} and an easy to grind workpiece.

6.4.3 Estimating Grinding Temperatures

The thermal model to estimate grinding temperatures is from Section 3.3, but now some of the quantities used in the calculations are available from calculation.

Repeating the expression for the composite properties of the grinding wheel,

$$k_s = \varepsilon \bullet k_f + (1-\varepsilon) \bullet k_g \tag{3.19a}$$

$$c_s = \varepsilon \bullet c_f + (1-\varepsilon) \bullet c_g \tag{3.19b}$$

$$\rho_s = \varepsilon \bullet \rho_f + (1-\varepsilon) \bullet \rho_g \tag{3.19c}$$

$$\kappa_s = \frac{k_s}{\rho_s \bullet c_s} \tag{3.19d}$$

The subscript "s" refers to the surface of the wheel, "f" to the grinding fluid that fills the voids, and "g" refers to the grits. The porosity, ε is used to average the properties of the abrasives and the grinding fluids. One way to make this calculation is in terms of the space available between the grits and the thermal boundary layer. This approach, idealized in Figure 6.9a, is used by Lavine [Lavine88]. For this idealization, the grits of diameter d_g, computed from Eq (6.2), are assume to be packed together and the depth into the wheel is defined by the average thermal boundary layer depth $\bar{\delta}$

$$\bar{\delta} = 2.1\sqrt{\frac{\kappa_s\, l_c}{v_s}} \tag{6.17a}$$

is used to estimate the porosity,

$$\varepsilon = 1 - \frac{4}{3}(\bar{\delta}\, K^2)\left(\frac{3\, d_g}{2} - \bar{\delta}\right) \tag{6.17b}$$

K is the circumferential spacing of active grit, first used in Eq (6.4). Another approach, based on the histogram P(x) of the measured profile of a grinding wheel, is illustrated in

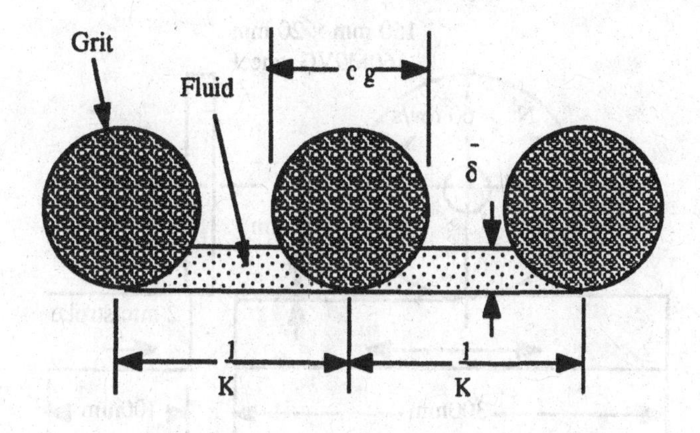

Figure 6.9a Geometric and thermal estimate of porosity.

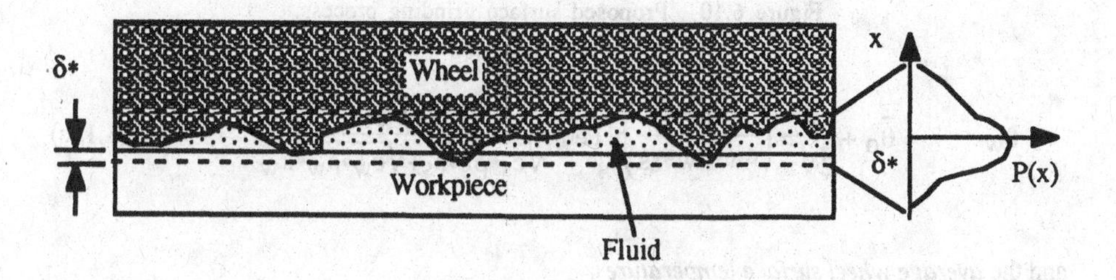

Figure 6.9b Empirical and mechanics estimate of porosity.

Figure 6.9 Ways to estimate porosity.

Figure 6.9b. This method determines δ^*, the depth into the abrasive needed to support the normal load F_n, and uses this mechanics based depth to estimate the porosity,

$$\varepsilon = 1 - \int_{\infty}^{\delta^*} P(x)\, dx \qquad (6.17c)$$

Both methods give values near one for porosity, meaning that in theory there is a lot of room for fluid in the grinding zone [Can*etal*89].

The Peclet number used both to determine if it is appropriate to ignore conduction through the grind zone area and in the partition of energy between the wheel and workpiece is

$$Pe_s = \frac{l_c \bullet v_s}{\kappa_s} = \frac{l_c \bullet v_s \bullet c_s \bullet \rho_s}{k_s} > 20 \qquad (3.17b)$$

and depends on the contact length l_c estimated from Eqs (6.3) for both surface and cylindrical grinding. For creep feed grinding where *a* is large and v_w is small compared to conventional grinding, Eq (6.3a or c) should probably be used to estimate the geometric contact length. Estimating the *average workpiece surface temperature*

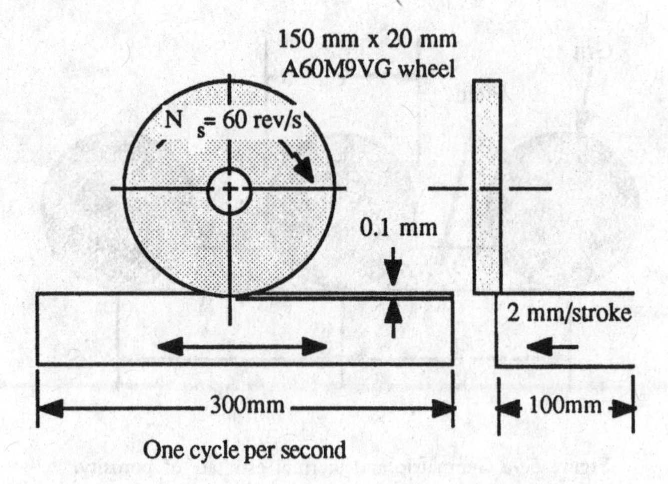

Figure 6.10 Proposed surface grinding process.

$$\bar{\theta}_w = \bar{\theta}_o + \frac{4}{3\sqrt{\pi}}\left(\frac{q_g}{c_w \bullet \rho_w \bullet v_s}\right)\sqrt{Pe_w}\frac{\sqrt{c_w \bullet \rho_w \bullet k_w}}{\sqrt{c_s \bullet \rho_s \bullet k_s}+\sqrt{c_w \bullet \rho_w \bullet k_w}} \quad (3.15e)$$

and the *average wheel surface temperature*

$$\bar{\theta}_s = \bar{\theta}_o + \frac{4}{3\sqrt{\pi}}\left(\frac{q_g}{c_s \bullet \rho_s \bullet v_s}\right)\sqrt{Pe_s}\frac{\sqrt{c_s \bullet \rho_s \bullet k_s}}{\sqrt{c_s \bullet \rho_s \bullet k_s}+\sqrt{c_w \bullet \rho_w \bullet k_w}} \quad (3.15f)$$

requires estimating the average heat flux $q_g = (F_t \bullet v_s)/(l_c \bullet b)$. The term $(F_t \bullet v_s)$ is the power at the spindle, which can be estimated by any of the methods in Section 6.4.1. If a grinding machine is instrumented with a power meter, these actual measurements can be used, after correcting for losses and transmission efficiencies, because like machining, the power to feed the wheel is negligible.

Example 6.1 A Proposed Surface Grinding Setup

Figure 6.10 shows a setup that is proposed for surface grinding a 4340 plate with dimensions of 200 mm x 300 mm x 50 mm and that has a Brinnell hardness of 300 kg/mm^2. The ground surface should have an R_a surface roughness of less than 0.1 μm and the average workpiece temperature should be below 100 °C to assure that instantaneous temperatures are below the workpiece recrystalization temperature. The surface grinder provides 5 kW at the spindle and rotates at 60 rev/s.

Check the process plan to determine if the power at the spindle will be exceeded, the average workpiece temperature is less than 100 °C and the theoretical surface roughness is less than the limit.

Begin by estimating the theoretical surface roughness using the analogy between slab milling and surface grinding. This means determining s_g from Eq (6.4), which in turn means finding the spacing between the active number of grits. Assume that the grits are tightly packed together, but only every tenth grit is actually cutting, so using Eq (6.2) to estimate the grit diameter

$$K \approx \frac{1}{10\,d_g} = \frac{60}{10\,(25.4\text{ mm/in})(0.6)} = 0.394 \text{ active grits/mm.}$$

The wheel speed is $v_s = \pi$ (150 mm)(10^{-3} m/mm)(60 rev/s) = 28.3 m/s. The workpiece speed can be estimated from Eq (6.7b). Since one cycle of the surface grinder is two strokes, one forward and one back, the 1 cycle per second means that $t_s(v_w)$ = 0.5 s/stroke so

$$v_w = \frac{10^{-3}\text{ m/mm}\left(150\text{ mm}\sqrt{\frac{4(0.1\text{mm})}{(150\text{ mm})}\left(1-\frac{(0.1\text{mm})}{(150\text{ mm})}\right)} + 300\text{ mm}\right)}{0.5\text{ s/stroke}} = 0.615\text{ m/s}$$

The feed per grit is s_g = (0.615 m/s)÷((0.394 active grits/mm)(28.3 m/s)) = 0.055 mm/active grit, and from Eq (5.2n), where the wheel radius is used instead of the milling cutter radius, the theoretical R_a roughness is

$$R_a = \frac{(0.055\text{ mm})^2(10^3\ \mu\text{m/mm})}{18\sqrt{3}(150\text{ mm})} = 0.001\ \mu\text{m.}$$

Clearly this is less than the 0.1 μm limit. Remember the approximations used in estimating the grit spacing, so the measured roughness would probably be higher, but it is unlikely that it would exceed 0.1 μm.

Next estimate the power using the specific grinding energy method. The only data available in Table 5.2 for 4340 with a hardness of $H_B \approx 300$ is for a large swarf, E_g = 29.78 J/mm^3. The material removal, neglecting wheel wear is Z_w = (0.1 mm)(0.615 m/s)(10^3 mm/m)(2 mm) = 123.1 mm^3/s, so the estimated power needed at the spindle is

$$P = (29.78\text{ J/mm}^3)(123.1\text{ mm}^3\text{/s})(10^{-3}\text{ kW/J/s}) = 3.67\text{ kW} = (F_t\, v_s)$$

which is less than the power rating for the spindle.

The contact length is one of the quantities needed to estimate the average workpiece temperature. The contact length from Eq (6.3a) is 3.8734 mm and from Eq (6.3b) l_c = 3.8730 mm; nearly the same. As a result the estimated heat flux at the interface is

$$q_g = \frac{3.67\text{ kW}}{(2\text{ mm})(3.8730\text{ mm})} = 0.473\text{ kW/mm}^2$$

While the porosity could be calculated using Eqs (3.19) and Eqs (6.17), assume that ε = 0.90, and the thermal properties for the aluminum oxide abrasive and the 4340 are the same as those in Example 3.2. The fluid in this case is a water based emulsion that has approximately the same properties as water. These properties are listed in Table 6.5.

From this data, the composite properties at the wheel surface are:

$$k_s = 0.95(0.613\text{ W/m-K}) + (1\text{-}0.95)(6.74\text{ W/m-K}) = 1.226\text{ W/m-K}$$

Table 6.5 Thermal Properties of Abrasive, Workpiece and Grinding Fluid

Property	Abrasive AlO_2	Workpiece 4340	Fluid Water Based Emulsion
Density (kg/m^3)	3800	7800	997
Heat Capacity (J/g-K)	0.77	0.574	4.18
Thermal Conductivity (W/m-K)	6.74	38.9	0.613

$$c_s = 0.95(4.18 \text{ J/g-K}) + (1\text{-}0.95)(0.77 \text{ J/g-K}) = 3.84 \text{ J/g-K}$$

$$\rho_s = 0.95(997 \text{ kg/m}^3) + (1\text{-}0.95)(3800 \text{ kg/m}^3) = 1277 \text{ kg/m}^3$$

so the Peclet number at the surface of the wheel is $Pe_s = 43{,}800 > 20$ and for the workpiece $Pe_w = 12{,}600$. If the ambient temperature is $\bar{\theta}_0 = 30$ °C, then the average workpiece temperature from Eq (3.15e) is

$$\bar{\theta}_w = 30\ °\text{C} + \frac{4}{3\sqrt{\pi}}\left(\frac{0.473 \text{ kW/mm}^2}{(0.574 \text{ J/g-°K})(7800 \text{ kg/m}^3)(28.3 \text{ m/s})}\right)\sqrt{12600}$$

$$\times \frac{\sqrt{(0.574\text{J/g-K})(7800\text{kg/m}^3)(38.9\text{W/m-K})}}{\sqrt{(3.84\text{J/g-K})(1277\text{kg/m}^3)(1.226\text{W/m-K})}+\sqrt{(0.574\text{J/g-K})(7800\text{kg/m}^3)(38.9\text{W/m-K})}}$$

$$= 75.8\ °\text{C}$$

This calculation indicates that there will not be any problem with the workpiece, based on the 100 °C criterion. However, if the specific grinding energy is above 60 J/mm^3, which is possible since h_{eq} is only 0.002 mm, the estimated temperature exceeds 100 °C.

Example 6.2 Planning a Cylindrical Grinding Cycle.

The outer diameter of the inner race of a bearing is produced by external cylindrical plunge grinding using a wheel that has the form of the track for the balls dressed into the wheel. For planning purposes, this contour can be neglected and the process idealized as in Figure 6.11 . The same type of wheel and spindle used in Example 6.1 is used in this example: with $D_s = 150$ mm , $w = b = 10$ mm, $P_{max} = 5$ kW and $N_s = 60$ rev/s so $v_s = 28.3$ m/s. The workpiece material in this case is 52100 steel that has been hardened to H_B =

500 and the specific grinding energy for this material is 55 J/mm^3. The contact, machine, and workpiece stiffnesses are 10 kN/mm, 100 kN/mm and 20 kN/mm, respectively.

Estimate the radial feedrate v_r that will use all the available spindle power if $\mu_g = 0.5$. Then estimate the grinding and sparkout times, assuming that the radial grinding allowance is 0.2 mm and the roundness tolerance of 0.01 mm can be used to determine sparkout.

The easiest solution, the one that will be given, neglects wheel wear and assumes that the diameter reduction of 2x(0.0.2 mm)/50 mm = 0.008 is small. Then the radial feedrate can be estimated by using Eq (6.14b) for Z_w and finding the v_{f0} that uses all the spindle power,

$$v_{f0} = \frac{5\ \text{kW}(10^3\ \text{W/kW})}{\pi(55\ \text{J/mm}^3)(10\ \text{mm})(50\ \text{mm})} = 0.058\ \text{mm/s}.$$

The grinding time is simply the time it will take to advance the wheel radially by an amount equal to the grinding allowance , or

$$t_g = \frac{0.2\ \text{mm}}{0.058\ \text{mm/s}} = 3.456\ \text{s}.$$

The sparkout time calculation substitutes $\Delta r^* = 0.01$ mm for Δr in Eq (6.16e) and solves for (t-tg). The stiffnesses and the time constant are need for this calculation. The equivalent structural stiffness is

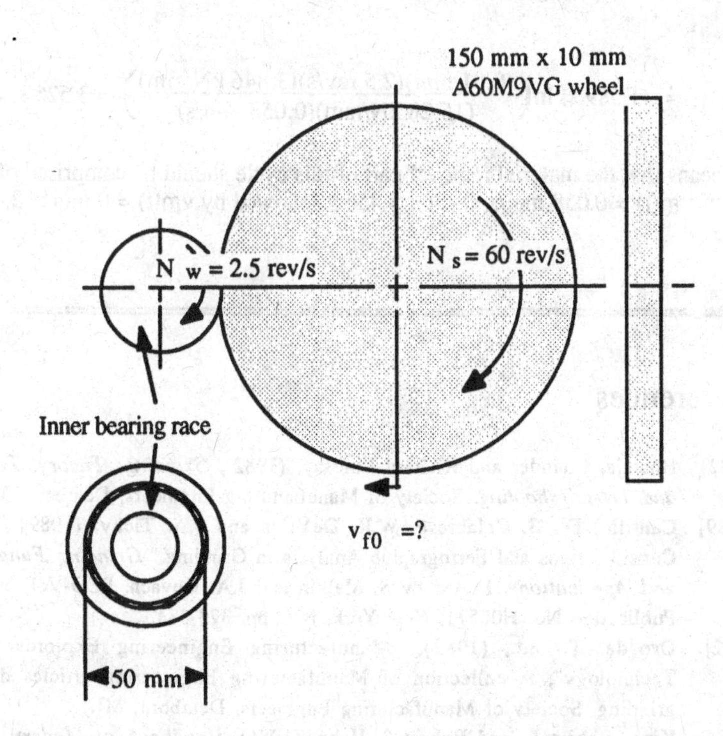

Figure 6.11 Idealization of plunge grinding for inner race of a bearing.

$$k_{eq} = \frac{1}{\left(\frac{1}{5\ \text{kN/mm}}\right)+\left(\frac{1}{100\ \text{kN/mm}}\right)+\left(\frac{1}{20\ \text{kN/mm}}\right)} = 3.846\ \text{kN/mm}$$

and the grinding stiffness from Eq (6.16a) is

$$k_g = \frac{(55\ \text{J/mm}^3)(10\ \text{mm})(0.393\ \text{m/s})}{0.5\ (28.3\ \text{m/s})} = 15.28\ \text{kN/mm},$$

from which the time constant is

$$\tau = \frac{(15.28\ \text{kN/mm})}{(2.5\ \text{rev/s})(3.846\ \text{kN/mm})} = 1.589\ \text{s}.$$

The sparkout time with this data is

$$t_{sparkout} = -\tau \ln\left(\frac{\Delta r^*\ N_w\ k_{eq}}{k_g\ v_{f0}}\right)$$

$$= -(1.589\ \text{s}) \ln\left(\frac{(0.01\ \text{mm})(2.5\ \text{rev/s})(3.846\ \text{kN/mm})}{(15.28\ \text{kN/mm})(0.058\ \text{mm/s})}\right) = 3.525\ \text{s}.$$

This means that the material removal part of this cycle should be comprised of a radial advance of $v_{f0}(t) = 0.058$ mm/s, $0 < t < 3.456$ s followed by $v_{f0}(t) = 0$ mm/s, $3.456\ \text{s} < t < 6.981$ s.

6.5 References

[BhaLin82] Bhateja, Chander and Richard Lindsay, (1982), *Grinding: Theory, Techniques and Troubleshooting*, Society of Manufacturing Engineers, Dearborn, MI.

[Can*etal*89] Cantillo, D., S. Calabrese, W.R. DeVries and J.A. Tichy, (1989), "Thermal Considerations and Ferrographic Analysis in Grinding," *Grinding Fundamentals and Applications*, Edited by S. Malkin and J.A. Kovach, PED-Vol. 39, ASME Publication No. H00571, New York, NY, pp. 323-334.

[Drozda82] Drozda, T., ed., (1982), "Manufacturing Engineering Explores Grinding Technology", A collection of Manufacturing Engineering articles discussing grinding, Society of Manufacturing Engineers, Dearborn, MI.

[KinHah86] King, Robert I., and Robert S. Hahn, (1986), *Handbook of Modern Grinding Technology*, Chapman and Hall, New York and London.

[Kiretal77] Kirk, J. A., Cardenas-Garcia, J.F., and Allison, C.R., (1977), "Evaluation of Grinding Lubricants - Simulation Testing and Grinding Performance," *Wear*, Vol.20, No.4, pp.333-339.

[KirCar77] Kirk, J.A., Cardenas-Garcia, J.F., (1977), "Evaluation of Grinding Lubricants-Simulation Testing and Grinding Performance," *Transactions of the American Society of Lubrication Engineers*, Vol. 20, pp.333-339.

[Lavine88] Lavine, Adrienne S., (1988), "A Simple Model for Convective Cooling During the Grinding Process," *ASME Transactions, Journal of Engineering for Industry*, Vol. 109, pp.1-6.

[LinBha82] Lindsey, Richard and Chander Bhateja, Ed.s, (1982), *Grinding-Theory, Technique and Troubleshooting*, SME.

[Malkin88] Malkin, S., (1988), "Grinding Temperatures and Thermal Damage," *Thermal Aspects in Manufacturing*,, PED-Vol. 30, ASME, New York, NY, pp. 145-156.

[Malkin89] Malkin, S., (1989), *Grinding Technology: Theory and Applications of Machining With Abrasives*, Ellis Horwood, Ltd. Chichester, England.

[Shaw90] Shaw, Milton C., (1990), "Cutting and Grinding: A Comparison," Fourth International Grinding Conference, Dearborn, MI, SME Paper #MR90-500, October 9-11.

[TicDeV89] Tichy, J.A. and W.R. DeVries, (1989), "A Model for Cylindrical Grinding Based on Abrasive Wear Theory," *Grinding Fundamentals and Applications*, Edited by S. Malkin and J.A. Kovach, PED-Vol. 39, ASME Publication No. H00571, New York, NY, pp. 335-347.

[Trietal77] Tripathi, K.C., Nicol, A.W., Rowe,G.W., (1977), "Observation of Wheel-Metal-Coolant Interactions in Grinding," *Transactions of the American Society of Lubrication Engineers*, Vol. 20, pp.249-256.

[Woodbu59] Woodbury, Robert S., (1959), *History of the Grinding Machine, A Historical Study in Tools and Precision Production*, The Technology Press, MIT, Cambridge, MA

6.6 Problems

6.1 For the two cylindrical grinding processes described below, determine the equivalent grinding wheel diameter D_{eq}, the equivalent chip thickness h_{eq}, and the material removal rate Z_w.

a. Internal cylindrical grinding of a workpiece to a 150 mm diameter is done using a 75 mm wheel that is 10 mm wide. There is a 0.5 mm grinding allowance. The wheel rotational speed is $N_s = 125$ rev/s, the workpiece rotational speed is $N_w = 2$ rev/s, the axial feed is 0.5 mm per revolution of the workpiece, and the radial depth increases 0.25 mm per axial stroke of the grinding wheel.

b. To produce a part for a bearing, cylindrical plunge grinding is used with a 200 mm wheel that is 25 mm wide and rotating at 7200 rev/min. The workpiece is the outer diameter of a bearing that is 20 mm wide and has to be ground to a 40 mm diameter. The workpiece is rotating at 150 rev/min, there is a 0.5 mm grinding allowance and the radial in-feed is 0.1 mm/rev of the workpiece.

6.2 Suppose that a number of pieces of high speed steel tool stock have to be ground flat using a surface grinder as in Figure 6.12.

a. Estimate the power required during a stroke of the workpiece table using the specific grinding energy method.

b. Assume that 95 percent of the power used in this grinding operation is converted to heat which flows into the workpiece through the contact area between the wheel and workpiece. Estimate the energy flux into the workpiece and the average workpiece temperature if the wheel is aluminum oxide and a water based emulsion with the same properties as in Example 3.2 is used.

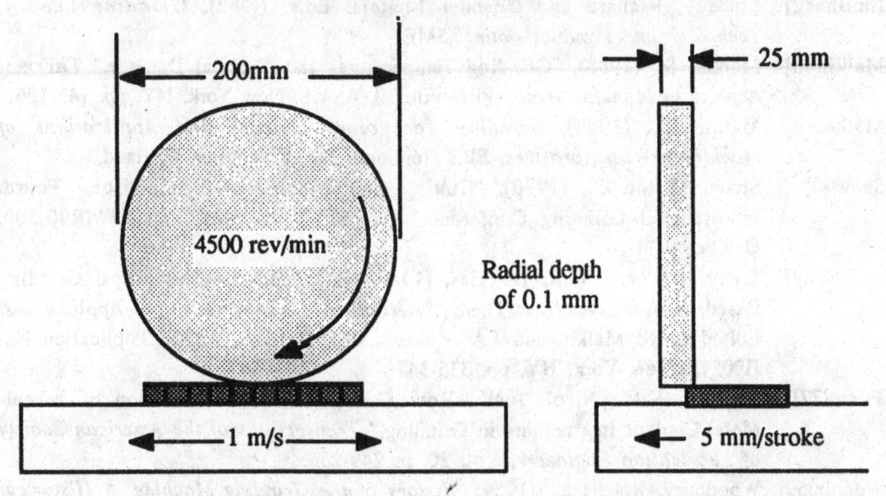

Figure 6.12 Surface grinding tool stock.

6.3 Cylindrical grinding will be the process used to remove the final 1.0 mm from the diameter of axles produced by a previous turning operation. The axles, some of the most expensive in the world, are made of 304 stainless steel and the spindles were set up on a grinder that has 4 kW available at the grinding wheel spindle. Other details for the grinding process are: the workpiece rotational speed is 2 rev/s, the radial feed is 0.25 mm/stroke and the axial feed is 2 mm per revolution of the workpiece. The grinding wheel code is 23A60-L5VBE, the wheel is 200 mm in diameter and 50 mm wide and is operated at a surface speed of 30 m/s.

a. Estimate the metal removal rate for this grinding operation.

b. Calculate the grinding equivalent chip thickness which is based only on the kinematics of the cut, the maximum chip thickness h_{max} which includes the abrasive size (Assume that the spacing between active grits is 10 grit diameters), and use the specific energy method to estimate the total power at the spindle for this grinding operation.

c. The grinding wheel-grinding fluid-workpiece combination results in a grinding ratio G = 1000. If the wheel will have to be dressed and trued when the radial wheel wear is 0.05 mm, how many axles can be made between dressing cycles? (Remember an axle needs 2 of these bearing surfaces and neglect any wear during sparkout that may occur on the return portion of a stroke.)

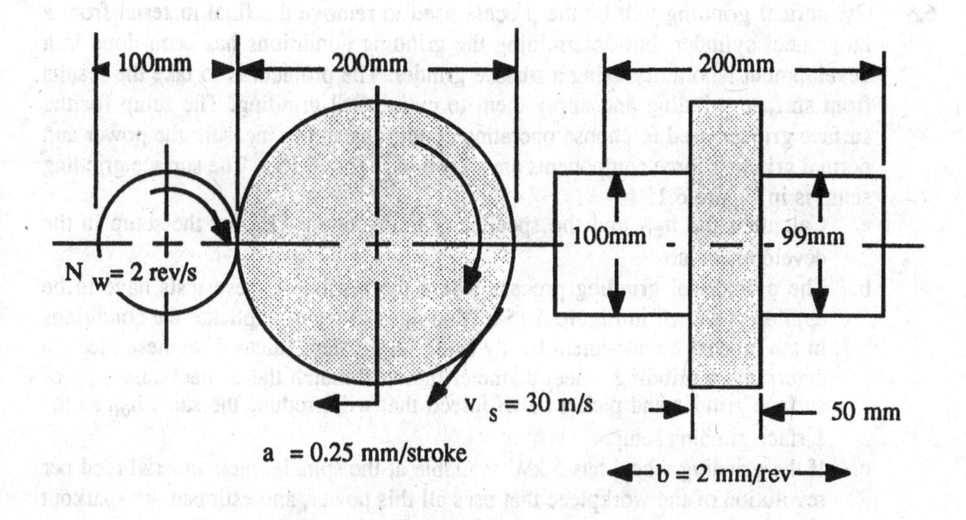

Figure 6.13 Traverse cylindrical grinding of one end of an axle.

6.4 A 100 mm diameter, 15 mm wide A80-M12V grinding wheel is used for plunge surface grinding as shown in Figure 6.14 . The equivalent stiffness of this setup has been measured as 500 MN/mm. The workpiece is a 100 mm x 250 mm piece of 4340 steel hardened to 450 Brinell. The work speed is 0.25 m/s, the rotational speed of the grinding wheel is 3600 rev/min and the radial depth set is at 0.20 mm.

a. Estimate the time to grind this workpiece considering the number of strokes so that the actual infeed is equal to half the thickness tolerance of ± 0.01 mm.

b. If the wheel has been conditioned to the point the $(F_t/F_n) = 0.5$, do a worst case analysis and estimate the spindle power on the first stroke.

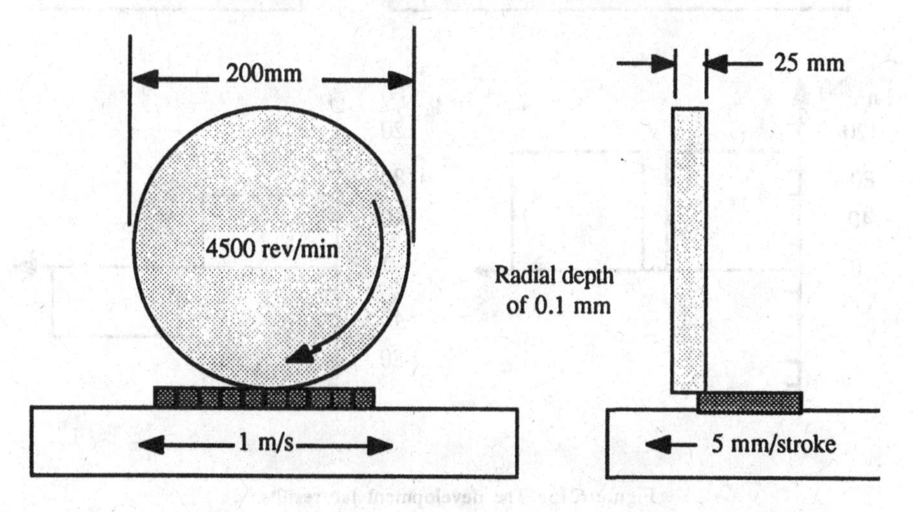

Figure 6.14 Plunge surface grinding for Problem 6.4.

6.5 Cylindrical grinding will be the process used to remove the final material from a large steel cylinder, but determining the grinding conditions has been done in a development laboratory using a surface grinder. The problem is to take the results from surface grinding and apply them to cylindrical grinding. The setup for the surface grinder used to choose operating conditions and to measure the power and normal grinding force components are shown in Figure 6.15 a. The surface grinding setup is in Figure 6.15 b.

a. Calculate the h_{eq} and the specific grinding energy E_c for the setup in the development lab.

b. The cylindrical grinding process where the results of these tests have to be applied is shown in Figure 6.15 a. You have to try to duplicate the conditions in the process development lab by using "equivalent" ideas. Use these ideas to determine a grinding wheel diameter that will match the contact curvature of surface grinder and pick a radial infeed that will produce the same h_{eq} as the surface grinding setup.

c. If the grinding wheel has 5 kW available at the spindle, pick an axial feed per revolution of the workpiece that uses all this power, and estimate the sparkout time for the radial force to reach 0.05 kN if the measured equivalent stiffness for the setup is K_{eq} = 1000 MN/mm.

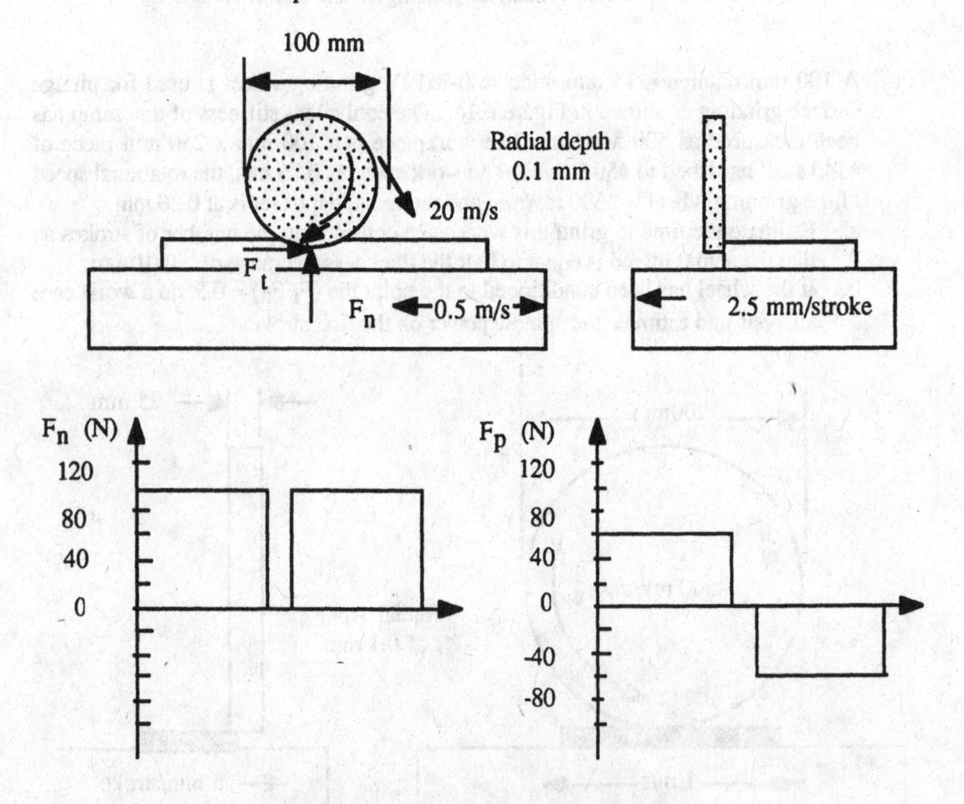

Figure 6.15a The development lab results.

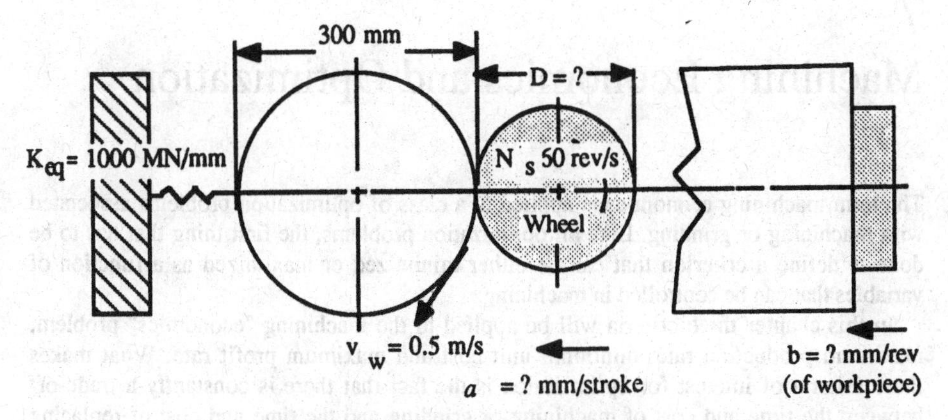

Figure 6.15b Proposed production setup.

Figure 6.15 Development lab and production grinding for problem 6.5.

7 Machining Economics and Optimization

The term machining economics is applied to a class of optimization problems associated with machining or grinding. Like all optimization problems, the first thing that has to be done is define a criterion that can be either minimized or maximized as a function of variables that can be controlled in machining.

In this chapter three criteria will be applied to the machining "economics" problem: maximum production rate, minimum unit cost and maximum profit rate. What makes these criteria of interest for optimization is the fact that there is constantly a trade-off between the time and cost of machining or grinding and the time and cost of replacing cutting edges or grinding wheels. The time and cost of machining or grinding can usually be reduced by operating at higher material removal rates, while the time and cost of replacing cutting edges or dressing grinding wheels increases at higher material removal rates. Machining economics applies optimization methods to the problem of selecting operating conditions. Quantitative methods can be used to select material removal rates are a compromise between times and costs that increase with material removal rates and those that decrease. Figure 7.1 shows the inputs constraints and methods used to determine cutting conditions based on these criteria.

Using turning as an example, the approach and solutions for the maximum production rate, minimum unit cost and maximum profit rate criteria will be developed. While other machining processes and grinding application will not be covered, these three criteria could be applied to them as well, as some of the references at the end of the chapter illustrate.

7.1 Terms and Notation for Turning Optimization

Before formulating the criteria that have to be minimized or maximized, some terms that are common to machining or grinding optimization have to be defined and the context of the optimization needs to be considered. The machining or grinding situations considered under the heading of machining economics are idealizations. The scope is quite narrow, the operation of a single machine tool, which may be part of a system, but the optimization is for the single machine and not the entire system. Associated with operating the machine tool are the functions of part loading, setup and unloading, tool changes to replace worn cutting edges, and tending to the normal operation of the machine tool. Because machining economics focuses on operating a single machine tool the terms and coefficients are on a per unit basis, e.g., unit times and costs for an operation or unit costs for tools.

The part loading and unloading time is denoted by a unit time per part t_{load} (s/unit). This may be the time it takes an operator to fixture and setup a workpiece, or the time for a robot/palletizing system to do the same thing. It is assumed that when this part of the manufacturing system is in operation no machining can be done, i.e., it doesn't allow staging the work on a separate table or spindle. Where does this data come from? For manual operations, t_{load} can be found from work standards, time and motion studies,

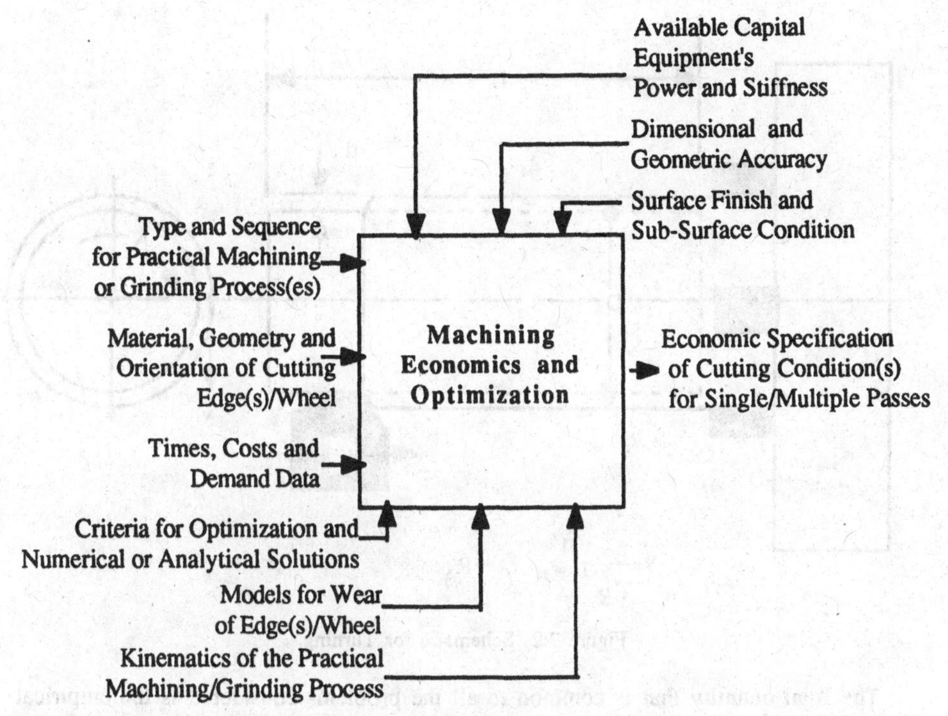

Figure 7.1 The Inputs, Constraints, Outputs and Mechanisms considered for the optimization of machining and grinding.

or estimated from experience with similar workpieces. For automated systems with pallet or robot part handling, this can be determined from the programmed cycle time.

Another associated time is that required to change a worn out cutting edge, t_{ch} (s/edge). The emphasis is on a single edge, so if there are N_t cutting edges on a milling cutter, the total time to change this cutter is assumed to be ($N_t \cdot t_{ch}$). As was the case with part loading and unloading, it's assumed that when cutting edges are being changed, no material is being removed. This time does not take into account changes in the cutting edge because a new type of tool is needed, e.g., changing from an end mill to a drill. Obtaining a value for the time to change a cutting edge, like the time to load and unload a workpiece, often relies on work standards or time and motion studies. If the change is done on a machining center where a new cutter can be mounted from a tool magazine, then t_{ch} is the time it takes to make the tool change divided by the number of cutting edges.

The time spent cutting and wearing out the cutting edge is denoted by t_c (s/unit). This time is a function of the machining process considered, and examples were given in Chapters 5 and 6 for computing these times for single machining or grinding passes. The general way that t_c is computed is by dividing the length of the pass by the feed rate v_f, i.e.,

$$t_c = \frac{\text{length of cut}}{\text{feed rate}} .$$

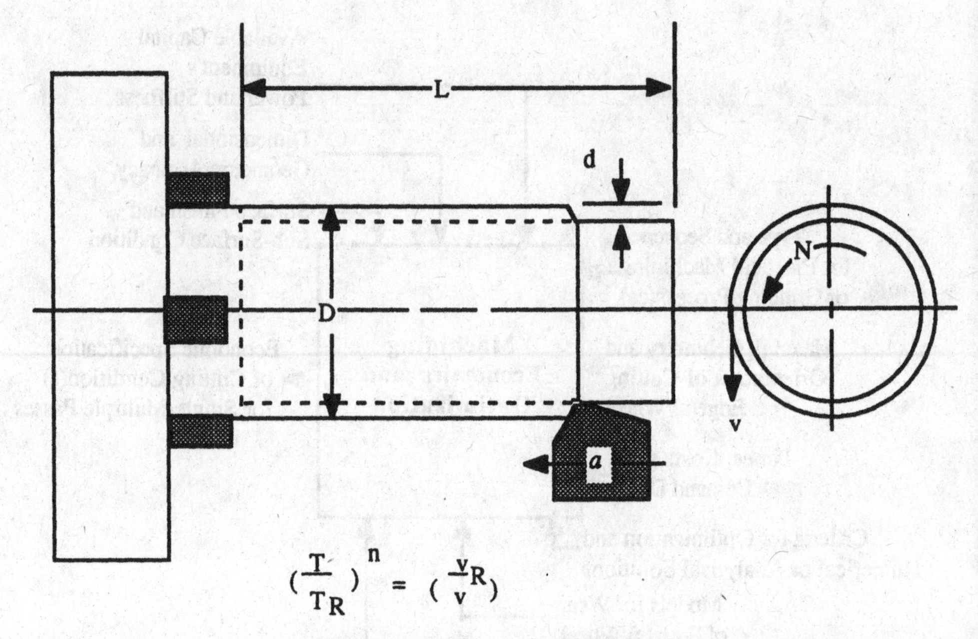

Figure 7.2 Schematic for Turning.

The final quantity that is common to all the problems considered is the empirical expression for the life of a cutting edge or grinding wheel. This chapter uses Taylor's tool life equation from Chapter 4,

$$\left(\frac{T}{T_R}\right)^n = \left(\frac{v_R}{v}\right). \tag{4.2a}$$

While the general tool life expression of Eq (4.2c) includes the other cutting conditions, namely, depth of cut and feed, examples in this chapter only use this simple form. The rationale for using Eq (4.2a), other than it simplifies things, goes like this: The depth of cut usually is chosen based on the geometry of a finished part or how many passes are needed based on power constraints. As a process planning variable, the feed is usually selected for two reasons: either surface finish for a final cut, or to limit excessive forces for a roughing cut. This leaves the cutting speed to be selected, and is the reason that this simpler tool life expression with cutting speed as the single variable is adequate in most cases. (As an aside, it also turns out that if a multi-variable optimization with constraints on power, force, and finish is done, the solution is usually such that the feed and depth are at constraint boundaries and the velocity is the free variable.)

Throughout this chapter the turning process in Figure 7.2 will be used. The optimization problems are all to determine the cutting speed v for a cut of length L. It is assumed that the tool with its useful life described by Eq (4.2a) has already been chosen, as have the feed and depth. Sections 7.2-4 use different criteria that give different solutions to this problem. A good review of these approaches can bc found in [DeVrie69].

7.2 Maximum Production Rate or Minimum Production Time

This is the oldest of the machining economics criteria, aimed at minimizing the unit production time [Gilber52]. To illustrate this criterion, a turning cut on an engine lathe will be used, where Figure 7.2 defines the terminology. It is assumed that two of the cutting conditions have already been specified: the depth of cut d, from the diameter reduction required, and the feed a, from either force or surface finish constraints.

The optimization problem is to choose a cutting speed v, to minimize the *average* unit production time, which is made up of the handling time, cutting time and the *average* time spent changing a cutting edge. These three components make up any maximum unit production rate problem, and are used to write the *average unit production time* $t_{prod}(v)$, in the form:

$$t_{prod}(v) = t_{load} + t_c(v) + t_{edge}(v) \tag{7.1}$$

t_{load} *Loading/unloading time* (s/unit) is assumed to be independent of v. It includes any time that is not a function of v, for example, the time to return the cutting edge to its home position or the time to remove a finished workpiece and set up a new piece of stock.

$t_c(v)$ *Time spent making a cut* (s/unit) is a function of the length of cut and the feed rate. This term is a function of v because of the relationship between the feedrate v_f, and v is determined by the feed. How this term is calculated depends on the particular machining process.

$t_{edge}(v)$ *Fraction of time to change a cutting edge* (s/unit) is made up of two parts: the time to change a single edge t_{ch} (s/edge), and the fraction of the life of a single cutting edge used up machining a single workpiece (edges/unit). This fraction is the ratio of the time that the edge is removing material, $t_c(v)$ for turning, to the tool life determined from Taylor's equation.

For the case of turning in Figure 7.2, the expression for the cutting time is:

$$t_c(v) = \frac{L}{v_f} = \frac{L}{a \bullet N} = \frac{L \bullet \pi \bullet D}{a \bullet v} \tag{5.10}$$

Because a single cutting edge is expected to last for several parts, the time spent changing a cutting edge has to be prorated. This is done using the ratio of the cutting time in Eq. (5.10) and the tool life in Eq. (4.2a):

$$t_{edge}(v) = t_{ch}\left(\frac{\text{Time spent machining one unit}}{\text{Useful life of one edge}}\right) = t_{ch}\left(\frac{t_c(v)}{T}\right) \tag{7.2a}$$

$$= t_{ch}\left(\frac{t_c(v)}{T_R\left(\frac{v_R}{v}\right)^{1/n}}\right) = t_{ch}\left(\frac{\pi \bullet L \bullet D}{a \bullet T_R\, v_R^{\ 1/n}}\right) v^{(1/n\,-\,1)} \tag{7.2b}$$

Combining these terms, the expression for the average unit production time is

$$t_{prod}(v) = t_{load} + \left(\frac{\pi \cdot L \cdot D}{a}\right) v^{-1} + t_{ch}\left(\frac{\pi \cdot L \cdot D}{f \cdot T_R \, v_R^{\,1/n}}\right) v^{(1/n - 1)} \qquad (7.3)$$

Equation (7.3) can always be evaluated to estimate the average unit production time. *The average production rate is simply the inverse of the average unit production time.* As can be seen by the form of the equation, the first term is constant, but the second term corresponding to the cutting time, decreases with increasing cutting speed. The optimization problem arises because of the last term in Eq (7.3), which represents the average time spent changing cutting edges. Because the Taylor exponent n, is always between zero and one, the third term is a function of $v^{(1/n - 1)}$ and increases with v, so that Eq. (7.3) has a unique minimum. Finding this minimum is the same as maximizing the unit production rate. The solution is obtained by differentiating Eq. (7.3) with respect to v, setting the result equal to zero, and solving for the v that gives the *cutting speed for maximum production rate*, v_{max};, i.e.,

$$v_{max} = \frac{v_R}{\left[(1/n - 1)\dfrac{t_{ch}}{T_R}\right]^n} \qquad (7.4)$$

The average unit production time is found by evaluating Eq.(7.3) at v_{max}. Similarly, the tool life at this cutting speed can be determined by evaluating the tool life at v_{max}.

A few comments about the solution for the cutting speed for maximum production rate:

- To increase, v_{max}, i.e., increase the production rate, the time to change a cutting edge t_{ch}, should be reduced. Ways to do this include: preset tooling that can quickly be dismounted and remounted; automation, like tool changers; or easily accessible mountings and readily identifiable tools for manual changes.
- Selection of cutting tool materials, as evidenced by v_R and n in Eq. (7.4), can increase v_{max}. Exactly what the effect will be is difficult to predict, since usually an increase in v_R, which is desirable, is accompanied by an increase in n. It is safe to say that changing from a high speed steel tool material to a ceramic would result in a significant increase in v_{max} and in the production rate.

7.3 Minimum Unit Cost

When cost is no object, deadlines have to be met, or the turning operation is a bottleneck in an overall process plan, the maximum production rate/minimum production time criterion gives the appropriate solution. Ways to increase the production rate were suggested like automation, quick change tooling, or new tool materials.

In reality, these improvements usually have a cost associated with them, either in terms of the operating rate or the direct expense associated with the tooling. In a company both of these costs are difficult to obtain because either they are not known precisely for a single operation or they are known but are considered part of the competitive advantage/disadvantage of a firm and usually are not publicized. However, these costs form the basis for the minimum unit production cost criteria.

The same problem considered in Section 7.2 is used to develop the objective function for optimization, an *average unit production cost* model $C_{prod}(v)$, or

$$C_{prod}(v) = C_{oper}\, t_{prod}(v) + C_{edge}\left(\frac{t_c(v)}{T}\right) \tag{7.5a}$$

$$= C_{oper} \cdot t_{load} + C_{oper}\left(\frac{\pi \cdot L \cdot D}{a}\right) v^{-1}$$

$$+ (C_{oper} \cdot t_{ch} + C_{edge}) \cdot \left(\frac{\pi \cdot D \cdot L}{a \cdot T_R \cdot v_R{}^{1/n}}\right) v^{(1/n - 1)} \tag{7.5b}$$

Because $C_{prod}(v)$ predicts the unit production cost of an operation ($/unit), the major difference between Eq (7.5) and the unit production time model Eq (7.3), is the introduction of the costs C_{oper} and C_{edge}.

C_{oper} *The total operating cost rate for the process* includes the rates for: direct labor; indirect costs due to labor, i.e., fringe benefits, mandatory contributions for unemployment insurance, etc.; the indirect cost of engineering and other staff and support functions; plus the cost of equipment, technology and the depreciation of these items. For this simple model, all of these costs ($/s) are lumped together and are applied to any time that requires an operator or the machine tool.

C_{edge} *The most clearly identified expense that can be charged against the process* of turning, viz., the expendable tooling, is included in this cost rate. It is the cost of replacing a cutting edge ($/edge), but excludes the time to install the edge, which is covered under operating costs. This cost can be estimated several ways, depending on the type of tools used. For example if the cutting edges are ground, then this cost is estimated by dividing the cost of the original tool stock by the number of re-grinds possible, and then adding the cost of re-grinding an edge. If, on the other hand, indexable inserts are used, the cost of a tool holder has to be spread out over the number of inserts the tool holder can be used on. Then to this unit cost must be added the cost of a single insert divided by the number of useable cutting edges on it, to give the unit cost of an edge.

Comparing Eq. (7.5) with Eq. (7.3), one conclusion is that if there isn't any cost associated with tooling ($C_{edge} = 0$), then these criteria only differ by a constant C_{oper}, so the solutions for minimum unit cost and maximum production rate are the same. What makes these models different when $C_{edge} \neq 0$ is the cost of a cutting edge, which is prorated in the same way as the time to change a cutting edge, viz., the cutting time is divided by the tool life. This means another term involving $v^{(1/n - 1)}$ so this term's weight increases by a factor $(1 + C_{edge}/t_{ch}\, C_{oper})$, shifting the solution for minimum cost to the left of the one for maximum production rate. Solving for the v that minimizes Eq (7.5) gives the *cutting speed for minimum unit cost* , v_{min}:

$$v_{min} = \frac{v_R}{\left[(1/n - 1)\, t_{ch}\left(1 + \frac{C_{edge}}{t_{ch} \cdot C_{oper}}\right)\right]^n} \tag{7.6a}$$

$$= \frac{v_{max}}{\left[1 + \frac{C_{edge}}{t_{ch} \cdot C_{oper}}\right]^n}. \tag{7.6b}$$

This solution indicates two things:

- First, v_{min} is always less than or equal to v_{max}. This is because n is positive and the denominator in Eq (7.6b) is always greater than one.
- When the cost of an edge C_{edge}, gets greater than the cost of changing an edge ($t_{ch} \cdot C_{oper}$), v_{min} drops way below v_{max} because the denominator is much greater than one.

Example 7.1 - Comparing the Unit Production Rate and Unit Production Cost Criteria

A large number of units require a turning cut 100 mm long that reduces the diameter of a piece of stock from 35 to 30 mm. Based on force considerations, a feed of 0.5 mm/rev has been selected. The particular lathe that will be used for this purpose has a chargeable rate of $180/hr ($0.05/s) that includes all direct and indirect costs. In addition, the following data is available from either vendors, time and motion studies or good engineering estimates.

- Inserts will be used that have 4 useable edges at a cost of $20/insert. The cost of the tool holder is negligible in this instance. The inserts have a Taylor tool life equation $\left(\frac{T}{60\text{ s/edge}}\right) = \left(\frac{v}{2\text{ m/s}}\right)^{-1/0.4}$.
- From time and motion studies, the average time for loading workpieces is 15 s/unit and the time to change a cutting edge is 10 seconds.

With this data, the values needed to determine the cutting speed for maximum production rate and minimum unit production time are: the tool edge changing time t_{ch} = 10 s/edge and the Taylor tool life parameters for the inserts, n = 0.4 and v_R = 2 m/s. Then substituting into Eq. (7.3),

$$v_{max} = \frac{2\text{ m/s}}{\left[(1/0.4 - 1)\,\frac{10\text{ s/edge}}{60\text{ s/edge}}\right]^{0.4}} = 3.48\text{ m/s}.$$

Once this value is known the estimated life of a cutting edge, the unit production time (which should be the minimum) and cost can be calculated from Eqs. (4.2a), (7.3) and (7.5). For example, the tool life at v_{max} = 3.48 m/s is

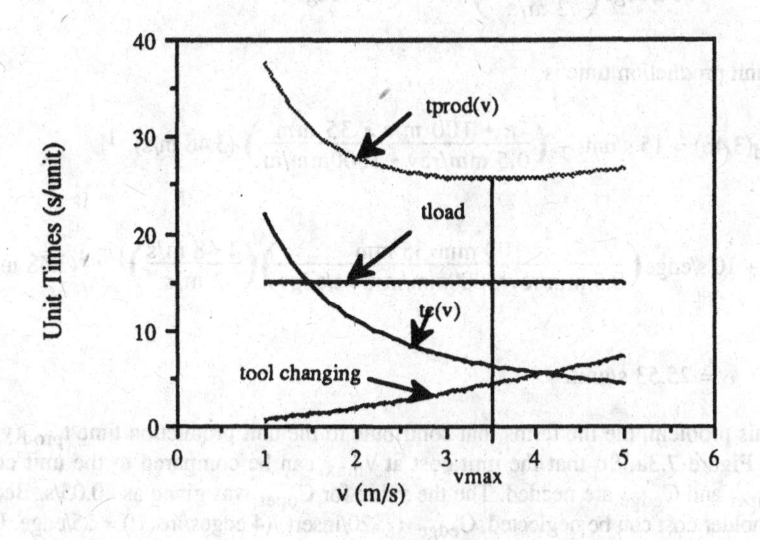

Figure 7.3a Unit production time

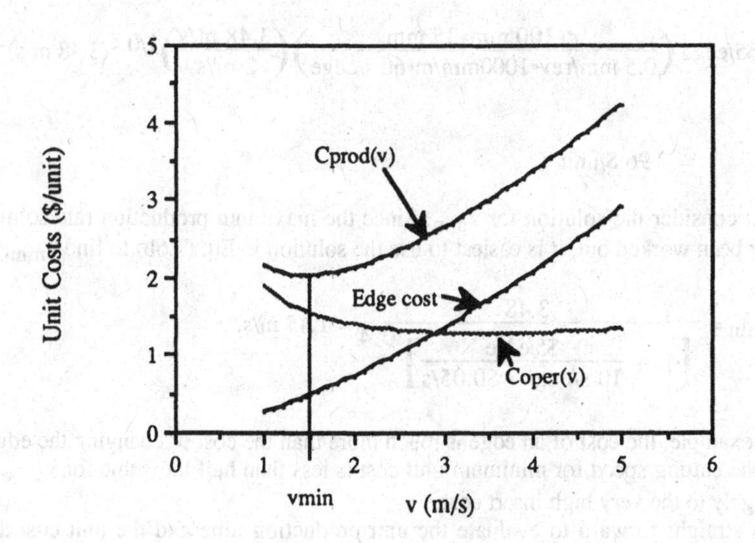

Figure 7.3b Unit production cost

Figure 7.3 Unit production time and cost for Example 7.1.

$$T = 60 \text{ s/edge}\left(\frac{3.48 \text{ m/s}}{2 \text{ m/s}}\right)^{-1/0.4} = 15 \text{ s/edge}$$

and the unit production time is

$$t_{prod}(3.48) = 15 \text{ s/unit} + \left(\frac{\pi \cdot 100 \text{ mm} \cdot 35 \text{ mm}}{0.5 \text{ mm/rev} \cdot 1000\text{mm/m}}\right)(3.48 \text{ m/s})^{-1}$$

$$+ 10 \text{ s/edge}\left(\frac{\pi \cdot 100 \text{ mm} \cdot 35 \text{ mm}}{0.5 \text{ mm/rev} \cdot 1000\text{mm/m} \cdot 60 \text{ s/edge}}\right)\left(\frac{3.48 \text{ m/s}}{2 \text{ m/s}}\right)^{1/0.4}(3.48 \text{ m/s})^{-1}$$

$$= 25.53 \text{ s/unit.}$$

For this problem, the the terms that contribute to the unit production time $t_{prod}(v)$, are given in Figure 7.3a. So that the unit cost at v_{max} can be compared to the unit cost at v_{min}, C_{oper} and C_{edge} are needed. The the value for C_{oper} was given as \$0.05/s. Because the tool holder cost can be neglected, C_{edge} = (\$20/insert)/(4 edges/insert) = \$5/edge. Using this cost data and the the unit production cost equation, Eq (7.5a)

$$C_{prod}(3.48) = \$0.05/\text{s } (25.53 \text{ s/unit})$$

$$+ \$5/\text{edge}\left(\frac{\pi \cdot 100 \text{ mm} \cdot 35 \text{ mm}}{0.5 \text{ mm/rev} \cdot 1000\text{mm/m} \cdot 60 \text{ s/edge}}\right)\left(\frac{3.48 \text{ m/s}}{2 \text{ m/s}}\right)^{1/0.4}(3.48 \text{ m/s})^{-1}$$

$$= 2.96 \text{ \$/unit}$$

Next consider the solution for v_{min}. Since the maximum production rate solution has already been worked out, it is easiest to use the solution in Eq. (7.6b) to find v_{min}, i.e.,

$$v_{min} = \frac{3.48}{\left[1 + \frac{\$5/\text{edge}}{10 \text{ s/edge} \cdot \$0.05/\text{s}}\right]^{0.4}} = 1.45 \text{ m/s.}$$

In this example, the cost of an edge is much more than the cost of changing the edge. As a result the cutting speed for minimum unit cost is less than half the value for v_{max}. This is due largely to the very high insert cost.

It is straight forward to evaluate the unit production time and the unit cost at v_{min}: $t_{prod}(1.45)$ = 31.34 s/unit $C_{prod}(1.45)$ = 2.02 \$/unit. As should be the case, the unit cost at v_{min} is less than it was at v_{max}, about two thirds the cost, and the unit production time is about twenty five percent larger. Looking at Figure 7.3b, the contribution of the operating cost and the expense of providing a cutting edge are shown, indicating how the minimum shifts to the left for the minimum unit production cost criterion.

7.4 Maximum Profit Rate

The minimum unit cost criterion assumes that it's always "good" to produce at the lowest possible unit cost, however this may not always be true. For a firm, division of a company or and individual entrepreneur, a goal is usually to maximize profit, or, if contractual obligations require that workpieces must be delivered even at a loss, then the goal becomes minimizing loss. A key point to understand when formulating a machining economics optimization problem is that even with low unit cost, the units still need a customer. If the selling price is above the unit cost, there is a profit, but if the cost is above what someone is willing to pay there will be a negative profit, a euphemism for a loss. In other words, unit cost and production rate can be affected by technology and engineering developments, but the units produced have to be bought by someone - a retail customer, a wholesaler or another operating division - at an agreed on price. It is the difference between price and cost that determines profit of loss.

The maximum profit rate criterion uses profit as the optimization criterion. The solution to this problem will be based on *the marginal principle in economics; the optimum production volume occurs where marginal cost equals marginal revenue.*

A brief review of introductory economics is the place to start, before considering the machining economics problem. The production volume u, is represented by the *supply curve* in the price-volume plane. For the idealized problems considered in this chapter, the *supply curve is vertical; the capacity exists to produce at any volume, independent of the price.* The customer side of the picture is represented by a *non-increasing demand curve D(u).* In a free market, the selling price is the point where D(u) and the supply curve intersect; *for a vertical supply curve the locus of selling prices is D(u).* Revenue R(u) is the total income from u units at their selling price D(u) or

$$R(u) \quad = D(u)u. \tag{7.7}$$

Associated with the production volume is a cost given by C(u). P(u) is the profit defined as the difference between revenue and cost,

$$P(u) \quad = R(u) - C(u). \tag{7.8}$$

With Eq (7.8) as the objective function, and the optimization problem is to find a u* that maximizes Eq (7.8) or

$$\frac{\partial P(u^*)}{\partial u} = \frac{\partial R(u^*)}{\partial u} - \frac{\partial C(u^*)}{\partial u} = 0 \text{ or} \tag{7.9a}$$

$$\frac{\partial R(u^*)}{\partial u} = \frac{\partial C(u^*)}{\partial u}. \tag{7.9b}$$

Equation (7.9b) is simply a mathematical statement of the marginal principle: *the optimum production volume occurs where marginal cost equals marginal revenue.*

Example 7.2 - Simple Examples of Applying the Marginal Principle

Let u denote the number of units to be supplied, and assume that the volume supplied is independent of price, i.e., a vertical supply curve. The problem is to determine u* to maximize profit. Associated with producing these units is a cost curve given by:

$$C(u) = C_0 + C_1 u + C_2 u^2 = 12 - 2 u + u^2. \quad (7.E1)$$

From market surveys, experience, and forecasting, a linearly decreasing demand curve was developed:

$$D(u) = D_0 - D_1 u = 12 - 2 u. \quad (7.E2a)$$

With the vertical supply curve, knowing the demand curve is equivalent to knowing the selling price, so the revenue curve from Eq (7.7) is,

$$R(u) = D(u) u = D_0 u - D_1 u^2 = 12 u - 2 u^2. \quad (7.E3)$$

The profit function is defined by Eq (7.8), in terms of the revenue curve Eq (7.E3) and the cost curve Eq. (7.E1),

$$P(u) = R(u) - C(u) = -12 + 14 u - 3 u^2 \quad (7.E4)$$

Applying the marginal principle, Eq (7.9b), to this problem means solving

$$D_0 - 2 D_1 u^* = C_1 + 2 C_2 u^* \text{ or}$$

$$u^* = \frac{D_0 - C_1}{2(C_2 + D_1)} \quad (7.E5)$$

$$= \frac{12 - (-2)}{2(1 + 2)} = 2.33 \text{ units, and } P(u^*) = 4.33.$$

Figure 7.4a shows this solution, along with the cost, demand, revenue, and profit curves.

As another quick example, suppose the cost curve stays the same, but the demand curve is lowered, i.e.,

$$D(u) = D_0 - D_1 u = 8 - 2 u. \quad (7.E2b)$$

The same solution in Eq (7.E5), is valid,

$$u^* = \frac{D_0 - C_1}{2(C_2 + D_1)} = \frac{8 - (-2)}{2(1 + 2)} = 1.67 \text{ units, but } P(u^*) = -3.67,$$

i.e., the negative profit or a loss is minimized. Figure 7.4b shows this solution.

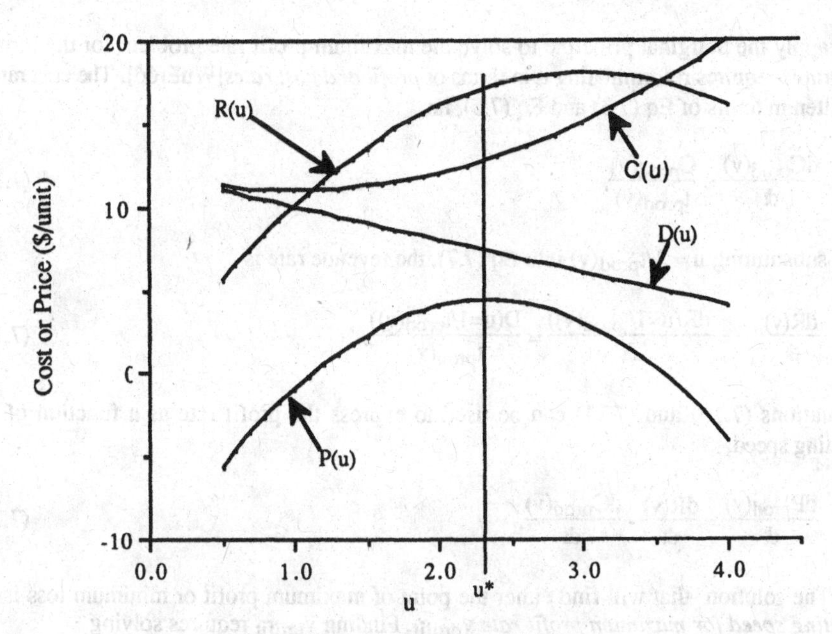

Figure 7.4a Revenue exceeding cost

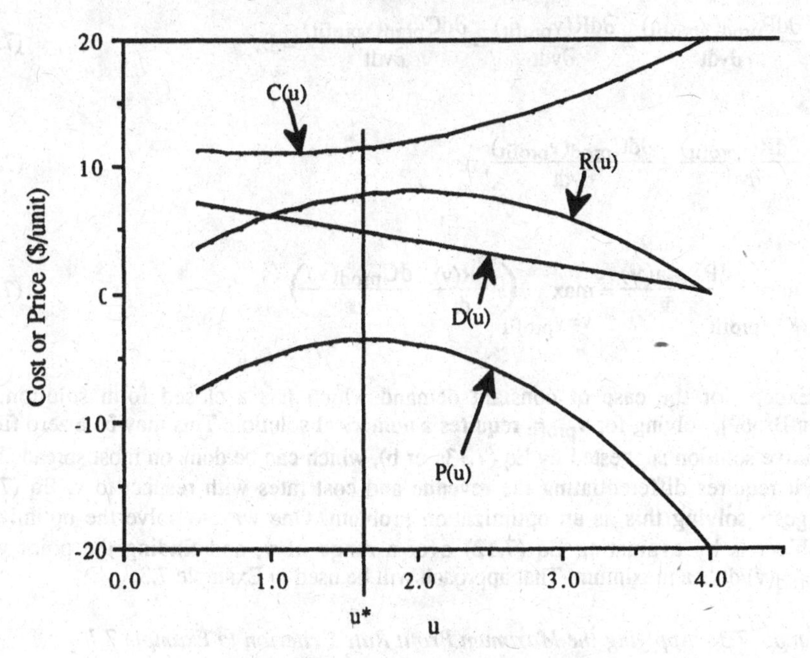

Figure 7.4b Cost exceeding revenue.

Figure 7.4 The maximum profit criterion in Example 7.2.

To apply the marginal principle to solve the maximum profit rate problem for the turning example requires reformulating it in terms of *profit and cost rates*[WuEr66]. The cost rate is written in terms of Eq (7.3) and Eq (7.5), i.e.

$$\frac{dC_{prod}(v)}{dt} = \frac{C_{prod}(v)}{t_{prod}(v)}. \tag{7.10}$$

By substituting $u = 1/t_{prod}(v)$ into Eq (7.7), the revenue rate is

$$\frac{dR(v)}{dt} = \frac{dR(u=1/t_{prod}(v))}{dt} = \frac{D(u=1/t_{prod}(v))}{t_{prod}(v)}. \tag{7.11}$$

Equations (7.10) and (7.11) can be used to express the profit rate as a function of the cutting speed,

$$\frac{dP_{prod}(v)}{dt} = \frac{dR(v)}{dt} - \frac{dC_{prod}(v)}{dt}. \tag{7.12}$$

The solution that will find either the point of maximum profit or minimum loss is the *cutting speed for maximum profit rate* v_{profit}. Finding v_{profit} requires solving

$$\frac{\partial dP_{prod}(v_{profit})}{\partial vdt} = \frac{\partial dR(v_{profit})}{\partial vdt} - \frac{\partial dC_{prod}(v_{profit})}{\partial vdt} = 0, \tag{7.13a}$$

$$\frac{\partial dR(v_{profit})}{\partial vdt} = \frac{\partial dC_{prod}(v_{profit})}{\partial vdt}, \text{ or} \tag{7.13b}$$

$$\max_{v=v_{profit}} \frac{dP_{prod}(v)}{dt} = \max_{v=v_{profit}} \left(\frac{dR(v)}{dt} - \frac{dC_{prod}(v)}{dt} \right). \tag{7.13c}$$

Except for the case of constant demand which has a closed form solution, c.f., [ArmBro69], solving for v_{profit} requires a numerical solution. This may be a zero finding iterative solution suggested by Eq (7.13a or b), which can be done on most spread sheets, but it requires differentiating the revenue and cost rates with respect to v. Eq (7.13c) suggests solving this as an optimization problem. One way to solve the optimization problem is by evaluating Eq (7.12) over a range of v, and finding the point where $dP_{prod}(v)/dt$ is a maximum. That approach will be used in Example 7.3.

Example 7.3 - Applying the Maximum Profit Rate Criterion to Example 7.1

The maximum profile rate criterion can be applied to the situation in Example 7.1. The specifications, cutting conditions, tools, times and cost for the turning problem will remain the same. Two demand curves will be introduced in this example: one that will make it possible to make a profit and the other will result in a selling price less than the unit cost. Determining demand curves is always difficult, and the ones in this example have been selected to illustrate different solutions.

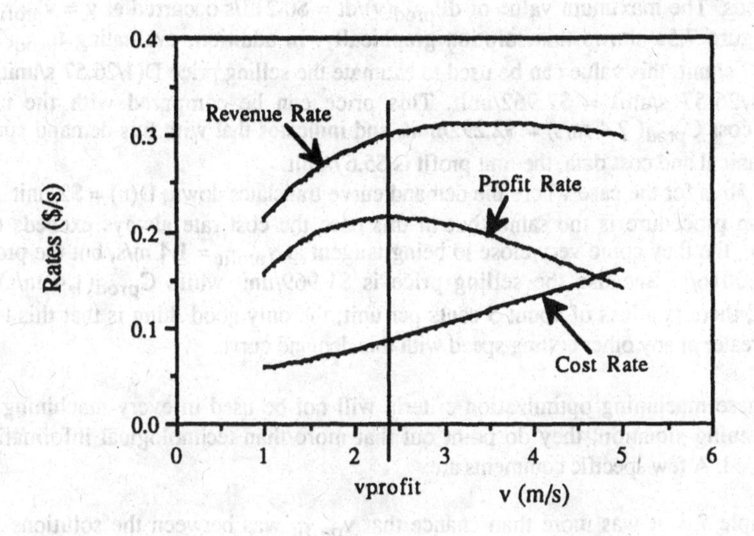

Figure 7.5a v_{profit} when the revenue rate exceeds the cost rate.

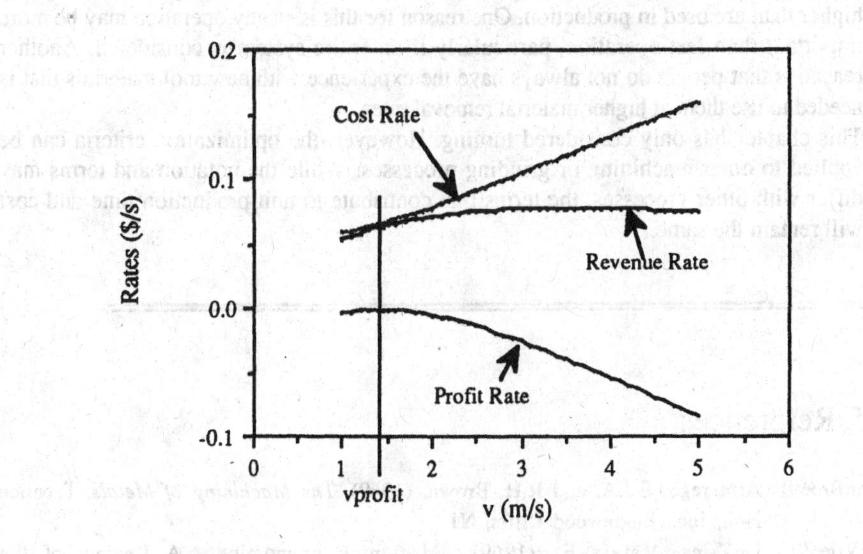

Figure 7.5b v_{profit} when the cost rate exceeds the revenue rate.

Figure 7.5 Maximum profit rate criteria for Example 7.3.

The first demand curve is: D(u) = \$8/unit - u. With this demand curve and the revenue and cost rate equations, a spread sheet was used to calculate Eq (7.12) over a range of cutting speeds. The maximum value of $dP_{prod}(v)/dt$ = \$0.213/s occurred at $v = v_{profit}$ = 2.4 m/s. Figure 7.5a shows this solution graphically. In addition, evaluating t_{prod}(2.4 m/s) = 26.57 s/unit, this value can be used to estimate the selling price D(1/26.57 s/unit) = \$8/unit - 1/26.57 s/unit = \$7.962/unit. This price can be compared with the unit production cost C_{prod}(2.4 m/s) = \$2.292/unit, and indicates that with this demand curve and the technical and cost data, the unit profit is \$5.67/unit.

Figure 7.5b is for the case where the demand curve translates down, D(u) = \$2/unit - u. The solution procedure is the same, but in this case the cost rate always exceeds the revenue rate; the they come very close to being tangent at v_{profit} = 1.4 m/s, but the profit rate is \$-0.0016/s. Because the selling price is \$1.969/unit while C_{prod}(1.4 m/s) = \$2.018/unit, there is a loss of about 5 cents per unit; the only good thing is that this loss would be greater at any other cutting speed with this demand curve.

While these machining optimization criteria will not be used in every machining or process planning situation, they do point out that more than technological information may be needed. A few specific comments are:

- In Example 7.3, it was more than chance that v_{profit} was between the solutions for v_{min} and v_{max}, when there was a region the profit rate was positive. This will always be true if the revenue rate curve is above the cost rate curve and there is a decreasing demand curve. Gilbert identified this *region bounded by v_{min} and v_{max} as the high efficiency or HiE region[Gilber52]*.
- Evaluating the cutting speeds using the criteria in this chapter usually results in speeds higher than are used in production. One reason for this is steady operation may be more important than fast operation, particularly if an entire system is considered. Another reason is that people do not always have the experience with new tool materials that is needed to use them at higher material removal rates.
- This chapter has only considered turning. However, the optimization criteria can be applied to other machining or grinding processes. While the notation and terms may differ with other processes, the terms that contribute to unit production time and cost will remain the same.

7.5 References

[ArmBro69] Armarego, E.J.A, and R.H. Brown, (1969), *The Machining of Metals*, Prentice Hall, Inc., Englewood Cliffs, NJ.

[DeVrie69] DeVries, Marvin F., (1969), "Machining Economics - A Review of the Traditional Approaches and Introduction to New Concepts", *American Society of Tool and Manufacturing Engineers Paper #MR69-279*, pp. 435-442.

[Drozda82] Drozda, T., ed., (1982), "Manufacturing Engineering Explores Grinding Technology", A collection of Manufacturing Engineering articles discussing grinding, Society of Manufacturing Engineers, Dearborn, MI.

[Gilber52] Gilbert, W.W., (1952), "Economics of Machining," in *Machining Theory and Practice*, American Society of Metals, Cleveland.

[Gilber62] Gilbert, W.W., (1962), "Economics of Machining," in *Machining With Carbides and Oxides*, ASTME/McGraw Hill, New York.

[WuEr66] Wu, S.M. and Ermer, D.S., (1966), "Maximum Profit as the Criterion in the Determination of the Optimum Cutting Conditions," *ASME Transactions, Journal of Engineering for Industry*, Series B, Vol. 88, pp. 435.

7.6 Problems

7.1. This problem looks a lot like Problem 5.4, but while the setup is the same it is aimed at comparing the three machining economics criteria for the turning operation shown in Figure 7.6. It is assumed that the cutting speed is the only cutting condition that has to be chosen, since the depth is determined from the geometry and the feed of 0.25 mm/rev was chosen based on force and power limits. The cutting edge is assumed to be a carbide with a Taylor tool life equation given by: $\left(\frac{T}{60 \text{ s/edge}}\right) = \left(\frac{1.5 \text{ m/s}}{v}\right)^{(1/0.25)}$. The applicable times, costs, and linear demand curve for u units are: t_{load} = 30 s/unit, t_{ch} = 0 s/edge, C_{oper} = \$0.05/s, C_{edge} = \$2/edge, D(u) = 25 -0.001 u(\$/unit).

a. Use the maximum production rate criterion to determine: v_{max}, $t_{prod}(v_{max})$ and $C_{prod}(v_{max})$.

b. Use the minimum cost rate criterion to determine: v_{min}, $t_{prod}(v_{min})$ and $C_{prod}(v_{min})$.

c. Use the maximum profit rate criterion to estimate: v_{profit}, $t_{prod}(v_{profit})$ and $C_{prod}(v_{profit})$. There are several ways to do this, e.g., Newton's method, evaluate and plot either the difference between the revenue and cost rates looking for a maximum or minimum or find the difference between the marginal rates looking for a zero, or use the iterate function on a spread sheet. Choosing the correct range for v will be important, but you should know the approximate region. Three decimal places of precision in your answer is plenty.

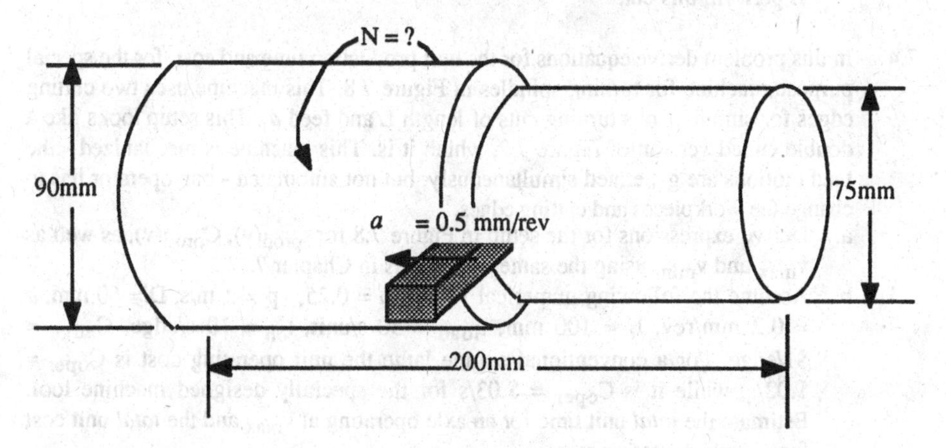

Figure 7.6 Cut for comparing three machining economics criteria and technology changes in Problems 7.1 and 7.2.

7.2 This problem is aimed at using Problem 7.1 as a baseline for comparing the effects of technological changes on unit production time and cost. Repeating the data in Problem 7.1: L = 200 mm, b = 7.5 mm, a = 0.25 mm/rev, t_{load} = 30 s/unit, t_{ch} = 10 s/edge, C_{oper} = \$0.05/s, C_{edge} = \$2/edge and the Taylor equation for carbide is $\left(\frac{T}{60 \text{ s/edge}}\right) = \left(\frac{1.5 \text{ m/s}}{v}\right)^{(1/0.25)}$

a. Suppose that either because of a time and motion study or because of a technological investment, the time to load and unload workpieces is reduced to t_{load} = 10 s/unit. For this change, evaluate v_{max}, $t_{prod}(v_{max})$, v_{min} and $C_{prod}(v_{min})$:

b. Suppose that to save money, rather than use a carbide cutting tool, high speed steel cutting edges will be used so that the following data: C_{edge} = \$1/edge, and $\left(\frac{T}{60 \text{ s/edge}}\right) = \left(\frac{0.5 \text{ m/s}}{v}\right)^{(1/0.10)}$ should be used to evaluate v_{max}, $t_{prod}(v_{max})$, v_{min} and $C_{prod}(v_{min})$.

c. Suppose that a new machine tool has been purchased that allows a higher feed, a = 0.35 mm/rev, but this also raises the operating cost to C_{oper} = \$0.06/s. Use this new data to evaluate v_{max}, $t_{prod}(v_{max})$, v_{min} and $C_{prod}(v_{min})$.

7.3 This problem is to select two cutting conditions for turning, viz., the cutting speed v and the feed a, based on the minimum unit cost criterion and the surface finish. Once these are known, estimate the power required for the cut. The same cutting tool and workpiece as in Problem 5.8 will be used, viz., a new carbide cutting tool with a signature of 0°, 10°, 5°, 5°, 15°,45°, 0.5 mm and a Taylor tool life specified by n = 0.25 and v_R = 2 m/s. The workpiece is 2024-T4 aluminum that is 200 mm long and has a diameter of 110 mm. The depth of cut b = 5 mm. This conditions are documented in Figure 7.7.

a. Select the largest feed so that $R_a \leq 0.75$ μm.

b. Based on time and motion studies, it takes 60 seconds to unload, setup and load a workpiece and it takes 15 seconds to change a cutting edge. The rate charged for all costs excluding tooling is \$0.01/sec and the cost of tooling is \$5/edge. Determine the cutting speed for minimum unit cost.

c. For the cutting conditions specified, estimate the power required at the spindle to perform this cut.

7.4 In this problem derive equations for the unit production time and cost, for the special purpose machine for turning spindles in Figure 7.8. This machine uses two cutting edges for simultaneous turning cuts of length L and feed a . This setup looks like a double ended version of Figure 7.2, which it is. This machine is mechanized - the feed motions are generated simultaneously, but not automated - one operator has to change the workpieces and cutting edges.

a. Derive expressions for the setup in Figure 7.8 for $t_{prod}(v)$, $C_{prod}(v)$, as well as v_{max} and v_{min}, using the same notation as in Chapter 7.

b. Assume the following numerical values: n = 0.25, v_R = 1 m/s, D = 60 mm, a = 0.2 mm/rev, L = 100 mm, t_{load} = 30 s/unit, t_{ch} = 10 s/edge, C_{edge} = \$1/edge. For a conventional engine lathe the unit operating cost is C_{oper} = \$.02/s, while it is C_{oper} = \$.03/s for the specially designed machine tool. Estimate the *total* unit time for an axle operating at v_{max} and the *total* unit cost for an axle operating at v_{min}.

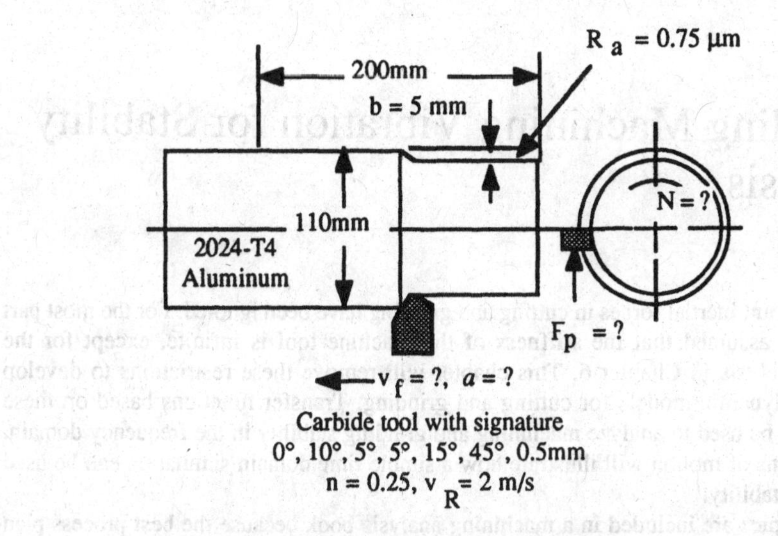

Figure 7.7 Turning operation for Problem 7.3.

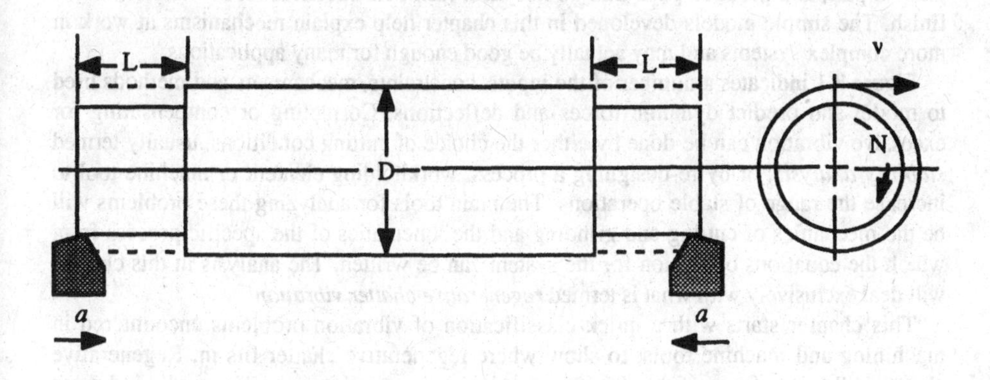

Figure 7.8 Double ended turning machine.

8
Modeling Machining Vibration for Stability Analysis

Up to this point inertial forces in cutting and grinding have been ignored. For the most part it has been assumed that the stiffness of the machine tool is infinite, except for the sparkout analysis in Chapter 6. This chapter will remove these restrictions to develop simplified dynamic models for cutting and grinding. Transfer functions based on these models will be used to analyze machining and grinding stability in the frequency domain. The equations of motion will illustrate how a simple time domain simulation can be used to analyze stability.

These topics are included in a machining analysis book because the best process plan based on static and kinematic analysis may not work because of excessive vibration. This vibration can be damaging to the machine tool, destroy the cutting edges or tooling used to make a part, and produce parts that do not meet technical specifications on dimension or finish. The simple models developed in this chapter help explain mechanisms at work in more complex systems and may actually be good enough for many applications.

Figure 8.1 indicates a number of the inputs, constraints, mechanisms and methods used to model and predict dynamic forces and deflections. Correcting or compensating for excessive vibration can be done by either the choice of cutting conditions, usually termed *stability analysis*, or by re-designing a process, workholding element or machine tool to increase the range of stable operations. The main tools for analyzing these problems will be the mechanics of cutting and grinding and the kinematics of the specific process from which the equations of motion for the system can be written. The analysis in this chapter will deal exclusively with what is termed *regenerative chatter vibration*.

This chapter starts with a quick classification of vibration problems encountered in machining and machine tools, to show where regenerative chatter fits in. Regenerative chatter will be the focus of the next two sections, concentrating on turning and grinding as applications. The topics covered in this chapter include: deriving the equations of motion, a simple numerical simulation, defining the stability boundaries of a particular machine tool using stability charts, and practical ways to identify and correct some machine vibration problems encountered in practice.

8.1 Classification of the Types of Vibration in Machining

Vibration in machining and grinding refers almost exclusively to the relative displacement between the cutting edge or grinding wheel and the new workpiece surface being generated by the process. There are other vibrations that are detrimental to the life of the machine tool, but for machining processes all these vibrations are of interest because they are potential causes of the vibration between the cutting edge and workpiece. A simple vibration classification will be used: vibrations independent of the material removal process and vibrations that occur because of the material removal process. Figure 8.2 schematically illustrates these types of machine tool vibrations.

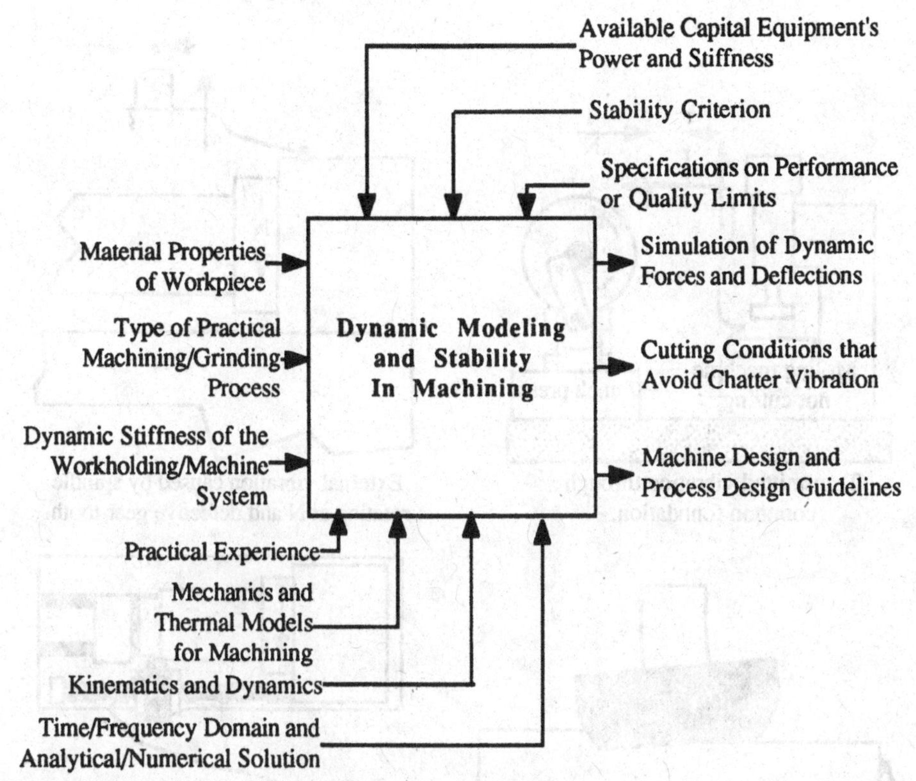

Figure 8.1 The Inputs, Constraints, Outputs and Mechanisms to analyze machine dynamics and stability.

8.1.1 Vibration Sources External To Chip Making

External or forced vibrations are external to the cutting process, and sometimes even the machine tool, but can cause vibrations between a workpiece and cutting edge or grinding wheel.

Transmitted vibrations have a source external to the machine tool and are usually modeled as impulses or shocks. From the source, illustrated in Figure 8.2 by the punch press, the vibrations are transmitted through the common foundation for the press and milling machine, to the structure of the machine tool, inducing a vibration between the workpiece and the cutter. The nature of this induced vibration, i.e., how fast it oscillates and how quickly it dissipates, depends largely on the machine tool system. The simplest characterization of the machine tool system will be the lumped parameter natural frequency ω_n and the damping ratio ζ. Separate foundations or ways to isolate the vibration source from the workpiece are solutions to this type of vibration, but providing a separate foundation is a significant cost that must be included in a capital expenditure budget. It is also true that if these obvious and significant external sources of vibration are not handled, the subsequent efforts to control vibration on the machine tool may have no effect.

Sources of vibrations due to the machine tool itself - not the process of making a chip - are termed *external forced* vibrations. As Figure 8.2 illustrates, these may be things like defective rolls or balls in a bearing, missing teeth on a gear, or unbalanced spindle or feed drive systems. Usually these vibrations are periodic because their source is often a rotating

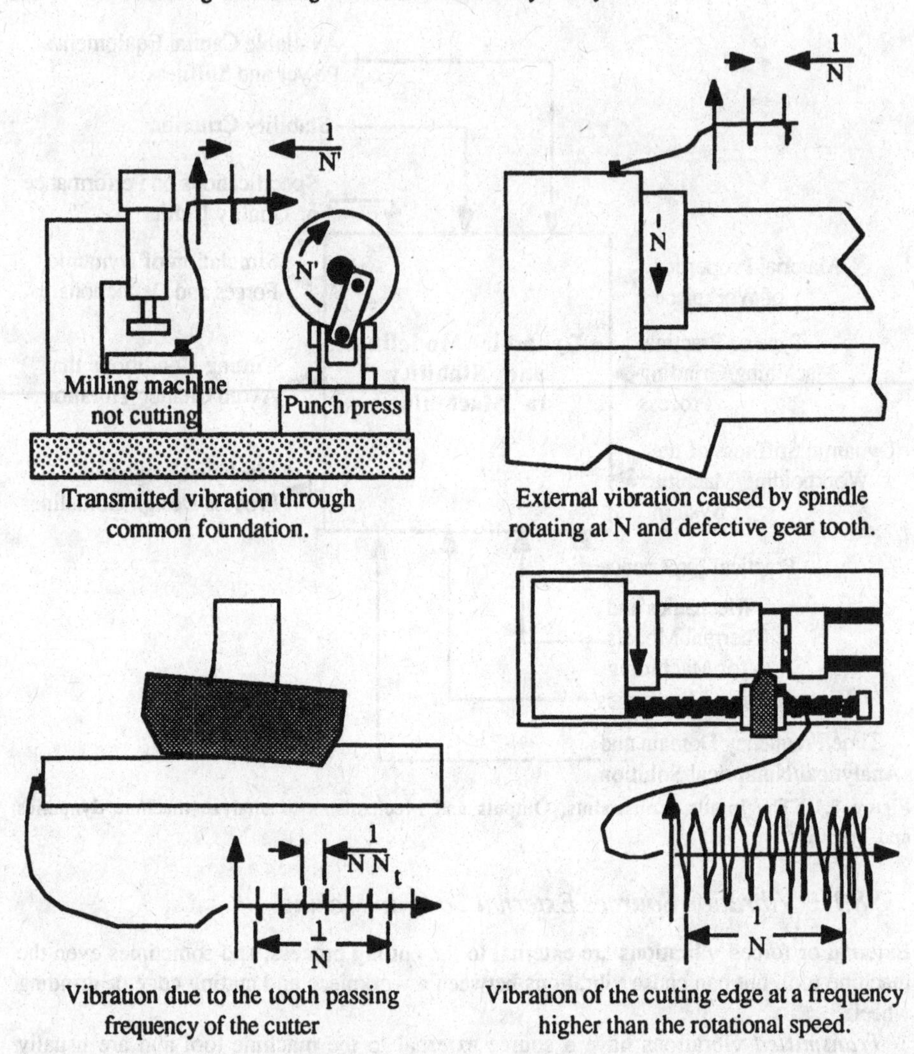

Figure 8.2 Illustrations of the types of vibration in machining. The top illustrations are due to sources not related to machining and the bottom two only occur when machining.

component of the machine tool. As a result, their frequency is usually an integer multiple of the rotational frequency of a drive or spindle. Rigid machine maintenance, on-machine diagnostics and good machine tool or component design are the ways that this source of vibration can be eliminated.

8.1.2 Vibration Sources Integral To Chip Making

The types of vibration in Section 8.1.1 can occur when a machine is not removing material. Ways to control or eliminate these types of vibrations are usually related to good machine tool design, siting the machine properly and maintenance. There are two important types of vibration that only occur during material removal, and they can be controlled by planning the process.

Internally forced vibration occurs because of the chip making process. The forces simulated in Example 5.3 for face milling are due to the chip making process. They are periodic forces that can excite a relative vibration between the machine tool and workpiece. On the scale of the contact between the chip and the cutting edge, discontinuous Type I chips or stick-slip friction on the rake face of a cutting edge can result in excitation between the tool and workpiece too. The frequency of these vibrations are either at the rotational speed of a cutter, workpiece or grinding wheel, at the tooth passing frequency ($N \bullet N_t$) in milling processes, or at a frequency that can be related to chip segmentation. Ways to control this type of vibration are related to understanding how cutting and grinding fluids can effect chip formation or knowing how to control the impact force by cutter design. Process planning solutions to these problems might be the specification of helical or unevenly spaced cutters, or choosing between climb and conventional milling to govern how the cutting edge should enter and exit a workpiece.

The final form of vibration is *self induced or chatter* vibration. It occurs because most cutting or grinding processes produce a new surface by removing material from a surface generated by a previous cutting edge or workpiece rotation. Undulations on the previously generated surface can cause vibrations to ***regenerate*** rather than die out. Self induced vibration can be controlled by selecting cutting conditions in a judicious way. This judicious way is often referred to as stability analysis. Both the kinematics and mechanics of chip formation serve as a forcing function that excites the dynamics of the machine tool-workpiece-workholding system. Stability analysis in Sections 8.2 and 8.3 deals with this form of machine tool vibration, how to model it, and how to select operating positions to avoid it.

8.2 Stability Analysis in Turning

Modeling in the time domain and stability analysis in the frequency domain will be illustrated using orthogonal turning as an example.

8.2.1 Modeling Cutting, Instantaneous Chip Thickness and the Machine Tool in Turning

As a starting point for modeling, Figure 8.3a shows a schematic of the turning process and Figure 8.3b shows the idealization of the cutting edge and motion of the system. Models in this chapter will have three parts: a model for the cutting process, a model for the equivalent stiffness of the machine-workpiece-fixturing system, and a model for the instantaneous feed taking into account undulations in the workpiece surface. In Section 8.3 on grinding, undulations in the wheel surface will be added to this kinematic model.

Starting with a model for the cutting process, a linearized representation assumes that the thrust or feeding force $F_q(t)$ is proportional to the cross section of the chip normal to the velocity vector v given by $A_c(t) = b\ h(t)$. The proportionality constant is referred to as the *specific cutting pressure* k'_c, and this empirical approach was used in Example 5.3 to estimate forces in face milling. Different ways to calculate k'_c are based on empirical models or models of the mechanics of chip forming. Table 8.1 gives examples of some of these ways in terms of quantities introduced in previous chapters. Then the model for the feed force in the cutting process is

Table 8.1 Ways to Estimate k'_c and k_p for Orthogonal Machining		
$k'_c = \frac{F_q}{A_c}$	$k_p = \frac{F_p}{F_q}$	Based on
$\frac{\sin(\beta-\gamma_o)\ \tau_s}{\cos(\phi_o+\beta-\gamma_o)\sin(\phi_o)}$	$\frac{1}{\tan(\beta-\gamma_o)}$	Eq (3.10e)
$E_c\bullet[K_c(h_{avg})\bullet K_r(\gamma)\bullet K_w(VB)]$ $\bullet[C_c(h_{avg})\bullet C_r(\gamma)\bullet C_w(VB)]$	$\frac{1}{[C_c(h_{avg})\bullet C_r(\gamma)\bullet C_w(VB)]}$	Eq.s (5.3a and c)
$B_q\left(\frac{a\ avg}{a\ R}\right)^{(a_q-1)} \times\left(\frac{b}{b_R}\right)^{(e_q-1)} a_R b_R$	$\frac{B_p}{B_q}\left(\frac{a\ avg}{a\ R}\right)^{(a_p-a_q)} \times\left(\frac{b}{b_R}\right)^{(e_p-e_q)}$	Table 5.1 $F_q=B_q\left(\frac{a}{a\ R}\right)^{a_q}\left(\frac{b}{b_R}\right)^{e_q}$ $F_p=B_p\left(\frac{a}{a_R}\right)^{a_p}\left(\frac{b}{b_R}\right)^{e_p}$

$$F_q(t) = (k'_c\ b)\ h(t) \tag{8.1a}$$

$$= (k_c)\ h(t) \tag{8.1b}$$

The second form of this model in Eq (8.1b) uses what is termed the *cutting stiffness* k_c, which is simply (k'_c b). k_c is referred to as a stiffness because it is the proportionality constant between the chip thickness h(t) and the feed force $F_q(t)$. In this chapter, all the modeling of the mechanics of the process is reduced to the simple expressions in Eq (8.1).

The model for the instantaneous chip thickness can be derived after looking at Figure 8.3b. The rotational speed N, is assumed to remain constant so (1/N) is the time delay between revolutions of the workpiece. From Eq (5.5) and Eq (5.8a) the relationships between the instantaneous feed rate $v_f(t)$, the instantaneous feed per revolution $a(t)$ and the

instantaneous uncut chip thickness h(t) are:

$$v_f(t) = a(t)\bullet N = \frac{h(t)}{\cos(\lambda)}\bullet N \tag{8.2a}$$

$$= h(t)\bullet N \qquad (\lambda = 0). \tag{8.2b}$$

This means that for orthogonal turning when $\lambda = 0$, the chip thickness and the *instantaneous feed* are the same. In the Z direction, the coordinates of the new surface are made up of the feed displacement and the vibration z(t), written as

$$Z(t) = z(t) - \int_0^t v_f(\tau)\ d\tau + Z(0). \tag{8.2c}$$

The most important expressions for modeling are based on the expressions for the instantaneous chip thickness,

$$h(t) = Z(t - 1/N) - Z(t) \tag{8.2d}$$

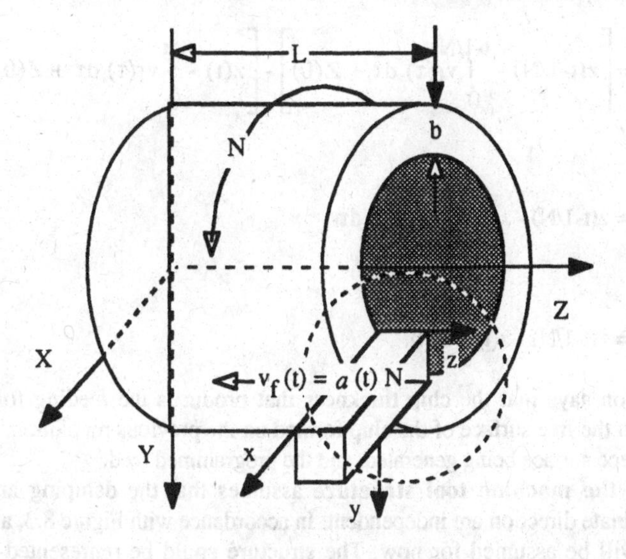

Figure 8.3a Model for turning.

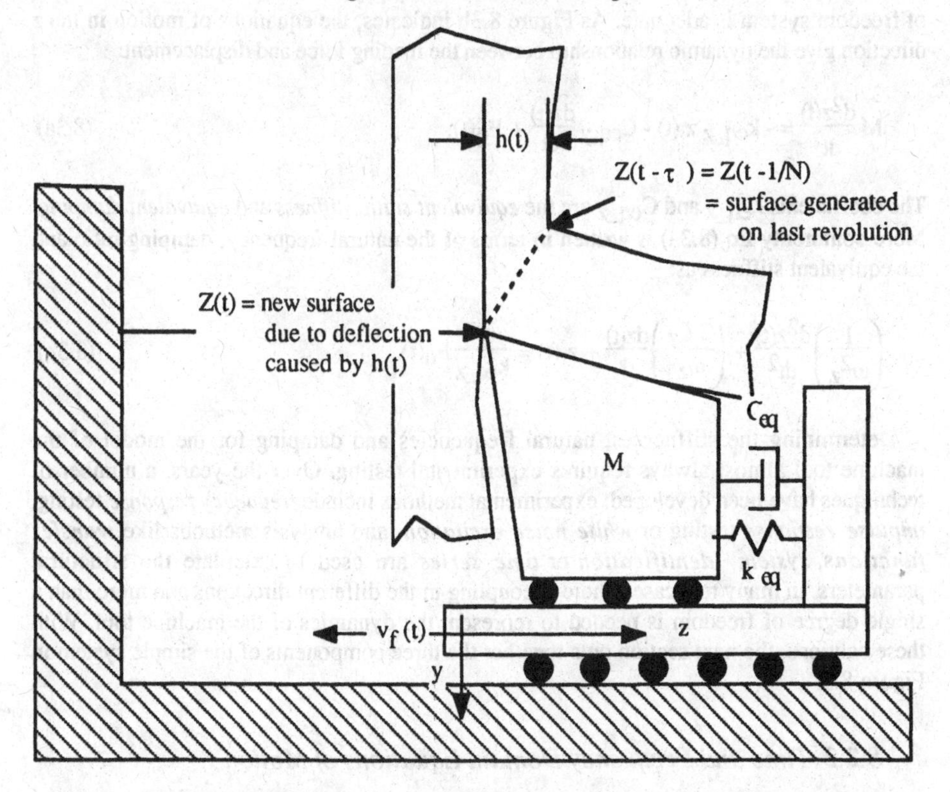

Figure 8.3b Idealization of single degree of freedom vibration model.

Figure 8.3 Idealizations for turning.

$$= \left[z(t-1/N) - \int_0^{t-1/N} v_f(\tau)\, d\tau + Z(0) \right] - \left[z(t) - \int_0^{t} v_f(\tau)\, d\tau + Z(0) \right]$$

$$= z(t-1/N) - z(t) - \int_{t-1/N}^{t} v_f(\tau)\, d\tau \tag{8.2e}$$

$$= z(t-1/N) - z(t) + a\,(t). \tag{8.2f}$$

This expression says that the chip thickness that produces the feeding force is due to: undulations on the free surface of the chip formed on the previous revolution, undulations on the workpiece surface being generated, and the programmed feed.

Modeling the machine tool structure assumes that the damping and stiffnesses in each coordinate direction are independent. In accordance with Figure 8.3, a single degree of freedom will be assumed for now. The structure could be represented by a transfer function of any order, but for illustration and for many practical situations the single degree of freedom system is adequate. As Figure 8.3b indicates, the equations of motion in the z direction give the dynamic relationship between the feeding force and displacement:

$$M \frac{d^2 z(t)}{dt^2} = - k_{eq\_z}\, z\,(t) - C_{eq,y} \frac{dz(t)}{dt} + F_q(t). \tag{8.3a}$$

The coefficients k_{eq_z} and C_{eq_z} are the *equivalent static stiffness* and *equivalent damping*. More commonly Eq (8.3a) is written in terms of the natural frequency, damping ratio and the equivalent stiffness as:

$$\left(\frac{1}{\omega^2_z}\right)\frac{d^2 z(t)}{dt^2} + \left(\frac{2\,\zeta_z}{\omega_z}\right)\frac{dz(t)}{dt} + z\,(t) = \frac{1}{k_{eq\_z}} F_q(t). \tag{8.3b}$$

Determining the stiffnesses, natural frequencies and damping for the model of the machine tool almost always requires experimental testing. Over the years, a number of techniques have been developed: experimental methods include *frequency response* testing, *impulse response* testing or *white noise excitation*, and analysis methods like *transfer functions*, *system identification* or *time series* are used to calculate the structural parameters. In many real cases, there is coupling in the different directions and more than a single degree of freedom is needed to represent the dynamics of the machine tool. With these cautions, the next section puts together the three components of the simple system in Figure 8.3.

8.2.2 Time and Frequency Domain Equations of Motion

Completing the equations of motion for this one degree of freedom system is as follows: The expression for the instantaneous uncut chip thickness Eq (8.2f), is used to eliminate h(t) from Eq (8.1), the model of the cutting process. Then the resulting force model is substituted in Eq (8.3b):

$$\left(\frac{1}{\omega^2_z}\right)\frac{d^2z(t)}{dt^2} + \left(\frac{2\,\zeta_z}{\omega_z}\right)\frac{dz(t)}{dt} + \left(\frac{k_c}{k_{eq\_z}} + 1\right)z\,(t) - \frac{k_c}{k_{eq\_z}}\,z(t\text{-}1/N) = \frac{k_c}{k_{eq\_z}}\,a\,(t)). \quad (8.4a)$$

This equation is nonlinear because of the term z(t-1/N). It is also apparent that in the ideal case when k_{eq_z} is very large, potential vibration problems are minimized because the "forcing functions" z(t-1/N) and *a*(t) are attenuated. Since the equivalent stiffness is assumed to be the serial combination of the stiffnesses of the machine tool, workpiece and workholding elements, k_{eq_z} is usually dominated by the most compliant element. This points out the importance of good machine tool design to assure that the machine tool is not the most compliant element in the system. Because the workpiece-fixture combination changes from part to part, their contribution to the equivalent stiffness becomes an important consideration in process planning when trying to make a machining setup resist chatter vibration. Because k_c, the cutting stiffness is in the numerator in several of the coefficients in Eq (8.4a), another way to control vibration is by reducing $k_c = (k'_c\ b)$, and in that way attenuate inputs to the system. Because the specific cutting pressure is primarily a function of the workpiece material and the contact interface conditions, usually there is a limit on what can be done to affect k'_c. As a result, b the width (or depth) of cut is one of the process planning variables that can be used to control chatter.

Stability analysis in the frequency domain will be done using transfer function models for Equation (8.1, 2 and 3). With s as the Laplace variable, the frequency domain models for the cutting process, chip thickness and machine tool-workpiece system are:

$$F_q(s) \quad = k'_c\, b\, h(s) = k_c\, h(s) \qquad (8.1c)$$

$$h(s) \quad = z(s)[e^{-s/N} - 1] + a\,(s) \qquad (8.2g)$$

$$\frac{z(s)}{F_q(s)} = \frac{1}{k_{eq\_z}\left[\left(\frac{s}{\omega_z}\right)^2 + 2\,\zeta_z\left(\frac{s}{\omega_z}\right) + 1\right]} = \frac{1}{k_{eq\_z}\, k_{dyn}(s)}. \qquad (8.3c)$$

The term $k_{eq\_z}\, k_{dyn}(s)$ is called the *dynamic stiffness*, and as Eq (8.3c) indicates, it is made up of the static stiffness and the transfer function for the dynamics between displacement of the structure and force. Solving these equations algebraically, the transfer function relating the feed *a*(s) to the vibration displacements z(s) is

$$\frac{z(s)}{a(s)} = \frac{k'_c\, b}{k_{eq\_z}\, k_{dyn}(s) + k'_c\, b\,[1 - e^{-s/N}]} \qquad (8.5a)$$

$$= \frac{\dfrac{k'_c\, b}{k_{eq\_z}}}{k_{dyn}(s) + \dfrac{k'_c\, b}{k_{eq\_z}}\,[1 - e^{-s/N}]} \qquad (8.5b)$$

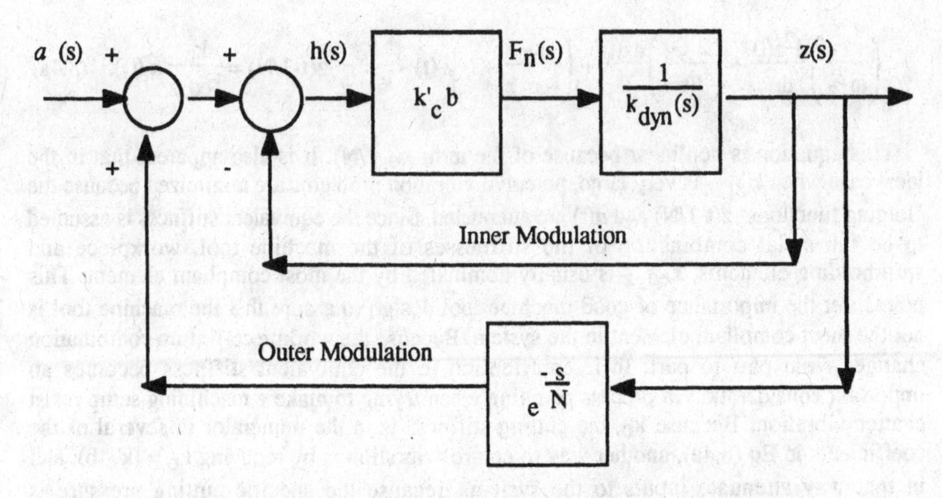

Figure 8.4 Feedback block diagram representation of Figure 8.3.

$$= \frac{\dfrac{k'_c b}{k_{eq\_z}}}{k_{dyn}(s) + \dfrac{k'_c b}{k_{eq\_z}}\left[\dfrac{s/N}{1!} - \dfrac{(s/N)^2}{2!} + \dfrac{(s/N)^3}{3!} \cdots\right]}. \tag{8.5c}$$

Eq (8.5) points out that the dynamic system has an infinite number of modes, so to compute the roots of the characteristic equation requires deciding where to truncate the power series in the denominator of Eq (8.5c). Also, the locus of these many roots is a function of cutting conditions, i.e., b and N.

The Nyquist stability criterion is easier to use with a transfer function representation of the non-linear system G(s), than calculating the roots of the characteristic equation. For stability, the criterion states that when the Laplace variable s is replaced by the complex frequency $j\omega$, the open loop transfer function $G(j\omega)+1$ must meet the requirement

$$G(j\omega) \quad > -1 = e^{-j\pi}, -\infty < \omega < \infty. \tag{8.6}$$

The model of cutting in Figure 8.3 was used to write Eqs (8.1c), (8.2g) and (8.3c). These equations can be used to construct the block diagram in Figure 8.4. Using block diagram reduction methods, Figure 8.4 can be used to find Eq (8.5). Note that Figure 8.4 has two feedback loops: one that represents undulations on the new surface termed the *inner modulation* and the outer loop with a delay of $e^{-s/N}$ that represents undulations on the surface from the previous revolution termed the *outer modulation*. For stable cutting conditions the undulations die out, and for unstable conditions $|z(t)|$ grows over time. For the idealized system in Figure 8.3, the Nyquist criterion indicates stable cutting conditions if

$$\frac{k_{eq\_z}}{k'_c\, b} \frac{k_{dyn}(j\omega)}{[1 - e^{-j\,\omega/N}]} > e^{-j\pi}, -\infty < \omega < \infty \text{ or} \tag{8.7a}$$

$$k_{dyn}(j\omega) > \frac{k'_c\, b}{k_{eq\_z}} \left[e^{j\pi} - e^{-j(\omega/N - \pi)}\right], -\infty < \omega < \infty, \text{ then } |z(\infty)| \to 0. \tag{8.7b}$$

This means that given the static stiffness and $k_{dyn}(j\omega)$ for a turning setup and the specific cutting pressure for a workpiece, Eq (8.7) can be evaluated to determine if the proposed cutting conditions, b and N, will cause chatter.

A graphical approach to this problem is to construct a *stability chart for a given machine tool.* This chart is the locus of (k'_c b) and N values that form the boundary between stable cutting and chatter. This chart can be constructed experimentally for a machine tool by covering a range of cutting stiffnesses and rotational speeds and determining where chatter begins. Evaluating Eq (8.7) as an equality and solving for (k'_c b) as a function of N is an analytical way to do the same thing. Either by experiment or calculation, a stability chart, with the characteristic lobed boundaries like that in Figure 8.5, is the result. This shape means that what may be chatter free operation at one cutting stiffness and speed may become unstable if the rotational speed is increased *or decreased.*

Because of this problem, the *global stability limit* can be defined: *the maximum cutting stiffness which gives stable operation for all rotational speeds.* Merritt's solution to the stability problem used this approach [Merrit65], based on the maximum value of the right hand side of Eq (8.7) occurring when $e^{-j(\omega/N - \pi)} = 1$ so

$$Re[k_{dyn}(j\omega)] > \frac{2\,(k'_c\, b)_{global}}{k_{eq\_z}}, -\infty < \omega < \infty. \tag{8.7c}$$

The global stability limit $(k'_c b)_{global}$, indicated by the dashed line in Figure 8.5, is a conservative value with lower material removal rates than are possible by operating between the lobes. One area of research is to try to move into these lobes using a control system that detects chatter while machining, to account for the limitations on the precision with which a stability chart can be specified.

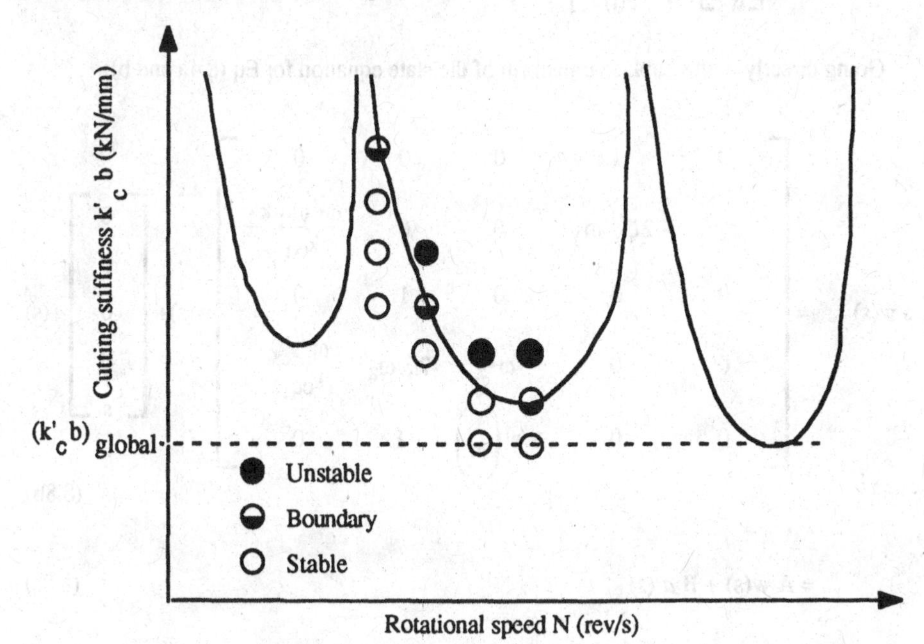

Figure 8.5 Stability chart for a lathe.

8.2.3 Two Degree of Freedom System in State Variable Form

An idealized two degree of freedom model for the machine tool introduces compliance and damping in the y or cutting force direction of Figure 8.3. This can be characterized by the static stiffness k_{eq_y}, natural frequency ω_y and damping ratio ζ_y. With this idealization, the y and z modes of the machine tool are uncoupled when not cutting. Because the force in the y-direction depends on the instantaneous chip thickness defined in the z-direction, the two modes of vibration are coupled during cutting. The force in the y-direction is the power component given by $F_p(t) = (k_p k_c)\, h(t)$, so the equation of motion in the y direction is:

$$\left(\frac{1}{\omega^2_y}\right)\frac{d^2y(t)}{dt^2} + \left(\frac{2\,\zeta_y}{\omega_y}\right)\frac{dy(t)}{dt} + y\,(t) - \frac{k_p k_c}{k_{eq\_y}}\,(z(t\text{-}1/N) - z(t)) = \frac{k_p k_c}{k_{eq\_y}}\, a\,(t). \qquad (8.4b)$$

Because z(t) and z(t-1/N) appear in Eq (8.4b), this shows that the two vibration modes are coupled by the cutting process model.

Equations (8.4a and b) are the solutions for the coupled motion of the cutting edge. Sometimes for time domain analysis or because of the availability of software for performing a computer simulation, it is convenient to use state variables to represent these equations of motion. This approach is a standard way to write the equations of motion, so standard solutions can be used to solve the equations. For example, for this two degree of freedom model, the vector of state variables can be defined as:

$$\mathbf{w} = \begin{bmatrix} w1 \\ w2 \\ w3 \\ w4 \\ w5 \end{bmatrix} = \begin{bmatrix} y(t) \\ dy(t)/dt \\ z(t) \\ dz(t)/dt \\ h(t) \end{bmatrix} \qquad (8.8a)$$

Going directly to the Laplace transform of the state equation for Eq (8.4a and b):

$$s\,\mathbf{w}(s) = \begin{bmatrix} 0 & 1 & 0 & 0 & 0 \\ -\omega^2_y & -2\zeta_y\,\omega_y & 0 & 0 & \dfrac{\omega^2_y k_p k_c}{k_{eq\_y}} \\ 0 & 0 & 0 & 1 & 0 \\ 0 & 0 & -\omega^2_z & -2\zeta_z\omega_z & \dfrac{\omega^2_z k_c}{k_{eq\_z}} \\ 0 & 0 & s\,e^{-\left(\frac{s}{N}\right)} & -1 & 0 \end{bmatrix} \mathbf{w}(s) + \begin{bmatrix} 0 \\ 0 \\ 0 \\ 0 \\ s \end{bmatrix} a\,(s) \qquad (8.8b)$$

$$= \mathbf{A}\,\mathbf{w}(s) + \mathbf{B}\,a\,(s) \qquad (8.8c)$$

The stability of the system while it is cutting is determined by the characteristic equation. In Laplace form, the characteristic equation is given implicitly by $|s\mathbf{I} - \mathbf{A}| = 0$ A time domain simulation can also be used to analyze stability. Standard packages are available to solve Eq (8.8), c.f., [MATRIX91]. Writing your own numerical integration software for state equations is also easy, since their solution is simply multiple applications of the integration of a first order differential equation. Theory says that if Eq (8.7) predicts unstable behavior, eventually the numerical integration will become unstable, so this is a computational way of mapping out a stability chart. But remember, the simulation is only as good as the specifications on the stiffnesses, dynamic coefficients, and cutting pressures supplied.

Example 8.1 Single Degree of Freedom Stability Analysis for Turning

The purpose of this example is to illustrate the link between the global stability analysis of Eq (8.7c) and a time domain numerical simulation. The simulation will be for a single degree of freedom system represented in state variable form and simulated numerically. The C procedure to implement the simulation is given in the Appendix.

For the simulation the state vector is

$$\mathbf{w} = \begin{bmatrix} w1 \\ w2 \\ w3 \end{bmatrix} = \begin{bmatrix} z(t) \\ dz(t)/dt \\ Z(t) \end{bmatrix},$$

where Z(t) is given by Eq (8.2c) and the chip thickness is determined by Eq (8.2d).

For this example, the following conditions are assumed: For the workpiece, the specific cutting pressure $k'_c = 1.5\ \mathrm{kN/mm^2}$. The lumped parameter representation of the machine tool is modeled with the static stiffness in the axial direction $k_{eq\_z} = 10$ kN/mm, the natural frequency $f_z = 20$ Hz or $\omega_z = 125.7$ rad/s and the damping ratio is $\zeta_z = 0.05$. Cutting conditions are a rotational speed of N= 10 rev/s, a feed of $a = 0.1$mm, with two depths of cut: b = 1 mm and 4 mm. With these values, Eq (8.7c) to determine global stability becomes:

$$\mathrm{Re}\left[\left(\frac{j\omega}{\omega_z}\right)^2 + 2\zeta_z\left(\frac{j\omega}{\omega_z}\right) + 1\right] = \left[1 - \left(\frac{\omega}{\omega_z}\right)^2\right] > \frac{2\,(k'_c\, b)_{global}}{k_{eq\_z}}, \quad -\infty < \omega < \infty.$$

The frequency limits used to check for global stability are 50% above the rotational frequency of the the workpiece, so for b = 1 mm the global stability region is bounded by,

$$\left[1 - \left(\frac{\omega}{125.7\ \mathrm{rad/s}}\right)^2\right] > \frac{2\,((1.5\ \mathrm{kN/mm^2})(1\ \mathrm{mm}))_{global}}{10\ \mathrm{kN/mm}}$$

$$= 0..30, \qquad 62.9\ s^{-1} < \omega < 62.9\ s^{-1}$$

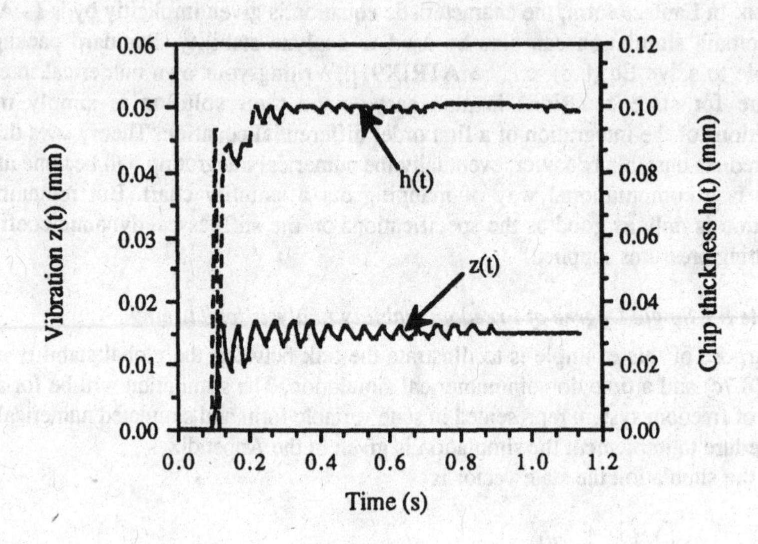

Figure 8.6a Stable cut with b = 1 mm.

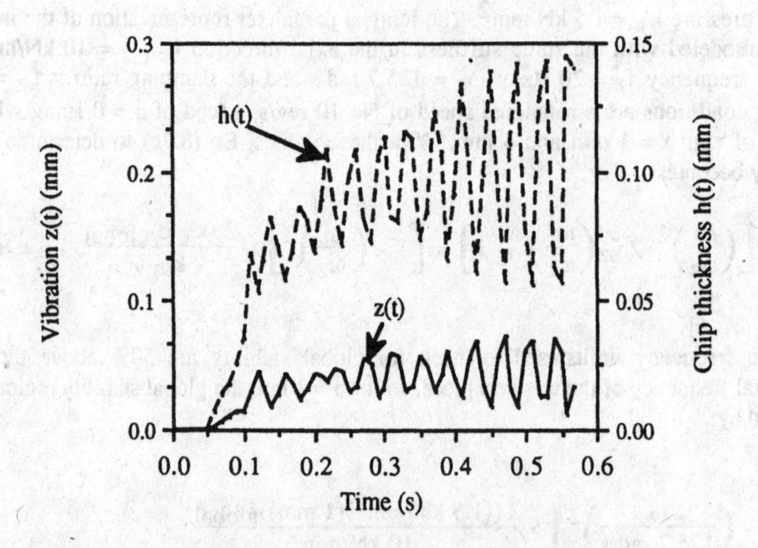

Figure 8.6b Unstable cut with b = 4 mm.

Figure 8.6 Simulation result for Example 8.1 using code in the Appendix.

which is globally stable. When b = 4 mm, then the limits are

$$\left[1-\left(\frac{\omega}{125.7\ \text{rad/s}}\right)^2\right] > \frac{2\,((1.5\ \text{kN/mm}^2)(4\ \text{mm}))_{\text{global}}}{10\ \text{kN/mm}}$$

$$= 1.2, \qquad -62.9\ s^{-1} < \omega < 62.85\ s^{-1}$$

so even when $\omega = 0$, the global stability condition is not satisfied.

Using the computer simulation, these conclusions can be seen in the time domain. Figure 8.6a shows a simulation using the case where b = 1 mm where the vibration z(t) and the instantaneous chip thickness are plotted. As can be seen, both z(t) and h(t) have transients after entering the cut, but these transients die out; a characteristic of stable cutting. If the depth of cut is increased to 4 mm, then z(t) and h(t) look like Figure 8.6b. The transients do not die out. In fact they grow because an increasing undulation on the surface generated on a previous revolution is in phase with a decrease on the current revolution, so large chip thicknesses result, causing larger forces, and larger out of phase undulations. This process is what is meant by regeneration. Also note that z(t) in both parts of Figure 8.6 have an average value that is not zero. This is the deflection of the cutting edge that could lead to geometric errors that must be within the tolerance limits on a part.

8.3 Stability Analysis of Cylindrical Plunge Grinding

When analyzing sparkout in Section 6.4.2 cylindrical plunge grinding was used to develop the expression for estimating sparkout time. There the equivalent stiffness k_{eq} and grinding stiffness k_g were introduced. The same idealization in Figure 6.7 is modified in Figure 8.7 to analyze the stability of cylindrical plunge grinding. The procedures are similar to those in Section 8.2 for analyzing turning, but begin directly with transfer functions

8.3.1 Modeling the Structure, Kinematics, Material Removal and Wear in Plunge Grinding

The motion of the center of the workpiece is modeled along a line normal to the contact area that joins the centers of the wheel and workpiece. Two things make modeling grinding more complicated than turning: wear of the wheel cannot be neglected, and there are two rotating elements that can "regenerate" - the workpiece and the grinding wheel. A particularly good reference on grinding stability is given by Srinivasan in [KinHah86].

The grinding machine structure is modeled as a transfer function between the normal grinding force and the structural displacement like Eq (8.3c) for turning

$$\frac{x(s)}{F_n(s)} = \frac{1}{k_{eq}\,k_{dyn}(s)}. \tag{8.9}$$

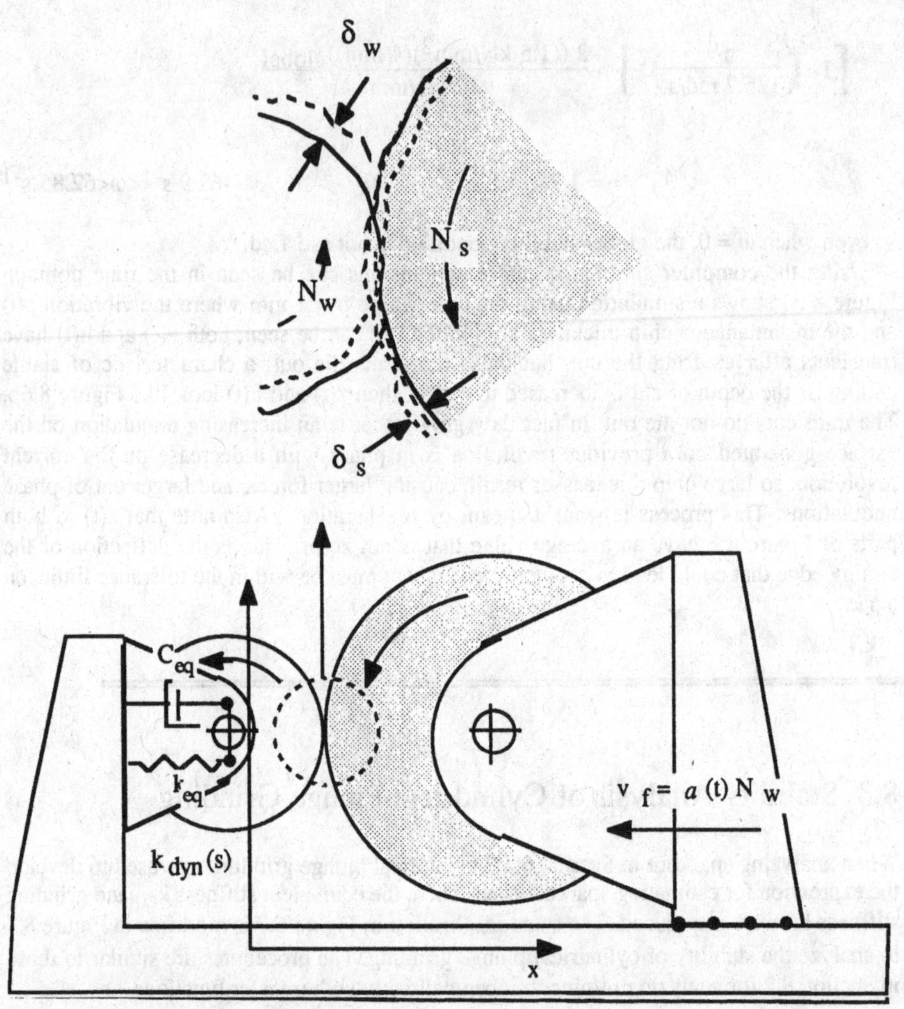

Figure 8.7 Idealization for analyzing chatter in cylindrical plunge grinding.

Here k_{eq} is the equivalent static stiffness of the structure used in the sparkout analysis, and $k_{dyn}(s)$ represents the structural dynamics. While higher order dynamic stiffnesses can be used, in Figure 8.7 it is idealized as a single degree of freedom system with natural frequency ω_n and damping ratio ζ.

Material removal is modeled as it was for the sparkout analysis with the *grinding pressure* k'_g or the grinding stiffness $k_g = (k'_g b)$. Ways to estimate k'_g and k_g are given in Table 8.2. The contact zone tends to act as an ideal low pass filter for undulations on the workpiece surface at frequencies above $\omega > (2\pi\, v_w / l_c)$. This filter will be represented by $H_w(s)$, so the transfer function that relates the normal grinding force to radial undulations on the workpiece $\delta_w(s)$ is

$$\frac{\delta_w(s)}{F_n(s)} = \frac{H_w(s)}{k_g\left(1 - e^{\frac{-s}{N_w}}\right)} = \frac{H_w(s)}{k'_g\, b\left(1 - e^{\frac{-s}{N_w}}\right)} \tag{8.10}$$

A positive undulation increases the radius and tends to reduce the instantaneous feed.

Wheel wear is modeled in much the same way. The contact zone also acts like an ideal low pass filter $H_s(s)$, but this time it filters out force variations at frequencies $\omega > (2\pi\, v_s / l_c)$ that cause radial undulations on the wheel. The undulations in the wheel that affect the radial depth are related to the reaction normal force by

$$\frac{\delta_s(s)}{F_n(s)} = \frac{H_s(s)}{k_g G \frac{v_s}{v_w}\left(1 - e^{\frac{-s}{N_s}}\right)} = \frac{H_s(s)}{k'_g G \frac{v_s}{v_w}\, b\left(1 - e^{\frac{-s}{N_s}}\right)} \,. \tag{8.11}$$

Note that the terms in the denominator of Eq (8.11) (grinding ratio G, the ratio of the tangential speeds and the grinding stiffness) might be termed the wheel stiffness.

The kinematic constraints and contact stiffness are the final elements in modeling Figure 8.7. If the radial feed set in a plunge grinding operations is a (s) then the structural vibration x(s), and the radial undulations in the workpiece $\delta_w(s)$ and wheel $\delta_s(s)$ all reduce the actual wheel depth as given by

$$d(s) = a(s) - x(s) - \delta_w(s) - \delta_s(s). \tag{8.12a}$$

Variations in d(s) generate normal force variations given by:

$$\frac{F_n(s)}{d(s)} = k_l \tag{8.12b}$$

where k_l is the *contact stiffness*. Most of the stiffnesses are difficult to obtain, but it is particularly difficult to find values for the contact stiffness, so Table 8.2 lists an empirical expression that has been used successfully.

Table 8.2 Ways to Estimate k'_g and k_l for Cylindrical Grinding

$k'_g = \frac{F_n}{A_c}$	$k'_l = \frac{F_n}{A_c}$	Based on
	$k_0 \left(\frac{D_{eq}}{D_{eq\_R}}\right)^{-0.25} \times \left(\frac{F_n b_R}{F_{n\_R} b}\right)^{0.75}$	[SnoBro69]
$\frac{E_c\, v_w}{\mu\, v_s}$		Eq (6.15b)
$\frac{v_w}{\Lambda_w}$		Eq (6.15b)

8.3.2 Transfer Function and Stability Analysis

The transfer functions from the previous section can be used to develop the relationship between one of the inputs in process planning, the radial feed *a*(s), and the structural vibration x(s) as in Section 8.2. Figure 8.8 shows a block diagram representation of the equations in Section 8.3.1.

Using standard techniques of block diagram reduction, the transfer function represented by Figure 8.8 is

$$\frac{x(s)}{a(s)} = \frac{\dfrac{1}{k_{dyn}(s)}}{\dfrac{k_{eq}}{k_l} + \dfrac{k_{eq}}{k_g}\dfrac{H_w(s)}{\left(1 - e^{\frac{-s}{N_w}}\right)} + \dfrac{k_{eq}\, v_w}{G\, k_g\, v_s}\dfrac{H_s(s)}{\left(1 - e^{\frac{-s}{N_s}}\right)} + \dfrac{1}{k_{dyn}(s)}} \tag{8.13a}$$

$$= \frac{1}{k_{dyn}(s)\left[\dfrac{k_{eq}}{k_l} + \dfrac{k_{eq}}{k_g}\dfrac{H_w(s)}{\left(1 - e^{\frac{-s}{N_w}}\right)} + \dfrac{k_{eq}\, v_w}{G\, k_g\, v_s}\dfrac{H_s(s)}{\left(1 - e^{\frac{-s}{N_s}}\right)}\right] + 1} \tag{8.13b}$$

In this form, Eq (8.6)'s representation of the Nyquist criterion is

$$k_{dyn}(j\,\omega)\left[\frac{k_{eq}}{k_l} + \frac{k_{eq}}{k_g}\frac{H_w(j\,\omega)}{\left(1 - e^{\frac{-j\,\omega}{N_w}}\right)} + \frac{k_{eq}\, v_w}{G\, k_g\, v_s}\frac{H_s(j\,\omega)}{\left(1 - e^{\frac{-j\,\omega}{N_s}}\right)}\right] > e^{-j\pi} \tag{8.14a}$$

or in terms of the compliance

$$\frac{1}{k_{dyn}(j\,\omega)} > -\left[\frac{k_{eq}}{k_l} + \frac{k_{eq}}{k_g}\frac{H_w(j\,\omega)}{\left(1 - e^{\frac{-j\,\omega}{N_w}}\right)} + \frac{k_{eq}\, v_w}{G\, k_g\, v_s}\frac{H_s(j\,\omega)}{\left(1 - e^{\frac{-j\,\omega}{N_s}}\right)}\right]. \tag{8.14b}$$

Equations (8.14a and b) could be used to construct a stability chart, but it would be three dimensional with the overall process stiffness represented as a function of N_w and N_s. For grinding, stability boundaries like those described by Eq (8.7c) for turning are

$$\mathrm{Re}\left(\frac{1}{k_{dyn}(j\,\omega)}\right) > -\left[\frac{k_{eq}}{k_l} + \frac{k_{eq}}{2\,k_g} + \frac{k_{eq}\, v_w}{2\,G\,k_g\,v_s}\right]. \tag{8.14c}$$

This means that the same type of checks on operating variables, particularly wheel and workpiece speeds, can be made for a proposed grinding setup if the data on stiffnesses is available.

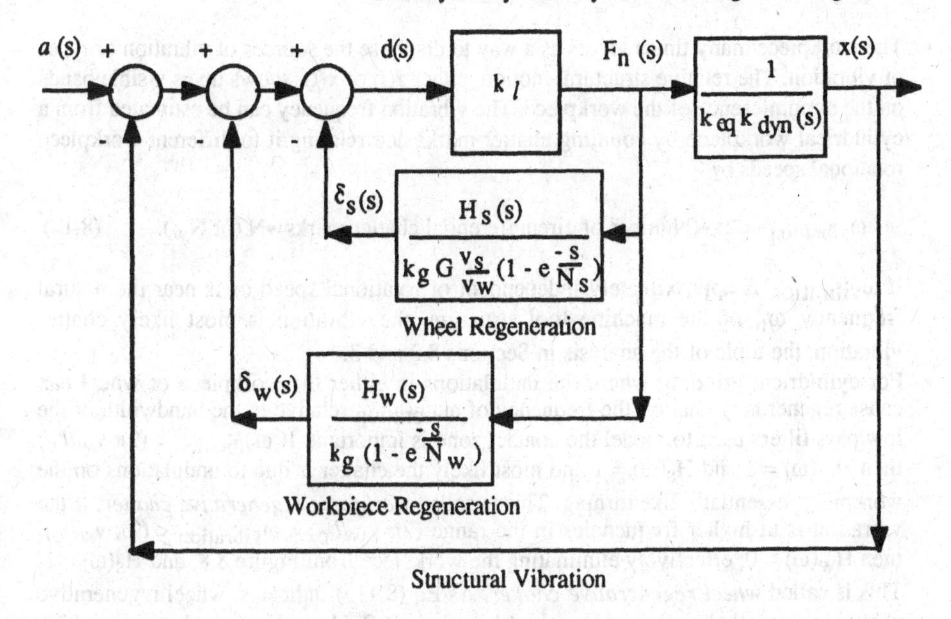

Figure 8.8 Block diagram of plunge cylindrical grinding.

8.4 Some Practical Ways to Diagnose and Minimize Chatter

Chatter stability for cutting and grinding were both analyzed with transfer functions used to represent the equations of motion. From these equations, as well as practical experience, ways to diagnose and correct machining or grinding chatter problems can be developed.

- While only the cutting and grinding conditions of rotational speed and width show up in the equations to determine stability, indirectly the feed also has an effect. Both k'_c and k'_g in Tables 8.1 and 8.2 increase as h_{avg} or h_{eq} decrease, and these measures of chip or swarf size are both direct functions of feeds. In other words, for all other things being equal, cuts at a high feed are more stable than those at low feed. In turning, for example, many times when the cutting edge first enters the cut it may begin to chatter and stabilize as the the full chip thickness develops.
- Cutting edges with small primary cutting edge angles and nose radii are more stable against chatter than if κ_r or R are large. Equation (8.2a) explains part of the effect of κ_r in light of the previous point, namely increasing the side cutting edge angle decreases the chip thickness which increases the specific cutting stiffness. A large nose radius, typical in finishing operations, has the same effect because most of the cutting is done on the nose. Equation (5.8b) shows the chip width is inversely proportional to $\cos(\kappa_r)$, so there is the added effect of increasing the cutting stiffness by increasing b. As expected from cutting mechanics, increasing the rake angle tends to reduce the cutting pressure and the susceptibility to chatter; this in part explains why grinding, with it's large negative effective rakes angles and thin swarf is susceptible to chatter. Conclusions on cutting edge geometry will also be valid for milling.

- The workpiece many times serves as a way to diagnose the sources of vibration or types of vibration. The relative structural motion, either z(t) or x(t), shows up as visible bands on the circumference of the workpiece. The vibration frequency can be estimated from a cylindrical workpiece by counting chatter marks and relating it to different workpiece rotational speeds by

$$\omega_{vibration} = 2\pi \bullet (\text{Number of circumferential chatter marks}) \bullet N \text{ (or } N_w). \qquad (8.13)$$

If $\omega_{vibration}$ is approximately independent of rotational speed or is near the natural frequency ω_n of the machine tool structure, the vibration is most likely chatter vibration, the topic of the analysis in Sections 8.2 and 3.
- For cylindrical grinding where the undulations in either the workpiece or wheel can cause regenerative chatter, the frequency of $\omega_{vibration}$ relative to the bandwidth of the low pass filters used to model the contact zone is important. If $\omega_{vibration} < (2\pi\, v_w/l_c)$ then $H_w(\omega) = 1$ and $H_s(\omega) = 1$ and most likely the chatter is due to undulations on the workpiece, essentially like turning. This is called *workpiece regenerative chatter*. If the vibration is at higher frequencies in the range $(2\pi\, v_w/l_c) < \omega_{vibration} < (2\pi\, v_s/l_c)$, then $H_w(\omega) = 0$, effectively eliminating the workpiece from Figure 8.8, and $H_s(\omega) = 1$. This is called *wheel regenerative chatter*. As Eq (8.11a) indicates, wheel regenerative chatter can usually be affected by a wheel-workpiece-fluid combination that gives a high G ratio, by reducing the workpiece speed relative to the wheel speed, or truing the wheel to remove undulations on the surface.
- The static and dynamic stiffnesses k_{eq} and $k_{dyn}(s)$ place limits on stability. They are equivalent stiffnesses comprising more than the machine tool stiffness. They include the workpiece itself, one of the reasons machining thin ribs is difficult, and fixturing setups between the machine tool and the workpiece. Because these elements are between the machine tool and the cutting edge, the weakest link will dictate stiffness and stability. As a result, a poorly fixtured compliant workpiece on the best machine tool money can buy will still chatter!

8.5 References

[Cook59] Cook, Nathan H., (1959), "Self-Excited Vibrations in Metal Cutting," *ASME Transactions, Journal of Engineering for Industry,* Series B, Vol. 81, pp. 183-186.

[DenHar34] DenHartog, J.P., (1934), *Mechanical Vibrations,* McGraw-Hill Book Co., New York and London.

[DeVEva84] DeVries, W. R., and M. S. Evans, (1984), "Computer Graphics Simulation of Metal Cutting," *CIRP Annals,* Vol. 33/1, pp. 15-18.

[HanTob74] Hanna, N.H. and S.A. Tobias, (1974), "A Theory of Nonlinear Regenerative Chatter," *ASME Journal of Engineering for Industry,* February, pp. 274-?.

[KinHah86] King, Robert I., and Robert S. Hahn, (1986), *Handbook of Modern Grinding Technology,* Chapman and Hall, New York and London.

[KoeTlu70] Koenigsberger, F., and J. Tlusty, (1970), *Machine Tool Structures, Vol. 1,* Pergamon Press,Oxford.

[Koenig78] Koenigsberger, F., (1978), *Testing Machine Tools: For the Use of Machine Tool Makers, Users, Inspectors, and Plant Engineers, Eighth Revision* Pergamon Press,Oxford.

[KolDeV91] Kolarits, Francis M., and Warren R. DeVries, (1991), "A Mechanistic Dynamic Model of End Milling for Process Controller Simulation," *ASME Transactions, Journal of Engineering for Industry*, Vol. 113, No. 2, May, pp.176-183.

[KniTob69] Knight, W.A. and S.A. Tobias, (1969), "Torsional Vibrations and Machine Tool Stability" *Proceedings of the Tenth Machine Tool Design and Research Conference*.

[MATRIX91] MATRIXx, (1991), Integrated Systems, Inc., Santa Clara, CA.

[Merrit65] Merritt, H.E., (1965), "Theory of Self-Excited Machine Tool Chatter, Contribution to Machine Tool Chatter Research - 1", *ASME Journal of Engineering for Industry*, November, Vol. 87, pp. 447-54.

[SmiTlu91] Smith S., and Tlusty, J., (1991), "An Overview of Modelling and Simulation of the Milling Process," *ASME Transactions, Journal of Engineering for Industry*, Vol. 113, No. 2, May, pp. 169-175.

[SnoBro69] Snoeys, R., and D. Brown, (1969), "Dominating Parameters in Grinding Wheel and Workpiece Regenerative Chatter," *Proceedings of the Tenth International Machine Tool Design and Research Conference*, pp. 325-48.

[SutDeV86] Sutherland, J. W., and R. E. DeVor, (1986), "An Improved Method for Cutting Force and Surface Error Prediction in Flexible End Milling Systems," *ASME J. of Engineering for Industry*, Vol. 108, November, pp. 269-279.

[Thomso80] Thomson, R.A., (1980), "A General Theory for Regenerative Stability," *Proceedings of the Eighth North American Manufacturing Research Conference*, University of Missouri-Rolla, May 18-21, pp. 377-387.

[Tlusty65] Tlusty, J., (1965), "A Method of Analysis of Machine Tool Stability," *Proceedings of the Sixth Machine Tool Design and Research Conference*.

[Tobia65] Tobias, Stephen Albert, (1965), *Machine Tool Vibration*, Authorized translation by A.H. Burton, John Wiley, New York.

[Weck84] Weck, Manfred, (1984), *Handbook of Machine Tools, Volume 1 - Types of Machines, Forms of Construction and Applications, Volume 2 - Construction and Mathematical Analysis, Volume 3 - Automation and Controls, Volume 4 - Metrological Analysis and Performance Tests*, John Wiley & Sons, New York.

[WelSmi70] Welbourn, D.B., and J.D. Smith, (1970), *Machine-Tool Dynamics; An Introduction*, University Press, Cambridge, England.

8.6 Problems

8.1 Suppose that the data given below is used to determine the stability of cutting conditions used to turn tubes. This problem is to show how cutting conditions, rather than machine tool design, affect stability. Use the criterion based on $\left|\frac{z(\omega)}{a(\omega)}\right| = 1$ to define stability at the damped natural frequency $\omega = \omega_n \sqrt{1-\zeta^2}$. Assume the following: for the machine tool-tool holder combination the static equivalent stiffness is k_{eq} = 10 kN/mm and the measured natural frequency and damping for this machine tool are f_n=85 Hz and ζ=0.01 respectively. You will have to calculate the cutting stiffness k_c, based on the mechanics of orthogonal cutting where the friction coefficient between the chip and rake face μ=1, the shear angle prediction equation to use is Lee and Shaffer's, the tool signature is 0°, 20°, 10°, 10°, 0°, 0°, 0 mm, and to account for work hardening assume that $\tau_s = 65 \text{ MPa } (h_{avg})/1\text{mm})^{-0.1}$.

a. If the wall thickness of the tube is 5 mm and the rotational speed of the workpiece N=400 rev/min, determine the stability of the cutting conditions if the relative feedrate $a = 0.1$ mm/rev and 0.4 mm/rev.

b. The table given below is to provide 4 points on a stability chart for $a = 0.4$ mm/rev.

A Four Point Stability Chart		
b=6 mm	$\left\|\frac{z(\omega)}{a(\omega)}\right\| =$	$\left\|\frac{z(\omega)}{a(\omega)}\right\| =$
b=5 mm	$\left\|\frac{z(\omega)}{a(\omega)}\right\| =$	$\left\|\frac{z(\omega)}{a(\omega)}\right\| =$
N(rev/min)	200	400

8.2 For a turning process the stability chart shown in Figure 8.9 applies. To use it, the first thing that needs to be done is to estimate the specific cutting pressure k'_c or the cutting stiffness k_c. Do this by assuming that the workpiece material is 4340 steel that has a hardness of 210 Brinell. The specific cutting energy correction factors for power and feeding force are: $K_c = 0.9$, $K_r = 1$, $K_w = 1.3$, $C_c = 1$, $C_r = 1$, $C_w = 3$.

a. If the depth of cut is b=5 mm what range(s) of rotational speeds will result in stable cutting.

b. What depth of cut will be stable for all rotational speeds, or at least the range of speeds covered by this chart?

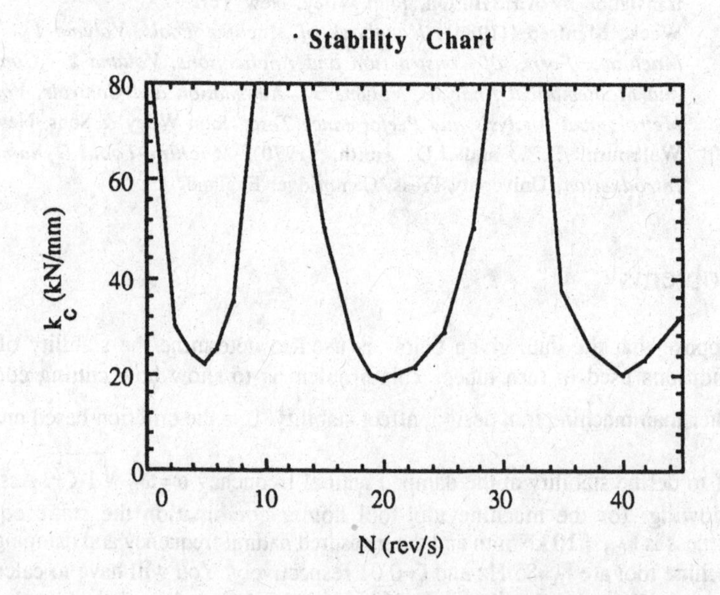

Figure 8.9 Experimental stability chart for Problem 8.2.

Bibliography

[AbbFir33] Abbot, E.J. and F.A. Firestone, (1933), "Specifying Surface Quality," *Mechanical Engineering*, Vol. 55, pp. 569-572.

[Abb*etal*33] Abbot, E.J., S. Bousky and D.E. Williamson, (1938), "The Profilometer," *Mechanical Engineering*, Vol. 60, pp. 205-216.

[Albrec60] Albrecht, P., (1960), "New Developments in the Theory of the Metal Cutting Process - Part I: The Ploughing Process in Metal Cutting," *ASME Transactions, Journal of Engineering for Industry*, Series B, Vol. 82, pp. 348-358.

[Albrec61] Albrecht, P., (1961), "Self Induced Vibrations in Metal Cutting," ASME Paper No. 61-WA-195.

[AllStu84] Allocca, John A., and Allen Stuart, (1984), *Transducers: Theory and Applications*, Reston Publishing Co., a Prentice-Hall Company, Reston, VA.

[AmeMac79] Edited by American Machinist, (1979), Metalcutting: Today's Techniques for Engineers and Shop Personnel, American Machinist/McGraw-Hill, New York.

[Ams*etal*84] Amstead, B. H., Ostwald, P. F., and Begeman, M. L., (1984), *Manufacturing Processes*, John Wiley, New York.

[And*etal*85] Andrew, Colin, Trevor D. Howes and Tom R.A. Pearce, (1985), *Creep Feed Grinding*, Holt, Rinehart and Winston, New York and London.

[Archar53] Archard, J.F., (1953), "Contact and Rubbing of Flat Surfaces," *Journal of Applied Physics*, Vol. 24, pp. 981-988.

[ArmBro69] Armarego, E.J.A, and R.H. Brown, (1969), *The Machining of Metals*, Prentice Hall, Inc., Englewood Cliffs, NJ.

[ASME52] American Society of Mechanical Engineers (ASME), (1952), *Manual on Cutting of Metals, Second Edition*, The American Society of Mechanical Engineers.

[ASTE50] American Society of Tool Engineers (ASTE), (1950), *Practical Design of Manufacturing Tools, Dies, and Fixtures*, First Edition, McGraw-Hill.

[ASTME60] American Society of Tool and Manufacturing Engineers, (1960) *Metal Cutting Research Reports, (1954-1960)*, ASTME Research Fund, Detroit, MI.

[ASTME67] American Society of Tool and Manufacturing Engineers, (1967), *Handbook of Industrial Metrology*, Prentice Hall, Englewood Cliffs, NJ.

[BS113472] British National Standard BS1134, (1972), "Assessment of Surface Texture," British Standards Institute, London.

[Bab*etal*85] Babin, T. S., J. M. Lee, J. W. Sutherland, and S. G. Kapoor, (1985), "A Model for End Milled Surface Topography," *Proc. 13th North American Manufacturing Research Conf.*, Berkeley, CA, May 1985, pp. 362-368.

[BabSut86] Babin, T. S., J. W. Sutherland, and S. G. Kapoor, (1986), "On the Geometry of End Milled Surfaces," *Proc. 14th NAMRC*, Minneapolis, MN, May 1986, pp. 168-176.

[BhaLin82] Bhateja, Chander, Lindsay, Richard, (1982) *Grinding Theory Techniques and Troubleshooting, Society of Manufacturing Engineers*, Dearborn, MI.

[BhaHam69] Bhattacharyya, Amitabha, and Inyong Ham, (1969), *Design of Cutting Tools: Use of Metal Cutting Theory*, American Society of Tool and Manufacturing Engineers, Dearborn, MI.

[BhaLin82] Bhateja, Chander and Richard Lindsay, (1982), *Grinding: Theory, Techniques and Troubleshooting*, Society of Manufacturing Engineers, Dearborn, MI.

[Bjørke89] Bjørke, Øyvund, (1989), *Computer Aided Tolerancing, Second Edition*, ASME Press, New York.

[Briks96] Briks, A.A., (1896) *Cutting of Metals* (Rezaniye Metallov).

[BoeKal88] Boerma, J., and Kals, H., (1988), "FIXES, a System for Automatic Selection of Set-Ups and Design of Fixtures," *Annals of the CIRP*, Vol. 37/1, pp. 443-446.

[Bollin72] Bollinger, J., (1972), "Computer Control of Machine Tools", *CIRP Annals*, Vol. 21, No. 2.

[BolDuf88] Bollinger, John G. and Duffie, Neil A., (1988), *Computer Control of Machines and Processes*, Addison Wesley.

[BooKni89] Boothroyd, Geoffrey, and Winston A. Knight, (1989), *Fundamentals of Machining and Machine Tools, Second Edition*, Marcel Dekker, Inc., New York and Basel.

[Boston41] Boston, Orlan W., (1941), *Metal Processing*, John Wiley & Sons, New York.

[Boston43] Boston, Orlan W., (1943), *A Bibliography on Cutting of Metals: 1864-1943*, The American Society of Mechanical Engineers, New York.

[BowTab71] Bowden, F.P., and D. Tabor, (1971 printing), *The Friction and Lubrication of Solids, Part 1*, Oxford University Press, London.

[Boyes82] Boyes, W.E., (1982), *Jigs and Fixtures, Second Edition*, SME, Dearborn, MI.

[Box*etal*78] Box, George E.P., William G. Hunter, J. Stuart Hunter, (1978), *Statistics for Experimenters: An Introduction to Design, Data Analysis and Model Building*, Wiley, NY.

[Bursta65] Burstall, Aubrey F., (1965), *A History of Mechanical Engineering*, The MIT Press, Canbridge, MA.

[Can*etal*89] Cantillo, D., S. Calabrese, W.R. DeVries and J.A. Tichy, (1989), "Thermal Considerations and Ferrographic Analysis in Grinding," *Grinding Fundamentals and Applications*, Edited by S. Malkin and J.A. Kovach, PED-Vol. 39, ASME Publication No. H00571, New York, NY, pp. 323-334.

[CarStr88] Carroll, J.T., and J.S. Strenkowski, (1988), "Finite Element Models of Orthogonal Cutting With Application to Single Point Diamond Turning, " *International Journal of Mechanical Science*, Vol. 30, No. 12., pp. 899-920.

[CarJae59] Carslaw, H.S., and J.C. Jaeger, (1959), *Conduction of Heat in Solids, Second Edition*, Oxford University Press, London.

[Centne64] Centner, R., (1964), *Final Report on the Development of the Adaptive Control Technique for a Numerically Controlled Milling Machine*, USAF Technical Documentary Report ML-TDR-64-279.

[Chang90] Chang, Tien-Chien, (1990), *Expert Process Planning for Manufacture*, Addison-Wesley, Reading, MA.

[Chiria90] Chiriacescu, Sergiu T., (1990), *Stability in the Dynamics of Metal Cutting*, Elsevier Science Publishers, New York.

[Chou*etal*89] Chou, Y-C., V. Chandru and M.M. Barash, (1989), "A Mathematical Approach to Automatic Configuration of Machining Fixtures: Analysis and Synthesis," *ASME Transactions, Journal of Engineering for Industry,* Vol. 111, No. 4, pp. 579-584.

[Cincin46] The Cincinnati Milling Machine Co., (1946), *A Treatise on Milling and Milling Machines, Section Two*, Cincinnati, OH.

[CIRPUn75] "CIRP Unifided Terminology," (1975), Ed. J. Peters, Provisional Edition 1975-01-01, CIRP Secretariat, Paris.

[CIRPUn86] "CIRP Unifided Terminology," (Edition 1986), CIRP Secretariat, Paris.

[CogDeV86] Cogdell, J. David, and Warren R. DeVries, (1986), "Asperity Peak Curvature Changes During Lubricated Sliding," *CIRP Annals*, Vol. 35/1, pp. 401-404.

[Col*etal*86] Colbert, John L., Roland J. Menassa and Warren R. DeVries, (1986), "A Modular Fixture for Prismatic Parts in an FMS," *Proceedings of the Fourteenth North American Manufacturing Research Conference*, University of Minnesota, pp. 597-602, May 28-30.

[Col*etal*78] Colwell, L. V., J. C. Mazur and W. R. DeVries, (1978), "Analytical Strategies for Automatic Tracking of Tool Wear," *Sixth North American Metalworking Research Conference Proceedings*, University of Florida, pp. 274-282, April 16-19.

[Cook59] Cook, Nathan H., (1959), "Self-Excited Vibrations in Metal Cutting," *ASME Transactions, Journal of Engineering for Industry,* Series B, Vol. 81, pp. 183-186.

[Cook66] Cook, Nathan H., (1966), *Manufacturing Analysis*, Addison-Wesley, Reading, MA.

[Cre*etal*57] Creveling, J.H., T.F. Jordan and E.G. Thomsen, (1957), "Some Studies of Angle Relationships in Metal Cutting," *ASME Transactions*, Vol. 79, pp.127-138.

[Datsko66] Datsko, J., (1966), *Materials Properties and Manufacturing Processes,* Joseph Datsko Consultants, Ann Arbor, MI.

[DanUls85a] Danai, K., and Ulsoy, A. G., (1985a), "A Model Based Approach to Tool Wear Estimation in Turning," *Proceedings of the Sensors 85 Conference*, Detroit, Michigan.

[DanUls85b] Danai, K., and Ulsoy, A. G., (1985b), "A Dynamic State Model for On-Line Tool Wear Estimation in Turning,"in Kannatey-Asibu, Jr., E., Ulsoy, A. G., and Komanduri, R. (eds.), *Sensors and Controls for Manufacturing*, ASME, New York.

[DanUls88] Danai, J., and Ulsoy, A. G., (1988), "Dynamic Modeling of Cutting Forces in Turning, Milling, and Drilling, from Experimental Data," in *Control Methods for Manufacturing Processes,* D. E. Hardt (Ed.), ASME, DSC Vol. 9, Dec., pp. 27-34.

[DanPak86] Daneshmend, L. K., and H. A. Pak, (1986), "Model Reference Adaptive Control of Feed Force in Turning," *ASME J. of Dynamic Systems, Measurement, and Control*, Vol. 108, No. 3, Sept., pp. 215-222.

[Decneu74] Decneut, A., (l974), "Verantwoorde Keuve Van De Slijpvoorwaarden," Dissertation, Catholic University of Leuven.

[DenHar34] DenHartog, J.P., (1934), *Mechanical Vibrations*, McGraw-Hill Book Co., New York and London.

[DeVetal91] DeVor, R.E., T.H. Chang, J.W. Sutherland, (1991) *Statistical Quality Design and Control: Contemporary Concepts and Methods*, MacMillan, New York, N.Y.

[DeVrie69] DeVries, Marvin F., (1969), "Machining Economics - A Review of the Traditional Approaches and Introduction to New Concepts", *American Society of Tool and Manufacturing Engineers Paper #MR69-279*, pp. 435-442.

[DeVetal83] DeVries, W. R., Raski, J. Z., and Ulsoy, A. G., (1983), "Microcomputer Applications in Manufacturing: A Senior Laboratory Course," *Proceedings of the International Computers in Engineering Conference*, Chicago.

[DeVrie79a] DeVries, W. R., (1979a), "Chatter Vibration Monitoring Using On-Line Identification of Discrete Second Order Autoregressive Time Series Models," *Seventh North American Metalworking Research Conference Proceedings*, University of Michigan, pp. 275-278, May 13-16.

[DeVrie79b] DeVries, W.R., (1979b), "Identification and Analysis of ARMA Time Series Models for Chatter Vibration," *Proceedings of the Tenth Annual Modeling and Simulation Conference*, University of Pittsburgh, pp. 429-435, April 25-27.

[DeVrie79c] DeVries, W.R., (1979c), "Autoregressive Time Series Models for Surface Profile Characterization," *CIRP Annals*, Vol. 28/1, pp. 437-440.

[DeVetal81] DeVries, W. R., J. Z. Raski and J. C. Mazur, (1981), "Investigation of Adaptive Exponential Smoothing Algorithms in Monitoring Tool Wear," *Ninth North American Manufacturing Research Conference Proceedings*, Pennsylvania State University, pp. 523-527, May 19-22.

[DeVrie81] DeVries,W. R., (1981), "Automatic Autoregressive Model Identification Applied to Surface Characterization," *Proceedings of the Twelfth Annual Modeling and Simulation Conference*, University of Pittsburgh, pp. 349-354, April.

[DeVrie82a] DeVries,W. R., (1982a), "The Surface Topography of Journals," *CIRP Annals*, Vol. 31/1, pp. 467-470.

[DeVrie82b] DeVries,W. R., (1982b), "A Three-Dimensional Model of Surface Asperities Developed Using Moment Theory," *ASME Transactions, Journal of Engineering for Industry*, Vol. 104, pp. 343-348.

[DeVEva84] DeVries, W. R., and M. S. Evans, (1984), "Computer Graphics Simulation of Metal Cutting," *CIRP Annals*, Vol. 33/1, pp. 15-18.

[DeVrLi85] DeVries, W. R., and Cheng-Jih Li, (1985), "Algorithms to Deconvolve Stylus Geometry From Surface Profile Measurements," *ASME Transactions, Journal of Engineering for Industry*, Vol. 107, pp. 167-174.

[Doebel83] Doebelin, Ernest O., (1983), *Measurement Systems: Applications and Design*,McGraw Hill Book Co., New York.

[Doyle85] Doyle, L. E., (1985), *Manufacturing Processes and Materials for Engineers*, Prentice-Hall, Englewood Cliffs.

[Drucke49] Drucker, D.C., (1949), "An Analysis of the Mechanisms of Metal Cutting," *Journal of Applied Physics*, Vol. 20., pp. 1013-1021.

[Drozda82] Drozda, T., ed., (1982), *Manufacturing Engineering Explores Grinding Technology, A collection of Manufacturing Engineering articles discussing grinding*, Society of Manufacturing Engineers, Dearborn, MI.

[Duncan74] Duncan, Acheson J., (1974), *Quality Control and Industrial Statistics, Fourth Edition*, Richard D. Erwin, Inc, Homewood, IL.

[Eggetal59] Eggleston, D.M., R. Herzog and E.G. Thomsen, (1959), "Observations on the Angle Relationships in Metal Cutting," *ASME Journal of Engineering for Industry*, Series B, Vol. 81, pp. 251-262.

[ErnMer41] Ernst, H. and M.E. Merchant, (1941), "Chip Formation, Friction, and High Quality Machined Surfaces," Surface Treatment of Metals, *Transactions of the American Society of Metals*, Vol. 29, pp. 299-378.

[Evans89] Evans, Chris, (1989), *Precision Engineering: An Evolutionary View*, Bedford, Cranfield.

[EvaDeV90] Evans, Mark S., and Warren R. DeVries, (1990),"Dynamic Model of Metal Cutting Suitable for Parameter Estimation From On-Line Measurements," *Proceedings of the 1990 ASME International Computers in Engineering Conference and Exposition*, Boston, Vol. 1, pp. 551-558, August 5-9.

[Fer*etal*85] Ferreira, P., Kocher, B., Liu, C., and Chandru, V., (1985), "AIFIX: An Expert System Approach to Fixture Design," *Computer-Aided/Intelligent Process Planning*, ASME PED-Vol. 19, pp.73-82.

[Flo*etal*84] Flom, D. G., Komanduri, R., and Lee, M., (1985), "High Speed Machining of Metals", *Annual Review of Material Science*, Vol. 14, pp 231-278.

[Fu*etal*84] Fu, H.J., R.E. DeVor and S.G. Kapoor, (1984), "A Mechanistic Model for the Prediction of the Force System in Face Milling Operations," *ASME Journal of Engineering for Industry*, Vol. 106, Feb., pp. 81-88.

[Fuj*etal*70a] Fujii, S., M.F. DeVries, and S.M. Wu, "An Analysis of Drill Geometry for Optimum Drill Design by Computer, Part I - Drill Geometry Analysis," *ASME Journal of Engineering for Industry*, Series B, Vol. 92, No. 3,. pp. 647-656.

[Fuj*etal*70b] Fujii, S., M.F. DeVries, and S.M. Wu, "An Analysis of Drill Geometry for Optimum Drill Design by Computer, Part II - Computer Aided Design," *ASME Journal of Engineering for Industry*, Series B, Vol. 92, No. 3,. pp. 657.

[FusSri89a] Fussell, B. K., and Srinivasan, K., (1989a), "An Investigation of the End Milling Process Under Varying Machining Conditions," *ASME Journal of Engineering for Industry*, Vol. 111, Feb., pp. 27-36.

[FusSri89b] Fussell, B. K., and Srinivasan, K., (1989b), "On-Line Identification of End Milling Process Parameters," *ASME J. of Engineering for Industry*, Vol. 111, Nov., pp. 322-330.

[GalKni86] Gallagher, C.C., and W.A. Knight (1986), *Group Technology Production Methods in Manufacture*, Ellis Horwood, Ltd. Chichester, England.

[Gat*etal*88] Gatlin, David, Warren R. DeVries, L.K. Lauderbaugh and J. Scott Reiss, (1988), "A Model for Cusped Surface Grinding Used in Designing a Robotic Die Finishing System," *Computer-Aided Design and Manufacture of Dies and Molds*, Edited by K. Srinivasan and W. R. DeVries, PED-Vol. 32, ASME Publication No. G00943, New York, NY, pp. 1-17.

[GeoMet82] Geo-Metrics II, (1982), "Dimensioning and Tolerancing," *ANSI/ASME Standard*, Y14.5M, pp. 35-52.

[Gilber52] Gilbert, W.W., (1952), "Economics of Machining," in *Machining Theory and Practice*, American Society of Metals, Cleveland.

[Gilber62] Gilbert, W.W., (1962), "Economics of Machining," in *Machining With Carbides and Oxides*, ASTME/McGraw Hill, New York.

[Hahn52] Hahn, R.S., (1952), "Metal Cutting Chatter and Its Elimination,"*ASME Transactions, Journal of Engineering for Industry*, Series B, Vol. 74, pp. 1073-1079.

[HanTob74] Hanna, N.H. and S.A. Tobias, (1974), "A Theory of Nonlinear Regenerative Chatter," *ASME Journal of Engineering for Industry*, February, pp. 274-?.

[Hatsch79] Hatschel, R.L., (1979), "Fundamentals of Drilling," *American Machinist*, Vol. 122, No. 2, pp.107-130.

[Hill50] Hill, R., (1950), *The Mathematical Theory of Placticity*, Oxford University Press, London.

[Hill54] Hill, R., (1956), "The Mechanics of Machining: A New Approach," *Journal of the Mechanics and Physics of Solids*, Vol. 3, pp. 47-53.

[HolMur83] Holzhauer, Wolfgang, and Murray, S.F., (1983), "Continuous Wear Measurements by On-Line Ferrography," *Wear*, Vol. 90, pp. 11-19.

[Hucks52] Hucks, H., (1952), "Plastizitatsmechanische Theorie der Spanbildung," *Werkstatt und Betrieb*, No. 1.

[HutYu90] Hutton, D.V., and Qinghuan Yu, (1990), "On the Effects of Built-Up Edge on Acoustic Emission in Metal Cutting," *ASME Transactions, Journal of Engineering for Industry*, Vol. 112, May, pp. 184-189.

[Iwa*etal*84] Iwata, K, K. Osakada and Y. Terasaka, (1984), "Process Modeling of Orthogonal Cutting by the Rigid-Plastic Finite Element Method," *ASME Transactions Journal of Materials and Technology*, Vol. 106, pp. 132-138.

[Jaeger42] Jaeger, J.C., (1942), "Moving Sources of Heat and the Temperatures of Sliding Contacts," *Proceedings of the Royal Society of New South Wales*, Vol. 76, pp.203-224.

[Jia*etal*88] Jiang, W., Wang, Z., and Cai, Y., (1988), "Computer-Aided Group Fixture Design," *Annals of the CIRP*, Vol. 37/1, pp 145-148.

[Kalpak82] Kalpakjian, S., Ed., (1982), *Tool and Die Failures Source Book*, American Society for Metals, Metals Park, OH.

[Kalpak84] Kalpakjian, S., (1984), *Manufacturing Processes for Engineering Materials*, Addison-Wesley, Reading, MA.

[KalJai80] Kalpakjian, Serope, and Sulekh Jain, Ed.s, (1980), *Metalworking Lubrication, Proceedings of the International Symposium on Metal Working Lubrication*, Century 2 Emerging Technologies Conferences, ASME Publication H00159, San Francisco, August 18-19.

[Kane82] Kane, George E., Ed., (1982), *Modern Trends in Cutting Tools*, SME, Dearborn, MI.

[Kassen69] Kassen, G., (1969), "Beschreibung der Elementaren Kinematik des Schleifuorganges," Dissertation T.H. Aachen.

[Kececi58] Kececioglu, D., (1958), "Force Components, Chip Geometry, and Specific Cutting Energy in Orthogonal and Oblique Machining of SAE-1015 Steel," *ASME Journal of Engineering for Industry*, Series B, Vol. 80, pp. 149-157.

[KenMet84] *Kennametal Metalcuttung Tools*, (1984), Kennametal Inc., Latrobe, PA.

[KenMil86] *Kennametal Milling Systems*, (1986), Kennametal Inc., Latrobe, PA.

[KinHah86] King, Robert I., and Robert S. Hahn, (1986), *Handbook of Modern Grinding Technology*, Chapman and Hall, New York and London.

[KinWhe66] King, Alan G., and W.M. Wheildon, (1966), *Ceramics in Machining Processes*, Academic Press, New York and London.

[Kiretal77] Kirk, J. A., Cardenas-Garcia, J.F., and Allison, C.R., (1977), "Evaluation of Grinding Lubricants - Simulation Testing and Grinding Performance," *Wear*, Vol.20, No.4, pp.333-339.

[KirCar77] Kirk, J.A., Cardenas-Garcia, J.F., (1977), "Evaluation of Grinding Lubricants-Simulation Testing and Grinding Performance," *Transactions of the American Society of Lubrication Engineers*, Vol. 20, pp.333-339.

[Klietal82a] Kline, W. A., DeVor, R. E., and Lindberg, J. R., (1982a), "The Prediction of Cutting Forces in End Milling with Application to Cornering Cuts," *Int. J. of Machine Tool Design and Research*, Vol. 22, No. 1, pp. 7-22.

[Klietal82b] Kline, W. A., DeVor, R. E., and Shareef, I. A., (1982), "The Prediction of Surface Accuracy in End Milling," *ASME J. of Engineering for Industry*, Vol. 104, Aug., pp. 272-278.

[KliDev83] Kline, W. A., and DeVor, R. E., (1983), "The Effect of Runout on Cutting Geometry and Forces in End Milling," *Int. J. of Machine Tool Design and Research*, Vol. 23, No. 2/3, pp. 123-140.

[KobTho60] Kobayashi, S. and E.G. Thomsen, (1960), "The Role of Friction in Metal Cutting," *ASME Transactions, Journal of Engineering for Industry*, Series B, Vol. 82, pp. 324-332.

[KobTho62a] Kobayashi, S. and E.G. Thomsen, (1962a), "Metal-Cutting Analysis - I: Re-Evaluation and New Method of Presentation of Theories," *ASME Transactions, Journal of Engineering for Industry*, Series B, Vol. 84, pp. 63-70.

[KobTho62b] Kobayashi, S. and E.G. Thomsen, (1962b), "Metal-Cutting Analysis - II: New Parameters," *ASME Transactions, Journal of Engineering for Industry*, Series B, Vol. 84, pp. 71-80.

[Kobetal60] Kobayashi, S., R.P. Herzog, D.M. Eggleston and E.G. Thomsen, (1960) "A Critical Review of Metal-Cutting Theories With New Experimenatl Data," *ASME Journal of Engineering for Industry*, Series B, Vol. 82, pp. 333-347.

[KoeTlu70] Koenigsberger, F., and J. Tlusty, (1970), *Machine Tool Structures, Vol. 1*, Pergamon Press,Oxford.

[Koenig78] Koenigsberger, F., (1978), *Testing Machine Tools: For the Use of Machine Tool Makers, Users, Inspectors, and Plant Engineers, Eighth Revision* Pergamon Press,Oxford.

[KolDeV89] Kolarits, F.M. and W.R. DeVries, (1989), "A Model of the Geometry of the Surfaces Generated in End Milling With Variable Process Inputs," *Mechanics of Deburring and Surface Finishing Processes*, Edited by R.J. Stango and P.R. Fitzpatrick, PED-Vol. 38, ASME Publication No. H00570, New York, NY, pp. 63-78.

[KolDeV90] Kolarits, Francis M., and Warren R. DeVries, (1990), "Adaptive Pole Placement Force Control of End Milling," *Proceedings of the 1990 American Control Conference*, San Diego, CA, Vol. 2/3, pp.1115-1120, May 23-25.

[KolDeV91] Kolarits, Francis M., and Warren R. DeVries, (1991), "A Mechanistic Dynamic Model of End Milling for Process Controller Simulation," *ASME Transactions, Journal of Engineering for Industry*, Vol. 113, No. 2, May, pp.176-183.

[KniTob69] Knight, W.A. and S.A. Tobias, (1969), "Torsional Vibrations and Machine Tool Stability" *Proceedings of the Tenth Machine Tool Design and Research Conferene.*

[Koeetal71] Koenig, W., *et. al.*, (1971), "A Survey of the Present State of High Speed Grinding," *CIRP Annals*, Volume 19/4.

[Komand85] Komanduri, R., (1985), "High Speed Machining", *Mechanical Engineering*, Vol. 107, No. 12, pp 64-76.

[KomDes82] Komanduri, R. and Desai, J.D., (1982), "Tool Materials for Machining", Technical Information Series #82CRD220, General Electric Company, Technical Information Exchange.

[KomSha75] Komanduri, R., and M.C. Shaw, (1975), "Formation of Spherical Particles in Grinding," *Phil. Mag.*, Vol. 32, pp. 711-724.

[KomErp90] Komvopolulos, K, and S.A. Erpenbeck, (1990), "Finite Element Modeling of Orthogonal Metal Cutting," *Computer Modeling and Simulation of Manufacturing Processes*, Edited by B. Singh *et al.*, MD-Vol. 20/PED-Vol. 48, ASME Publication No. G00552, New York, NY, pp. 1-24.

[Koren77] Koren, Y., (1977), "Computer Based Machine Tool Control", *IEEE Spectrum*, Vol. 14, March, pp 80-84.

[Koren83] Koren, Y., (1983), *Computer Control of Manufacturing Systems* , McGraw-Hill , New York.

[KorBen78] Koren, Y., and Ben-Uri, J., (1978), *Numerical Control of Machine Tools*, Khanna, Delhi.

[KraRat90] Krar, Stephen F., and E. Ratterman, (1990), *Superabrasives: Grinding and Machining With CBN and Diamond*, Gregg Division of McGraw-Hill, New York.

[Kronen54] Kronenberg, M., (1954) *Grundzuge der Zerspanungslehre* (2nd ed) Berlin: Springer-Verlag.

[LauUls88] Lauderbaugh, L. K., and A. G. Ulsoy, (1988), "Dynamic Modeling for Control of the Milling Process," *ASME J. of Engr. for Industry*, Vol. 110, Nov., pp. 367-375.

[LauUls89] Lauderbaugh, L. K., and A. G. Ulsoy, (1989), "Model Reference Adaptive Force Control in Milling," *ASME J. of Engr. for Industry*, Vol. 111, Feb., pp. 13-21.

[Lavine88] Lavine, Adrienne S., (1988), "A Simple Model for Convective Cooling During the Grinding Process," *ASME Transactions, Journal of Engineering for Industry*, Vol. 109, pp.1-6.

[LeClair82] LeClair, Steven R., (1982), "IDEF the Method, Architecture the Means to Improved Manufaccturing Productivity," SME Technical Paper #MS82-902, SME.

[Lee84] Lee, D., (1984), "The Nature of Chip Formation in Orthogonal Machining," *ASME Transactions Journal of Materials and Technology*, Vol. 106, pp. 9-15.

[LeeSha51] Lee, E.H., and B.W. Shaffer, (1951), "The Theory of Plasticity Applied to a Problem of Machining," *ASME Transactions, Journal of Applied Mechanics*, Vol. 73, pp.405-413.

[LeeHay87] Lee, J. D., and Haynes, L.S., (1987), "Finite Element Analysis of Flexible Fixturing System," *ASME Transactions, Journal of Engineering for Industry*, Vol. 109, pp. 579-584.

[LeeWoo90] Lee, Woo-Jong and T.C. Woo, (1990), "Tolerances: Their Analysis and Synthesis," *ASME Transactions, Journal of Engineering for Industry*, Vol. 112, May, pp. 113-121.

[LezSha90] Lezanski, P., and M.C. Shaw, (1990), "Tool Face Temperatures in High Speed Milling," *ASME Transactions, Journal of Engineering for Industry*, Vol. 112, May, pp. 132-135.

[LiDeV84] Li, C. J., and W. R. DeVries, (1984), "Computer Simulation to Assess the Effect of Stylus Geometry on Surface Metrology," *Proceedings of the Twelfth North American Manufacturing Research Conference*, Michigan Technological University, May 30-31 and June 1, pp. 418-423.

[Li87] Li, C.J., (1987), "Dynamic Modeling of Turning to Include Flank Wear and Non-Linearities Due to Dynamic and Kinematic Factors," Ph.D. Thesis, Department of Mechanical Engineering, Rensselaer Polytechnic Institute, May.

[LiDeV88] Li, Cheng-Jih, and Warren R. DeVries, (1988), "The Effect of Shear Plane Length Models on Stability in Machining Simulation," *Proceedings of the Sixteenth North American Manufacturing Research Conference*, University of Illinois, Urbana, IL, May 25-27, pp. 195-201.

[LiChen88] Li, Y.Y and Y. Chen, (1988), "Simulation of Surface Grinding," *ASME Transactions, Journal of Engineering for Industry*, Vol. 111, No. 1, pp. 46-53.

[LiuLia89] Liu, M., and S. Liang, (1989), "Model Reference Adaptive Control of Nonminimum-Phase Milling," *Proc. 17th NAMRC*, Columbus, OH, May pp. 266-273.

[LinBha82] Lindsey, Richard and Chander Bhateja, Ed.s, (1982), *Grinding-Theory, Technique and Troubleshooting*, SME.

[LoeSha54] Loewen, E.G., Shaw, M.C., (1954) "On the Analysis of Cutting Tool Temperatures," *ASME Transactions*, Vol. 76.

[Lud*etal*87] Ludema, Kenneth C., Robert M. Caddell and Anthony G. Atkins, *Manufacturing Engineering: Economics and Processes*, Prentice Hall, Inc. Englewood Cliffs, NJ, 1987.

[MTTF80] Machine Tool Task Force, (1980), *Technology of Machine Tools, Volume 1 - Executive Summary, Volume 2 - Machine Tool Systems Management and Utilization, Volume 3 - Machine Tool Mechanics, Volume 4 - Machine Tool Controls, Volume 5 - Machine Tool Accuracy*, Lawrence Livermore Laboratory, Livermore, CA.

[MachDa80] *Machining Data Handbook, Third Edition*, (1980), Vol.s 1 and 2, Machinability Data Center, Metcut Research Associates, Inc., Cincinnati, OH .

[Malkin88] Malkin, S., (1988), "Grinding Temperatures and Thermal Damage," *Thermal Aspects in Manufacturing*,, PED-Vol. 30, ASME, New York, NY, pp. 145-156.

[Malkin89] Malkin, S., (1989), *Grinding Technology: Theory and Applications of Machining With Abrasives*, Ellis Horwood, Ltd. Chichester, England.

[ManWil88] Mani, M., and Wilson, W., (1988), "Automated Design of Workholding Fixtures Using Kinematic Constraint Synthesis," *Proceedings of the Sixteenth North American Manufacturing Research Conference*, pp. 437-444.

[Martel41] Martellotti, M.E., (1941), "An Analysis of the Milling Process," *ASME Transactions*, Vol. 63, pp. 667.

[Martel45] Martellotti, M.E., (1945), "An Analysis of the Milling Process - Part II: Down Milling," *ASME Transactions*, Vol. 67, pp. 233.

[MasKor85] Masory, O., and Y. Koren, (1985), "Stability Analysis of a Constant Force Adaptive Control System for Turning," *ASME J. of Engineering for Industry*, Vol. 107, November, pp. 295-300.

[MasKor80] Masory, O., and Y. Koren, (1980), "Adaptive Control System for Turning," *Annals of the CIRP*, Vol. 29, No. 1, pp. 281-284.

[MATRIX91] MATRIXx, (1991), Integrated Systems, Inc., Santa Clara, CA.

[Menass89] Menassa, Roland J., (1989), "Synthesis, Analysis and Optimization of Fixtures for Prismatic Parts," Ph.D. Thesis, Rensselaer Polytechnic Institute.

[MenDeV88] Menassa, R.J., and W.R. DeVries, (1988), "Optimization Methods Applied to Selecting Support Positions in Fixture Design," *Proceedings of the USA-Japan Symposium on Flexible Automation*, Minneapolis, Vol. 1, pp. 475-482, July 18-20.

[MenDeV89] Menassa, Roland J. and Warren R. DeVries, (1989), "Locating Point Synthesis in Fixture Design," *CIRP Annals*, Vol. 38/1, pp. 165-170.

[Mercha44] Merchant, M.E., (1944), "Basic Mechanics of the Metal Cutting Process," *Journal of Applied Mechanics*, Vol. 66, pp. 168.

[Mercha45a] Merchant, M.E., (1945a), "Mechanics of the Metal Cutting Process," *Journal of Applied Physics, 16:5*, pp. 267-275.

[Mercha45b] Merchant, M.E., (1945b), "Mechanics of the Metal Cutting Process," *Journal of Applied Physics, 16:6* , pp. 318-324.

[Merrit65] Merritt, H.E., (1965), "Theory of Self-Excited Machine Tool Chatter, Contribution to Machine Tool Chatter Research - 1", *ASME Journal of Engineering for Industry*, November, Vol. 87, pp. 447-54.

[Mitrof66] Mitrofanov, S.P., (1966), *Scientific Principles of Group Technology, Parts 1-3*, Translated from Russian by E. Harris and edited by T.J. Grayson, National Lending Library for Science and Technology, Boston Spa, Yorkshire England.

[Moh*etal*88] Mohamed, Y., Elbestawi, M. A., and Liu, L., (1988), "Application of Some Parameter Adaptive Control Algorithms in Machining," in *Control Methods for Manufacturing Processes*, D. E. Hardt (Ed.), ASME, DSC Vol. 9, Dec., pp. 63-70.

[MonAlt91] Montgomery, D., and Y. Altintas, (1991), "Mechanism of Cutting Force and Surface Generation in Dynamic Milling," *ASME Transactions, Journal of Engineering for Industry*, Vol. 113., No. 2, May, pp. 160-168.

[Nakaya78] Nakayama, K., (1978),*The Metal Cutting in its Principles*, (in Japanese) Corona Publishing Co., Ltd., Tokyo, Japan.

[OkuHit61] Okushima, K. and Hitomi, K., (1961), "An Analysis of the Mechanics of Orthogonal Cutting and Its Application to Discontinuous Chip Formation," *ASME Transactions, Journal of Engineering for Industry*, Series B, Vol. 83, pp. 545-556.

[OkuHit64] Okushima, K. and Hitomi, K., (1964), "A Study of Economical Machining: An Analysis of the Maximum-Profit Cutting Speed", *The International Journal of Production Research*, Volume 3, No. 1, pp. 73-78.

[Olesto70] Oleston, N. O., (1970), *Numerical Control*, Wiley, New York.

[Ono61] Ono, K., (1961), "Analysis of the Grinding Force," *Bulletin of the Japan Society of Grinding Engineers*, No. 1.

[Oxley63] Oxley, P.B.L., (1963), "Mechanics of Metal Cutting for a Material With Variable Flow Stress," *ASME Transactions, Journal of Engineering for Industry*, Series B, Vol. 85, pp. 339-345.

[Oxley89] Oxley, P.B.L., (1989), *Mechanics of Machining: An Analytical Approach to Assessing Machinability*, Ellis Horwood Limited, Chichester, England.

[Paietal88] Pai, D.M., E. Ratterman and M.C. Shaw, (1989), "Grinding Swarf," *Wear*, Vol. 131, No. 2, pp. 329-339.

[PalOxl59] Palmer, W.B., and P.L.B. Oxley, "Mechanics of Orthogonal Machining," *Proceeding of the Institution of Mechanical Engineers*, Vol. 173, pp. 623.

[PanWu83] Pandit, S. M., and Wu, S. M., (1983), *Time Series and System Analysis with Applications*, Wiley, New York.

[Peklen57] Peklenik, J., (1957), "Ermittlung Vol Geometrisch en und Physikalischen Kenngrussen Fur Die Grundlagen Forschung des Schleifens," Dissertation T.H. Aachen .

[Peklen67] Peklenik, J., (1967-68), New Developments in Surface Characterization and Measurements by Means of Random Process Analysis," *Proceedings of the Institution of Mechanical Engineers*, Vol. 182, 3K, pp. 108-126.

[Peters84] Peters, J., (1984), "Contribution of CIRP Research to Industrial Problems in Grinding," *CIRP Annals, v33/2*, pp.451-468.

[Pet*etal*79] Peters, J., P. Vanherck and M. Sastrodinoto, (1979), "Assessment of Surface Topology Analysis Techniques," *CIRP Annals, v28/2*.

[Phadke89] Phadke, Madhav Shridhar, (1989), *Quality Engineering Using Robust Design*, Prentice Hall, Englewood Cliffs, NJ.

[Piispa37] Piispanen, Vaino, "Lastunmuodostumisen Teoriaa," *Teknillinen Aikakauslehti*, Vol. 27, p 315-322.

[Pollac76] Pollack, H.W., (1976), *Tool Design*, Reston Publishing Comp., Inc., Reston, Virginia.

[PreWil77] Pressman, R. S., and Williams, J. E., (1977), *Numerical Control and Computer-Aided Manufacturing*, John Wiley, New York.

[Reiche56] Reichenback, G.S. *et. al.*, (1956), "The Role of Chip Thickness in Grinding," *Transactions ASME*, Volume 78.

[Ruff78] Ruff, A.W., (1978), "Debris Analysis of Erosive and Abrasive Wear," *Proceedings of the International Conference on the Fundamentals of Tribology*, MIT Press, Cambridge, MA, June, pp.877-885.

[RamRma87] Ramanath, S., and M.C. Shaw, (1988), "Abrasive Grain Temperatures at the Beginning of a Cut in Fine Grinding," *ASME Transactions, Journal of Engineering for Industry*, Vol. 110, February, pp.15-24.

[Ram*etal*88] Ramanath, S., T.C. Ramaraj and M.C. Shaw, (1987), "What Grinding Swarf Reveals," *CIRP Annals*, Vol. 36/1, pp.245-247.

[Reason44] Reason, R.E., "Surface Finish and Its Measurement," (1944), *Journal of the Institution of Production Engineers*, Vol. 23, pp. 347-372.

[Rea*etal*44] Reason, R.E., M.R. Hopkins and R.I. Garrod, (1944), *Report on Measurement of Surface Finish by Stylus Methods*, Rank Taylor Hobson, Leicester.

[Roe26] Roe, Joseph Wickham, (1926), *English and American Tool Builders*, McGraw-Hill Book Company, New York and London.

[Roweetal90] Rowe, W.B., J.A. Pettit, M.N. Morgan and A.S. Lavine, (1990), "A Discussion of Thermal Models in Grinding," Fourth International Grinding Conference, Dearborn, MI, SME Paper #MR90-516, October 9-11.

[RusBro66] Russell, J.K., and R.H. Brown, (1966) "The Measurement of Chip Flow direction," *International Journal of Machine Tool Design and Research*, Vol. 6, pp. 129.

[Sata54] Sata, Toshio, (1954), "Friction Process on Cutting Tool and Cutting Mechanism," *Proceedings of the Fourth Japan National Congress for Applied Mechanics.*

[*SenRev*82] "Surface Geometry and Texture Measured Together," (1982), *Sensor Review*, Vol. 2, 2, April, pp. 59-63.

[Shaw*etal*53] Shaw, M.C., N.H. Cook and I. Finnie, (1953), "Shear Angle Relationships in Metal Cutting," *ASME Transactions*, No. 2, Vol. 75.

[Shaw84] Shaw, Milton C., (1984), *Metal Cutting Principles,* Clarendon Press, Oxford, England.

[Shaw90] Shaw, Milton C., (1990), "Cutting and Grinding: A Comparison," Fourth International Grinding Conference, Dearborn, MI, SME Paper #MR90-500, October 9-11.

[Shey87] Shey, J. A., (1987), *Introduction to Materials and Manufacturing Processess, Second Edition,* McGraw-Hill Book Company, New York.

[SmiTlu91] Smith S., and Tlusty, J., (1991), "An Overview of Modelling and Simulation of the Milling Process," *ASME Transactions, Journal of Engineering for Industry*, Vol. 113, No. 2, May, pp. 169-175.

[SmiTlu90] Smith S., and Tlusty, J., (1990), "Update on High-Speed Milling Dynamics," *ASME Transactions, Journal of Engineering for Industry*, Vol. 112, May, pp. 142-149.

[Spotts83] Spotts, M.F., (1983), *Dimensioning and Tolerancing for Quality Production*, Prentice-Hall, Inc., Englewood Cliffs, NJ.

[Snoeys74] Snoeys, R., (1974), "The Significance of Chip Thickness in Grinding," *CIRP Annals*, Volume 23/2.

[SnoBro69] Snoeys, R., and D. Brown, (1969), "Dominating Parameters in Grinding Wheel and Workpiece Regenerative Chatter," *Proceedings of the Tenth International Machine Tool Design and Research Conference*, pp. 325-48.

[Stable51] Stabler, G,V., (1951), "The Fundamental Geometry of Cutting Tools," *Proceedings of the Institution of Mechanical Engineers*, Vol. 165, p. 14.

[StepWu88a] Stephenson, D.A., and S.M. Wu, (1988a), Computer Models for the Mechanics of Three-Dimensional Cutting Processes-Part I: Theory and Numerical Method," *ASME Transaction, Journal of Engineering for Industry*, Vol. 110, February, pp. 33-37.

[StepWu88b] Stephenson, D.A., and S.M. Wu, (1988b), Computer Models for the Mechanics of Three-Dimensional Cutting Processes-Part II: Results for Oblique End Turning and Drilling," *ASME Transaction, Journal of Engineering for Industry*, Vol. 110, February, pp. 38-43.

[Stephe91a] Stephenson, D.A., (1991a), "Assessment of Steady State Metal Cutting Temperature Models Based on Simultaneous Infrared and Thermocouple Data," *ASME Transaction, Journal of Engineering for Industry*, Vol. 113, No. 2. May, pp. 121-128.

[Stephe91b] Stephenson, D.A., (1991b), "An Inverse Method for Investigating Deformation Zone Temperatures in Metal Cutting," *ASME Transaction, Journal of Engineering for Industry*, Vol. 113, No. 2. May, pp. 129-136.

[StuGoe75] Stute, G., and F. R. Goetz, (1975), "Adaptive Control System for Variable Gain in ACC Systems," *Proc. 16th Int. Machine Tool Des. and Res. Conf.*, Manchester, UK, Sept., pp. 117-121.

[SurTex78] *Surface Texture, Surface Roughness, Waviness and Lay*, (1978), American National Standard Institute B46.1-1978., Published by ASME, United Engineering Center, 345 East 47th Street, New York, NY 10017.

[Suther88] Sutherland, J. W., (1988), "A Dynamic Model of the Cutting Force System in the End Milling Process," in *Sensors and Controls for Manufacturing - 1988*, E. Kannatey-Asibu et al. (Eds.), ASME, PED Vol. 33, Dec., pp. 53-62.

[SutDeV86] Sutherland, J. W., and R. E. DeVor, (1986), "An Improved Method for Cutting Force and Surface Error Prediction in Flexible End Milling Systems," *ASME J. of Engineering for Industry*, Vol. 108, November, pp. 269-279.

[SutBab88] Sutherland, J. W., and Babin, T. S., (1988), "The Geometry of Surfaces Generated by the Bottom of an End Mill," *Proc. 16th NAMRC*, Urbana, IL, May pp. 202-208.

[Taguch86] Taguchi, Gen'ichi, (1986), *Introduction to Quality Engineering: Designing Quality Into Products and Processes*, English Translation by Asian Productivity Organization, Tokyo.

[Taguch87] Taguchi, Gen'ichi, (1987), *System of Experimental Design: Engineering Methods to Optimize Quality and Minimize Costs*, Don Clausing, Technical Editor, (English translation by Louise Wantanabe Tung), UNIPUB/Kraus International Publications, Dearborn MI and American Supplier Institute.

[Tay*etal*74] Tay, A.O., Stevenson, M.G. and Davis, G., (1974) "Using the Finite Element Method to Determine Temperature Distributions in Orthogonal Machining", *Proceedings of Institution of Mechanical Engineers*, Vol. 188, pp. 627-638.

[Taylor07] Taylor, Frederick Winslow, (1907), "On the Art of Cutting Metals," *ASME Transactions*, Vol 28.

[Taylor11] Taylor, Frederick Winslow, (1911), *The Principles of Scientific Management*, Harper, New York and London.

[TekUls90] Tekinalp, O. and A. Galip Ulsoy, (1990), "Effects of Geometric and Process Parameters on Drill Transverse Vibrations," *ASME Transactions, Journal of Engineering for Industry*, Vol. 112, May, pp. 189-194.

[Tho*etal*61] Thomsen, E.G., S. Kobayashi and M.C. Shaw, (1961), "Some Controlled Metal-Cutting Studies With Resulfurized Steel," *ASME Transactions, Journal of Engineering for Industry*, Series B, Vol. 83, pp. 513-522.

[Thomso80] Thomson, R.A., (1980), "A General Theory for Regenerative Stability," *Proceedings of the Eighth North American Manufacturing Research Conference*, University of Missouri-Rolla, May 18-21, pp. 377-387.

[Thomso86] Thomson, R.A., (1986), "Chatter Growth - Tests to Evaluate the Theory," *Modelling, Sensing and Control of Manufacturing Processes, PED Vol. 23 and DSC Vol. 4.*, ASME special publication.

[TicDeV89] Tichy, J.A. and W.R. DeVries, (1989), "A Model for Cylindrical Grinding Based on Abrasive Wear Theory," *Grinding Fundamentals and Applications*, Edited by S. Malkin and J.A. Kovach, PED-Vol. 39, ASME Publication No. H00571, New York, NY, pp. 335-347.

[TriCho51] Trigger, K.J., and Chao, B.T., (1951) "An Analytical Evaluation of Metal Cutting Temperatures," *ASME Transactions*, Vol. 73, pp. 57-68.

[Tri*etal*77] Tripathi, K.C., Nicol, A.W., Rowe,G.W., (1977), "Observation of Wheel-Metal-Coolant Interactions in Grinding," *Transactions of the American Society of Lubrication Engineers*, Vol. 20, pp.249-256.

[Tlusty65] Tlusty, J., (1965), "A Method of Analysis of Machine Tool Stability," *Proceedings of the Sixth Machine Tool Design and Research Conference.*

[Tlusty78] Tlusty, J., (1978), "Analysis of the State of Research in Cutting Dynamics," *CIRP Annals, 27:2*.

[Tlu*etal*78] Tlusty, J., A. Cowley, and M. A. A. Elbestawi, (1978), "A Study of an Adaptive Control System for Milling with Force Constraint," *Proc. 6th North American Metalworking Research Conf.*, Gainesville, FL, April pp. 364-371.

[TluIsm81] Tlusty, J. and F. Ismail, (1981), "Basic Non-Linearity in Machining Chatter," *CIRP Annals, 30:2*, pp. 299.

[TluIsm83] Tlusty, J., and Ismail, F., (1983), "Special Aspects of Chatter in Milling," *ASME J. of Vibration, Acoustics, Stress, and Reliability in Design*, Vol. 105, Jan. pp. 25-32.

[Tlu*etal*82] Tlusty, J., Ismail, F., Hoffmanner, A., and Rao, S., (1982), "Theoretical Background for Machining Tests of Machining Centers and Turning Centers," *Proc. 10th NAMRC*, Hamilton, Ontario, May, pp. 385-392.

[Tobia65] Tobias, Stephen Albert, (1965), Machine Tool Vibration, Authorized translation by A.H. Burton, John Wiley, New York.

[Tom*etal*85] Tomizuka, M., S-J. Zhang, J-H. Oh, and M-S. Chen, (1985), "Modeling of Metal Cutting Processes for Digital Control," *Proc. 13th North American Manufacturing Research Conference, Berkeley*, CA, May, pp. 575-580.

[Tom*etal*83] Tomizuka, M., J. H. Oh, and D. A. Dornfeld, (1983), "Model Reference Adaptive Control of the Milling Process," in *Control of Manufacturing Processes and Robotic Systems*, D. E. Hardt and W. J. Book (Eds.), ASME, New York, pp. 55-63.

[Trent77] Trent, Edward Moor, (1977), *Metal Cutting*, Butterworths, London and Boston.

[Ulsoy85] Ulsoy, A. G., (1985), "Applications of Adaptive Control Theory to Metal Cutting," in Donath, M. (ed), *Dynamic Systems: Modeling and Control*, ASME, New York.

[UlsDeV89] Ulsoy, A. Galip and Warren R. DeVries, (1989), *Microcomputer Applications in Manufacturing*, John Wiley and Sons, Inc.

[UlsHan85] Ulsoy, A. G., and Han, E., (1985), "Tool Breakage Detection Using a Multi-Sensor Strategy," *Proceedings of the IFAC Conference on Control Science and Technology for Development*, Beijing, China.

[Uls*etal*83] Ulsoy, A. G., Koren, Y., and Rasmussen, F., (1983), "Principal Developments in the Adaptive Control of Machine Tools", *ASME Journal of Dynamic Systems Measurement and Control* , Vol. 105, No. 2, pp 107-112.

[UlsTek84] Ulsoy, A. G., and Tekinalp, O., (1984), "Dynamic Modeling of Transverse Drill Bit Vibrations," *Annals of the CIRP*, Vol. 33/1.

[UsuHir78] Usui, E., and A. Hirota, (1978), "Analytical Prediction of Three Dimensional Cutting Process - Part 2: Chip Formation and Cutting Force With Conventional Single Point Tool," *ASME Transactions, Journal of Engineering for Industry*, Vol. 100, No. 2, pp. 229.

[vonTur70] von Turkovich, B.F., (1970), "Shear Stress in Metal Cutting," *ASME Journal of Engineering for Industry*, Vol. 92, pp. 151-157.

[vonTur74] von Turkovich, B.F., (1974), "Deformation Mechanics During Adiabatic Shear," *Proceedings of the 2nd North American Metalworking Research Conference - Supplement*, Madison, WI, pp. 682-690.

[VorTea81] Vorburger, T.V., and E.C. Teague, (1981), "Optical Techniques for On-Line Measurement of Surface Topography," *Precision Engineering*, Vol. 3, pp. 61.

[Weck84] Weck, Manfred, (1984), *Handbook of Machine Tools, Volume 1 - Types of Machines, Forms of Construction and Applications, Volume 2 - Construction and Mathematical Analysis, Volume 3 - Automation and Controls, Volume 4 - Metrological Analysis and Performance Tests*, John Wiley & Sons, New York.

[WelSmi70] Welbourn, D.B., and J.D. Smith, (1970), *Machine-Tool Dynamics; An Introduction*, University Press, Cambridge, England.

[Werner78] Werner, G., (1978), "Influence of Work Material on Grinding Forces," *CIRP Annals*, Volume 27/1.

[WhiArc70] Whitehouse, D.J., and J.F. Archard, (1970), "The Properties of Random Surfaces of Significance in Their Contact," *Proceedings of the Royal Society, v316A*, pp. 97-121.

[Whietal74] Whitehouse, D.J., P. Vanherck, W. deBruin, and C.A. vanLuttervelt, "Assessment of Surface Topology Analysis Techniques in Turning," *CIRP Annals*, Vol. 23/2, pp. 256.

[Woodbu59] Woodbury, Robert S., (1959), *History of the Grinding Machine, A Historical Study in Tools and Precision Production*, The Technology Press, MIT, Cambridge, MA

[Woodbu60] Woodbury, Robert S., (1960), *History of the Milling Machine, A Study in Technical Development*, The Technology Press, MIT, Cambridge, MA.

[Woodbu61] Woodbury, Robert A., (1961), *History of the Lathe to 1850, A Study in the Growth of a Technical Element of an Industrial Economy*, Nimrod Press Inc., Boston, MA.

[Wu64a] Wu, S.M., (1964a), "Tool-Life Testing by Response Surface Methodology - Part I," *ASME Transactions, Journal of Engineering for Industry*, Series B, Vol. 86, pp. 105-110.

[Wu64b] Wu, S.M., (1964b), "Tool-Life Testing by Response Surface Methodology - Part II," *ASME Transactions, Journal of Engineering for Industry*, Series B, Vol. 86, pp. 110-116.

[Wu*etal*77] Wu, S. M., M. F. DeVries and W. R. DeVries, (1977), "Analysis of Machining Operations by the Dynamic Data System Approach,"*Fifth North American Metalworking Research Conference Proceedings*, University of Massachusetts, pp. 219-223, May 23-25.

[WuEr66] Wu, S.M. and Ermer, D.S., (1966), "Maximum Profit as the Criterion in the Determination of the Optimum Cutting Conditions," *ASME Transactions, Journal of Engineering for Industry,* Series B, Vol. 88, pp. 435, November.

[ZhaKap91a] Zhang, G.M., and S.G. Kapoor, (1991a), "Dynamic Generation of Machined Surfaces, Part 1: Description of Random Excitation System," *ASME Transactions, Journal of Engineering for Industry,* Vol. 113, No. 2, May, pp.137-144.

[ZhaKap91b] Zhang, G.M., and S.G. Kapoor, (1991b), "Dynamic Generation of Machined Surfaces, Part 2: Construction of Surface Topography," *ASME Transactions, Journal of Engineering for Industry,* Vol. 113, No. 2, May, pp.145-153.

[Zorev66] Zorev, N.N., (1966), *Metal Cutting Mechanics,* translated from Russian by H.S.H. Massey and edited by Milton C. Shaw, Pergamon Press, Ltd., London.

[Zvory96] Zvorykin, K.A., (1896), "Work and Force Necessary for Separating a Metal Chip," (Rabota i Usiliye, Neobkhodimyye Dlya Otdeleniya Metallicheskoi Struzhki), *Tekhnicheski Sbornik i Vestnik Promyshlennosti.*

Appendix
Code for Simulation Examples

This appendix has the computer code, written in "C", for the two computer simulations in the text, Examples 5.3 and 8.1. Only the functions that are used to implement the simulations are included. In addition to these functions, the "front end" functions to enter, change, check and display simulation results are needed.

These functions are not *the way* to do the simulations, but *a way*. After all, like process planning, programming is a design activity with a number of alternative ways to accomplish the design objectives. Students are encouraged to copy and try this code, and then improve it. This is one way to learn the details of the methods and equations covered in this book. To assist in doing this, a number of the comments refer to equations in the text or the specific examples.

1 C Code for Example 5.3 Simulation of Face Milling

```
void    Face_Mill(Y0,X0,Xf,N_per_Rev,D,Lx,Ly,vf,Nt,Kappa,N,Kc)
/*
                            Arguments to Face_Mill
*/
float   Y0,X0;/*            Initial position of the center of the cutter */
float   Xf;/*               Final position in feed direction of cutter center */
int     N_per_Rev;/*        Increments in a simulation rotation */
float   D;/*                Diameter of the face mill (mm) */
float   Lx,Ly;/*            length and width of the workpiece (mm)      */
float   vf;/*               feedrate along the length (mm/s) */
int     Nt;/*               Number of cutting edges on the cutter */
float   Kappa;/*            Lead angle (degrees) */
float   N;/*                Rotational speed of the cutter (rev/s) */
float   Kc[3];/*            Cutting stiffness vector in Example 5.3 (kN/mm) */
{
float   DeltaTheta,DeltaX,X,x,sz,Theta,Theta0=0,a2,a1,Radius,
        ThetaEntry,ThetaExit,Theta2pi,Engaged,DeltaRev,Rev,hmax;
float   DeltaPosition[MAXTEETH],EdgePosition[MAXTEETH],h[MAXTEETH];
float   FTangent[MAXTEETH],FRadial[MAXTEETH],Fz[MAXTEETH];
float   Torque,FX,FY,FZ;
int     CuttingFlag[MAXTEETH], ithEdge, index;

FILE    *Face_Mill_Stream;
```

```
Face_Mill_Stream            =fopen("Face_Mill","a");
fprintf(Face_Mill_Stream,"\n__X(mm)____Rev___T(N-
        m)__FX(kN)__FY(kN)__FZ(kN)___F(kN)__Pc(kW)__Pf(kW)__");
/*
        Define the entry and exit angles
*/
Radius =D/2,    a2 =TopEdge(Y0,D,Ly),   a1 =BottomEdge(Y0,D,Ly);
if(a2<Ly)       {ThetaEntry =pi/2;
        }
else            {ThetaEntry =atan2(a2,Radius);
        }
if(a1>0)        {ThetaExit =-pi/2;
        }
else            {ThetaExit =atan2(-Y0,Radius);
        }
/*
        Initiaize the flags and the cutting edge angles.
*/
for (ithEdge = 0; ithEdge < Nt; ++ithEdge){
                CuttingFlag[ithEdge]        =!CUT;
                DeltaPosition[ithEdge]      =2.*pi*ithEdge/Nt;
                EdgePosition[ithEdge]       =0.;
                h[ithEdge]                  =0.;
                }
DeltaTheta      =2*pi/N_per_Rev,        DeltaRev        =1./N_per_Rev;
sz              =vf/(N*Nt),             DeltaX          =vf*DeltaTheta/N;
X               =X0,                    Rev             =0;
Theta2pi        =Theta0,                index           =0;
hmax            =sz*cos(Kappa*Deg_to_Rad);/*    Eq(5.12d) */
while(X<Xf){
/*
        Set the cutting edge angles; the function PlusMinusRadians keeps
        the angles used for the calculations between +/- pi.
*/
        for (ithEdge = 0; ithEdge < Nt; ++ithEdge){
                EdgePosition[ithEdge]
                =PlusMinusRadians(fmod(Theta2pi+DeltaPosition[ithEdge],2*pi));
/*
                Check if engaged and compute the chip thickness
*/
                if((EdgePosition[ithEdge]<ThetaEntry)
                        && (EdgePosition[ithEdge]>ThetaExit)){
                        Engaged         =X+Radius*cos(EdgePosition[ithEdge]);
                        if((Engaged<0) || (Engaged>Lx)){
                                CuttingFlag[ithEdge]        =!CUT;}
                        else{
                                CuttingFlag[ithEdge]        =CUT;}
                        }
                else{
```

```
			CuttingFlag[ithEdge]	=!CUT;}
		if(CuttingFlag[ithEdge] == CUT){
			h[ithEdge] =hmax*cos(EdgePosition[ithEdge]);}/*	Eq(5.13)*/
		else{
			h[ithEdge] = 0;}
/*

		Compute the Theta-R-Z forces at this position,
		they will be summed later.
*/

		FTangent[ithEdge]		=Kc[0]*h[ithEdge];
		FRadial[ithEdge]		=Kc[1]*h[ithEdge];
		Fz[ithEdge]			=Kc[2]*h[ithEdge];
		}
/*

	Sum the force components and estimate the torque and
	Cartesian components due to all the edges cutting.
*/

	Torque	=0.,	FX	=0.,	FY	=0.,	FZ	=0.;
		for (ithEdge = 0; ithEdge < Nt; ++ithEdge){
		if(CuttingFlag[ithEdge] == CUT){
			Torque	=Torque +Radius*FTangent[ithEdge];
			FX	=FX	+cos(EdgePosition[ithEdge])*FRadial[ithEdge]
				+sin(EdgePosition[ithEdge])*FTangent[ithEdge];
			FY	=FY	-sin(EdgePosition[ithEdge])*FRadial[ithEdge]
				+cos(EdgePosition[ithEdge])*FTangent[ithEdge];
			FZ	=FZ	+Fz[ithEdge];
			}

		}
		if(fmod(index,20)<=0){
			index	=0,	eraseplot();
			printf("\n__X(mm)____Rev___T(N-m)__FX(kN)__FY(kN)__
				FZ(kN)___F(kN)__Pc(kW)__Pf(kW)__");
			}
	printf("\n %6.2f %6.2f %6.2f %6.2f %6.2f
		%6.2f %6.2f %6.2f %6.3f ",X,Rev,Torque,FX,FY,FZ,
		sqrt(FX*FX+FY*FY),(Torque*2*pi*N*.001),(FX*vf*.001));
		index	=index+1;
	fprintf(Face_Mill_Stream,"\n%6.2f\t%6.2f\t%6.2f\t%6.2f\t%6.2f\t
		%6.2f\t%6.2f\t%6.2f\t%6.3f",X,Rev,Torque,FX,FY,FZ,
		sqrt(FX*FX+FY*FY),(Torque*2*pi*N*.001),(FX*vf*.001));
/*

	The last thing to do is move the cutter, both in translation
	and rotation. Note that the angle is decremented because with
	the right hand rule a clockwise rotation is negative.
*/

	X	=X+DeltaX;
	Rev	=Rev+DeltaRev, Theta2pi		=fmod(Theta2pi-DeltaTheta,2*pi);
	}

}
```

2 C Code for Example 8.1 Single Degree of Freedom Stability Analysis for Turning

```
void    Turning_Dynamics(Z0,Zf,KPrime,D,Lz,N,a,d,Keq,zeta,fn,
                         Steps_per_Rev,Nsteps,IOFlag,Stream)
/*

        Procedure to do a Turning Dynamics Simulation. The equations of
        motion are based on Eq (8.1a) - the cutting process model,
        Eq (8.2d) the instantaneous chip thickness, and the single degree
        of freedom machine structure in Eq (8.3b).

        A simple Euler (1st Order) numerical integration is used to solve the
        equations of motion. This single degree of freedom case uses a
        state variable representation. The states are:
        w1(t) = z(t) the displacement of the moving cutting edge,
        w2(t) = dz(t)/dt the relative velocity of the moving cutting edge, and
        w3(t) = Z(t) + z(t) the workpiece surface measured from the spindle.
        The time lag is handled by a buffer that keeps values from z(t)
        to z(t-1/N).

        Written in THINK'S LightspeedC, Copyright 1988 and run on a
        Macintosh SE.

        Warren R. DeVries                                    November, 1990

*/

float   Z0,Zf;/*        Initial and final cutter position (mm),
                                                   Z0 > Zf          */
float   KPrime;/*       Specific cutting pressure (kN/mm^2)*/
float   D,Lz;/*  Workpiece diameter and length (mm), Lz < Z0 */
float   N,a,d;/*        Cutting conditions: speed (rev/s),
                        feed (mm/rev) and depth (mm)      */
float   Keq,zeta,fn;/*  Equivalent static siffness (kN/mm),
                        damping ratio (dimensionless), and
                        natural frequency (Hz)      */

int     Steps_per_Rev; /*Number of times in a revolution that
                        data is logged, either in a file or
                        on the console     */
int     Nsteps;  /*     Number of steps in a revolution used for
                        numerical integration,
                        Nsteps >= Steps_per_Rev           */
```

```
int     IOFlag; /*      Flag for I/O 1 <=> console, 2 <=> file
                        pointed to by Stream and 3 <=> both       */
FILE    *Stream;/*      For file I/O        */
{
float   Kc,     w1, dw1dt,      w2, dw2dt, dw3dt,       w3,
        Z_lagged[MAXPOINT], vf, h,      Z;
float   Omegan,         Two_zeta_Omegan,        Omegan_Squared,
        Keq_over_Kc;
float   Delta_Time, Time,       Delta_Rev,      Rev;
int     Angle_Index,    Cut_Flag,       Printindex,     Screenindex;

if((IOFlag == 2) || (IOFlag == 3))      {
        fprintf(Stream,
"\n_____Rev\t_Time(s)\t__w1(mm)\tw2(mm/s)\t__w3(mm)\t___h(mm)\tdw2dt(g)\n");
        }
/*
        Calculation of the Coefficients and Initialization
*/
        Omegan                  =       2*pi*fn;
        Two_zeta_Omegan         =       2*zeta*Omegan;
        Omegan_Squared          =       Omegan*Omegan;
        Keq_over_Kc             =       Omegan_Squared*KPrime*d/Keq;
        vf                      =       N*a;

        for(Angle_Index = 0; Angle_Index<Nsteps; Angle_Index++){
                Z_lagged[Angle_Index]   =Lz;
        }
        Printindex      =Nsteps/Steps_per_Rev,  Screenindex     =0;
        w1      =0,     w2      =0,     w3      =Z0,    Z       =w3,
        h       =0,     Time    =0,     Rev     =0;
        Angle_Index = 0;
        Delta_Time      =1./(N*Nsteps), Delta_Rev       =1./Nsteps;
/*
        Start of the Main Computational Loop
*/
while(Z>Zf) {
/*
        Print and save the states
*/
        if(fmod(Angle_Index,Printindex) == 0)           {
                if((Screenindex == 0) && (IOFlag==1 || IOFlag==3))  {
                        eraseplot();
                        printf("_____Rev _Time(s) __w1(mm) w2(mm/s)
                                __w3(mm) ___h(mm) dw2dt(g)\n");
                }
                Screenindex     =fmod(Screenindex+1,20);
                if(IOFlag==1 || IOFlag==3)              {
                        printf("%8.2f %8.3f %8.3f %8.2f %8.2f %8.3f %8.3f\n",
                                Rev,Time,w1,w2,w3,h,dw2dt/9800);
```

```
                }
            if(IOFlag==2 || IOFlag==3)          {
                fprintf(Stream,
                "%8.2f\t%8.3f\t%8.3f\t%8.2f\t%8.2f\t%8.3f\t%8.3f\n",
                        Rev,Time,w1,w2,w3,h,dw2dt/9800);
                }
            }

    if(w3<Lz) {Cut_Flag     =CUT;}/*    Not cutting unless the edge tip
                                        is less than the length of
                                        the workpiece */
    else      {Cut_Flag     =!CUT;}
/*
    Compute the time derivatives of each state
*/
    dw1dt   =w2;

    if(Cut_Flag == CUT)     {h      =Z_lagged[Angle_Index]-w3;}
    else        {           h       =0.;}
    if(h < 0){              h       =0.;}/* Case of jumping out         */
    dw2dt   =-Omegan_Squared*w1    -Two_zeta_Omegan*w2    +Keq_over_Kc*h;

    dw3dt   =w2     -vf;
/*
    Euler integration of the state variables
*/
    w1      =w1     + dw1dt*Delta_Time;
    w2      =w2     + dw2dt*Delta_Time;
    w3      =w3     + dw3dt*Delta_Time;
/*
    Integrate some "convenience" variables
*/
    Z       =Z      -vf*Delta_Time;  /*     Cross feed position */
    Rev     =Rev    +Delta_Rev;      /*     Rotation counter */
    Time    =Time   +Delta_Time;     /*     Time variable */
    Z_lagged[Angle_Index]   =w3;/*          New surface & index  */
    Angle_Index     =fmod(Angle_Index+1,Nsteps);

    }
}
```

Index

9 780387 977287